# Cosmic Biology
## How Life Could Evolve on Other Worlds

D1232513

Louis Neal Irwin and Dirk Schulze-Makuch

# Cosmic Biology

## How Life Could Evolve on Other Worlds

 Springer

Published in association with
**Praxis Publishing**
Chichester, UK

Professor Louis Neal Irwin
The University of Texas at El Paso
El Paso
Texas
USA

Professor Dirk Schulze-Makuch
Washington State University
Pullman
Washington
USA

SPRINGER–PRAXIS BOOKS IN POPULAR ASTRONOMY
SUBJECT *ADVISORY EDITOR*: John Mason, M.B.E., B.Sc., M.Sc., Ph.D.

ISBN 978-1-4419-1646-4      e-ISBN 978-1-4419-1647-1
DOI 10.1007/978-1-4419-1647-1

Springer New York Dordrecht Heidelberg London

© Springer Science+Business Media, LLC 2011

All rights reserved. This work may not be translated or copied in whole or in part without the written permission of the publisher (Springer Science+Business Media, LLC, 233 Spring Street, New York, NY 10013, USA) except for brief excerpts in connection with reviews or scholarly analysis. Use in connection with any form of information storage and retrieval, electronic adaptation, computer software, or by similar or dissimilar methodology now known or hereafter developed is forbidden.
The use in this publication of trade names, trademarks, service marks, and similar terms, even if they are not identified as such, is not to be taken as an expression of opinion as to whether or not they are subject to proprietary rights.

Cover design: Jim Wilkie
Project copy editor: Dr John Mason
Typesetting: BookEns, Royston, Herts., UK

Printed on acid-free paper

Springer is part of Springer Science+Business Media (springer.com)

# Contents

*This book is dedicated to the people of the former Soviet Union, who opened the space age on 4 October 1957, to the people of the United States who first sent humans to another world on 20 July 1969, to the thousands of scientists and engineers of all nations whose hearts have been broken by the countless failed missions that inevitably occurred on the way to the spectacular successes of Luna, Venera, numerous Mars orbiters and rovers, Pioneer, Mariner, Viking, Magellan, and Hubble; and especially to the visionary government leaders, scholars, and technicians who funded, built, and sent forth the Voyager, Galileo, and Cassini-Huygens spacecraft to give us the incredible views we have of the other worlds in our cosmic neighborhood that may well hold our nearest neighboring forms of alien life.*

# Preface

This is a work of both science *and* imagination.

Science uses logic and demonstrable facts to reveal the nature of the material world. Imagination leads the search for facts in new directions, and stretches the limits of logical inference in creative ways. Both are needed to uncover the mysteries of the natural world.

Life is a dynamic manifestation of the material world – complex, incredibly variable, and constantly changing – but material nonetheless, and therefore subject to scientific study just as much as the stars that burn and the atoms that make them up. We strongly suspect that life extends as far as the stars do into the vastness of the cosmos. Unlike the stars that we can see with certainty, though, life lies hidden from view, of unknown extent, with properties that can be reasonably surmised but not yet confirmed on even our nearest planetary neighbor. Our science will reveal the nature of that life in due time. But where to look, what to expect, and how to interpret what we find, for now lies in the province of our imaginations.

So this is a book grounded in science but driven by imagination to predict the form that life *may have taken* on other worlds. It hasn't been easy to put a narrative together that genuinely attempts to stay well within the boundaries of scientific constraint, while at the same time stretching as far as we dare the limits of what ought to be possible. As scientists addicted to the comfort of known facts, we have had to grit our teeth with every use of the subjunctive case: *might be* instead of *is*, *may have been* instead of *was*, *could* instead of *did*, *ought* instead of *can*, *possibly* or *probably* instead of *certainly*, and so forth. But once we committed to writing a book about what we fully believe but of which we as yet have no certain knowledge, we had no choice. We have given the reader fair warning in the subtitle that this is ultimately a work of speculation, then we have forged ahead.

We wrote our first book, *Life in the Universe: Expectations and Constraints*, with the conviction that exobiology (morphing into NASA's astrobiology) needed a firm base of factual information, grounded in the physical sciences, with more than a token treatment of biology. We deliberately wrote a technical treatise, leaving to others the task of conveying a more accessible account for the general reader of life as it could exist in the rest of the universe. We made some pretty clear-cut predictions about the possibilities for life on other specific worlds, based on a logical extension of facts and principles. But we kept our predictions couched in technical language, and limited them simply to the existence of life, *per se*, without elaboration.

Encouraged by the success of *Life in the Universe*, we have decided that those predictions deserve full-blown elaboration, in a manner accessible to a reader without a higher education in science. Since everyone interested in the subject has a basic knowledge of our Solar System, we thought a case-history approach might be the best way to write an engaging narrative. This being a book of applied science, a certain amount of science can't be avoided. We have tried to provide this as painlessly as possible in the first three chapters – defining life precisely, reviewing the relevant essentials of chemistry and physics, and outlining the principles of evolutionary biology and ecology that most books on astrobiology avoid. Amateur and professional scientists alike can skip directly to chapter 4 without missing the basic message, though occasional reference back to the underlying principles and assumptions may prove helpful on occasion.

In chapter 4, we consider life on our home planet of Earth, not as we know it to be, but as an intelligent observer on another world could deduce it. We placed our hypothetical observer in the nearest solar system that we think could host a level of technology comparable to ours. By comparing what our alien scientists might infer about Earth, and why they would do so, we hope to illustrate *our* strategy for imagining how life could unfold on other worlds. Our exercise, beginning in chapter 5, takes a travelogue approach through our own Solar System, visiting first Mars, the planet with a climate and environment least different from Earth's, then Venus, as the other planet with Mars that resembled Earth most in its early history. We then consider the possibility of life in the unconventional habitats of the atmospheres of Venus and the gas giant planets. From there, we look at two exotic worlds circling Jupiter, Io and Europa (with a brief glance at Saturn's Enceladus). From there, our attention turns to the only other world besides Earth to host a nitrogen-rich atmosphere and abundance of organic chemicals – Saturn's giant moon, Titan – a featured attraction for every astrobiologist. But Neptune's captured satellite, Triton, and the biplanetary dwarfs, Pluto and Charon, may hold very different forms of life that we can't afford to disregard, so they will be our final destinations in the Solar System. Along the way, we'll look briefly at exoplanets and their capacity for hosting some forms of life. By the time we reach the outer reaches of our own Solar System, we will have concluded that we are very likely the only technologically-competent species in our region of the universe. We will therefore devote chapter 12 to an examination of why that might be, and what it says about the chances of our finding similar beings elsewhere (or of them finding us). Finally, we will conclude with a discussion in chapter 13 of the ultimate fate of life, on Earth in particular, but wherever it may arise in the universe.

Affirming the widely-held assumption that scientific truth holds true throughout the universe, our logical inferences are based on application of physicochemical principles on other worlds just as they apply on Earth. We likewise have no reason to doubt that the principles of evolutionary biology and ecology are just as universal. Therefore, they too have been applied without regard to the specific world on which we assume they operate.

On the other hand, we have made a few concessions to the uncertainty of our subject matter. Several terms, like *species* and all other taxonomic units, and specific subcellular structures or organ systems which have a specific reality for living organisms on Earth, cannot be assumed for life on other worlds, and therefore have been avoided. Even the broadest categories, like *plant* and *animal*, are usually replaced with *plant-like* or *animal-like*. For the same reason, while we are tempted to assume that cellular dimensions and structures obey some fairly universal constraints, we have felt uncomfortable assuming anything about higher-level cellular organization in alien biospheres, and so avoid referring to "multi-cellular" organisms except when speaking of known examples on Earth. We hope that any reader who finds this form of dissimilation annoying will grant us the freedom to be imprecise, for the sake of avoiding the implication of knowing what we do not.

We are honored to offer this work in the year of the 400th anniversary of Galileo's discovery of the four Jovian satellites. Two of them, Io and Europa, are featured chapters in this book – worlds on which life could exist today. Or it may not. We simply hope that, in setting forth the possibilities, we have imagined today what the science of tomorrow will be able to confirm or refute.

Louis Irwin thanks his wife, Carol, for her constant support and encouragement. Dirk Schulze-Makuch thanks his wife and kids for their continuous patience. Both of us acknowledge with deep appreciation our many colleagues for their insights and valuable discussions with us over the years.

Louis Neal Irwin and Dirk Schulze-Makuch
El Paso and Pullman
April, 2010

# Illustrations

# 1 Rare Earths and Life Unseen

## A strategy for cosmic biology

Speculation that alien life might be found on other worlds can be traced to ancient times. Copernicus, Kepler, and Galileo fed those thoughts by showing that the Earth is not the center of the universe. Once it was understood that our planet is just one of several rotating around a solar center, thoughts of other solar systems naturally arose. By 1584, Giordano Bruno was asserting that there must be "... countless suns, and countless earths all rotating around their suns." The next conceptual step – that on some of those "countless earths" there might be life – was easy to take.

By the 19th Century, the possibility of alien civilizations on Earth's nearest sibling planets, Venus and Mars, was taken seriously in the educated world. Spurred by the human propensity for both fear and wonder, ideas ranged from a war-like race of Martians bent on conquering Earth, to the prospect of a peaceful paradise flourishing beneath the clouds of Venus. Percival Lowell, an American astronomer, fueled speculation of intelligent life on Mars by claiming to see long canals, built perhaps to transport water from the poles to drying equatorial regions. This inspired science fiction writers to imagine a technologically advanced but dying, warring world, especially in the work of Edgar Rice Burroughs in the early 20th century. Speculations about life on Venus were generally more benign. A dense cloud cover enshrouded the planet in mystery, giving free reign to romantic images of a verdant biosphere. From Ray Bradbury's short stories depicting incessant rainfall, to Robert Heinlein's vision of a Carboniferous swamp, the Venus of science fiction most often was seen as a warm, dripping-wet, water world, capable perhaps of supporting a prolific alien flora and fauna.

As the space age dawned, reality set in. Robotic probes to our nearest neighbors doused all thoughts of technological civilizations or even widespread but unintelligent forms of life, as cameras landing on the surface of both Mars and Venus saw no evidence of either. In 1969, humans first set foot on the Moon, finding it as sterile on site as it appeared through the telescope. Since Mercury resembles the Moon in both topology and torrid fluctuations between extreme heat and cold, life seems equally improbable on that planet. As scientific data on the nature of the gas giants and their satellites accumulated, it became clear that Earth-like conditions were not to be found on any other world in our Solar System – that our planet, at least within a radius of four light years (the distance to our nearest star), is not only rare but unique.

L.N. Irwin and D. Schulze-Makuch, *Cosmic Biology: How Life Could Evolve on Other Worlds*,
Springer Praxis Books, DOI 10.1007/978-1-4419-1647-1_1,
© Springer Science+Business Media, LLC 2011

Against this revelation, admittedly discouraging for those who dream of finding sister systems of life beyond Earth, came the counterbalancing realization of just how vast the universe truly is. Maybe, except for Earth, *our* Solar System isn't all that inviting for life, but there must be solar systems galore in a universe that measures over 13 billion light years across. With evidence accumulating over the past decade that planets orbit other suns in our Galaxy, the assumption that solar systems pervade the universe seems reasonable. Even if Earth-like planets are rare, given enough solar systems, planets as inviting for life as Earth must exist elsewhere. Since we know that life exists on Earth, we're most likely to find it and recognize it on other worlds that resemble Earth. This is the logic of the search for Earth-like planets in other systems, and the terracentric search for life on those worlds. It fuels the drive to "follow the water." It forms the basis for seeking planets where liquid water and protective atmospheres exist, preferably on medium-sized, rocky planets with stable near-circular orbits with diurnal and seasonal cycles that are not too extreme. It supports the concept of the "stellar habitable zone," which presumes that life is restricted to those planets at just the right distance from their central star for Earth-like conditions to exist. It fits perfectly with the search for life in our own image.

The alternative logic is one that starts from a different premise – that life on Earth is not representative necessarily of life elsewhere, and that therefore the conditions that led to and sustain life on Earth need not be viewed as limitations [1,2]. A nuanced version of this alternative is that, in fact, the life we see around us as humans is not even typical of the vast majority of biomass on Earth; that the unseen microbial world beneath our feet, in our oceans, and to a considerable depth within the crust of our planet is the rule for life on most worlds, including our own, with the larger organisms that catch our eye simply being the exceptions that command our attention.

Peter Ward and Donald Brownlee, a paleontologist and astronomer respectively, at the University of Washington tried to merge these disparate models into a unified hypothesis in their influential book, *Rare Earth: Why Complex Life Is Uncommon in the Universe* [3]. They defined their Rare Earth Hypothesis as, "the paradox that life may be nearly everywhere but complex life almost nowhere." They argued that complex life is likely to arise only under conditions (implicitly, Earth-like conditions) that come about through a very special combination of circumstances. Since that combination on statistical grounds is likely to occur very rarely, complex life itself must be a rarity. Microscopic (by inference, "non-complex") life, on the other hand, is known to be highly adaptable to the most extreme environments. Since extreme environments, by terracentric standards, are common, then microscopic forms of "simple" life may well be common throughout the universe.

In the sections that follow, we review these two major points of view, then offer our own strategy for a systematic study of cosmic biology. But first, since history matters for the way things are today, we need to understand how all the potential habitats for alien life, from rare earths, to snowball planets, to gas giants, came into being. We will therefore start by providing a brief overview of the history of the universe and our Solar System.

## 1.1 How habitats come about

An assumption we make is that life cannot exist in the midst of an actively burning star, because the chemical reactions that make up living processes cannot proceed stably at the temperatures present in them. Therefore, cosmic biology is really planetary biology (meaning planets, planetoids, and their satellites). So we begin by considering where planets come from.

Science has not always assumed that the universe had a finite beginning. The visionary cosmologist, Fred Hoyle, long championed the idea that mass and energy are products of a steady, continuous, and endless process of generation throughout space. But increasingly scientists have come to the mind-bending view that the universe did indeed begin at a singular point in both time and space. The evidence for this is that the universe is expanding now, that the oldest objects in the universe are at its furthest reaches, and that remnants of the violent birth of the universe – what Fred Hoyle facetiously called the "Big Bang" – still lurk in the form of a background of electromagnetic radiation that would be predicted to be left over from such an event.

### 1.1.1 Genesis: A scientific story of creation
No myth or metaphor from any of the world's religions can match the audacity and incredibility of the currently favored scientific explanation for the origin of the universe. Until evidence to the contrary comes along, though, most scientists give grudging credence to the "Big Bang" theory, which envisions the emergence of all the matter and energy of the universe according to the following model [4].

Space and time did not exist before the universe was born. Our universe came into being about 13.7 billion years ago, when something emerged from nothing with such violence and rapidity that matter and energy were indistinguishable.

In the first microsecond of its existence, the universe expanded from a singular point to a volume of space 100 billion kilometers across at a temperature of 10 trillion °C. This explosive expansion of space enabled matter in the form of subatomic quarks to differentiate from pure energy and coalesce into baryons (protons and neutrons).

By the end of the first second of the universe, neutrons started changing into protons by emitting more electrons. Once the ratio of protons to neutrons reached about seven to one, they started combining into combinations of (mostly) two protons and two neutrons, the nuclei for future helium atoms. Nearly all the free neutrons were soaked up by this process within a couple of minutes, leaving a huge excess of free protons to form the nuclei for hydrogen atoms.

It took 200,000 years or more for the universe to cool to 2700 °C, a temperature at which the protons and neutron-proton aggregates could start

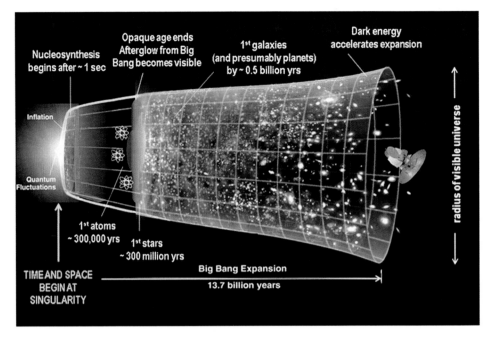

**Figure 1.1** Conceptual scheme for the "Big Bang" formation of the Universe. According to this model, all matter and energy in the universe had a singular point of origin in space and time. With expansion, energy and matter differentiated, and matter in time coalesced into galaxies consisting of stars, planets, and related objects (Modified from art by NASA/WMAP Science Team).

capturing the free flying electrons to form the first atoms of hydrogen and helium. As the dense fog of electrons condensed into newly formed atoms, the opaqueness cleared and photons could stream without interruption across the vastness of space, which now filled a volume of a hundred million light years.

By 300 million years into its life span, the universe was filled almost but not quite uniformly with mostly hydrogen, a little bit of helium, a tiny amount of lithium (the three smallest atoms), a lot of energy, and a whole lot of "dark matter" whose nature is still not understood today. Another fundamental force of nature, gravity, now worked on those discontinuities in the density of the "normal matter" in the universe. Clouds of hydrogen and helium began to collapse in regions where the gases were slightly more concentrated. When the clouds became massive enough, gravity sucked the atoms into a ball that gradually became compressed enough to force atoms of hydrogen to fuse together, forming helium and emitting energy as light and other forms of electromagnetic and particle energy. The first stars were born.

The stars of the early universe were massive by comparison with the average size of stars today. This meant more gravity, which meant a faster fusion at their cores. When all the hydrogen at the center of a star was consumed, the helium atoms started fusing into heavier elements like carbon, nitrogen, and oxygen.

**Figure 1.2** Star formation. Cool, dark clouds of gas and dust are just starting to condense in this region of our Milky Way Galaxy. There is a surprising amount of turmoil: the interstellar material is condensing into continuous and interconnected filaments glowing from the light emitted by new-born stars at various stages of development. The image was acquired by Herschel's spectral and photometric imaging receiver. Herschel is a European Space Agency mission with important participation from NASA (ESA/NASA/JPL-Caltech).

With heavier elements raising the density of their cores, the crushing gravity of the biggest stars caused the collapse of atoms down to their positively charged nuclei, which repulsed one another with such explosive force that the star blew apart. This first generation of stars thus spewed the heavier elements it had made into space, giving rise to metallic particles of dust as well as free-floating lighter elements to join the hydrogen and helium gas already in abundance. These clouds of dust and gas would eventually collapse into the next, more chemically enriched generation of stars and the cycle would repeat itself. In this way, the heavier elements that would give rise in time to all the worlds and all the living things in the universe were created.

### 1.1.2 How solar systems and planets form
The laws of angular momentum dictate that as matter collapses on itself, it must begin to spin. In many if not most cases, as the dust and gas fell into the

compacted gravity well of a protostellar mass, some of the matter would be kept by centrifugal force from falling into the central star. The matter spinning at the periphery would flatten into a disk, resulting in a spherical ball at the center with a smattering of material spinning around the central mass in a flat plane.

Thus compressed into a flattened space, dust particles, water, and other molecules that had formed in the interstellar void would collide with increasing frequency, often sticking together at the frigid temperatures that prevailed some distance from their growing sun at the center of their orbital trajectories. Discontinuities again played a role, with slightly heavier grains growing more massive slightly faster than those around them. The trend would continue for millions of years, leading to boulder-sized aggregates that coalesced into larger boulders, until masses of rock and ice called *planetesimals* had grown to several kilometers in diameter.

Planetesimals are large enough to generate gravitational attraction. The larger ones would pull in the smaller ones in a runaway process of accretion. At this point, the process would accelerate toward survival of a small number of planetesimals evolving toward *protoplanets* in a relatively short time. The larger the protoplanets became, the more fragments from the formative protostellar disk they would pull in.

**Figure 1.3** Exoplanet orbiting a red dwarf. This is an artist's conception of the smallest star known to host a planetary system. The dim red star, a red dwarf about the size of Jupiter named VB 10, is orbited by a planet of approximately the same size, about 20 light years away in the constellation Aquila (NASA/JPL-Caltech).

The newborn planets that were large enough became bombarded with so much matter and energy that their interiors melted, allowing the heaviest elements like iron and nickel to sink to the center. An outer crust of lighter metallic compounds like silicates formed a solid surface that underwent a high level of bombardment for millions of years more. This delivered both matter and energy to the growing planet's surface, while radioactive decay of the heavier elements in its interior generated heat from inside out.

Like the protostellar disk, the planetesimals that had collapsed to form the protoplanets were set spinning themselves, sometimes generating satellites that rotated around their central planet, as the planets rotated around their central star. As planets formed, they scooped up gases like hydrogen and nitrogen. The planet's size and temperature (depending on its distance from the central sun) dictated whether these light gases would evaporate back into space, be held as gases in an atmosphere, or be pulled with enough gravitational pressure to form a liquid shell around the planet's rocky core. The latter became characteristically massive "gas giants."

Our own Solar System was born about 4.6 billion years ago. The protostellar disk that gave rise to our local part of the universe generated a somewhat larger than average sun encircled by four rocky planets (Mercury, Venus, Earth, and Mars), four gas giants (Jupiter, Saturn, Uranus, and Neptune), and an assortment of smaller bodies including at least one binary dwarf planetary system (Pluto and Charon). All but the innermost two of our Solar System's planets have satellites of their own. The two largest giants, Jupiter and Saturn, also have the largest number of satellites, while the smaller inner planets have none (Mercury and Venus) to one (Earth) to two (Mars), perhaps because their parent planets had less starting material to begin with, or because keeping satellites in stable orbit that close to the immense gravitational pull of the Sun is more difficult.

Hundreds of other solar systems are now known to exist, and protostellar clouds can be seen in various stages of formation both near and far, suggesting a universe filled with recurrent and ongoing acts of creation.

## 1.1.3 Exoplanets

The first planets beyond our own Solar System were not discovered until 1992. Prior to that, we had no reason to doubt that the logic of our own system – smaller rocky planets and light atmospheres near the sun, with gas giants hosting dense atmospheres further out – was a general rule for the structure of solar systems everywhere. That assumption was dispelled with the discovery of gas giants larger than Jupiter orbiting their central stars much closer than the Earth orbits our Sun. Planets the size of our own rocky members, being much harder to detect, are just now being discovered, but enough have been found to suggest a far greater diversity of organizational structures for other solar systems, than our own "logical" planetary arrangement, or the rapidly revolving giant planets first found near their central suns, would suggest. This includes, for example, many

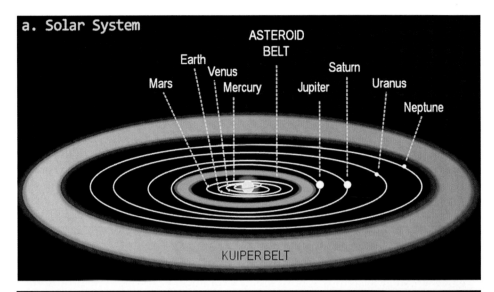

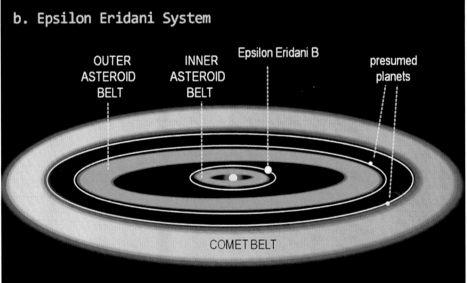

**Figure 1.4** Planetary formation. Schematic representation of two different planetary systems forming around their central stars (sizes and distances not to scale). (a) In our Solar System, four inner rocky planets and four larger gas giant planets orbit the Sun, with an asteroid belt between the inner and outer planets, and another comet forming belt beyond the furthest planet. (b) A smaller, younger star, Epsilon Eridani, a little over 10 light years away is surrounded by an inner asteroid belt with a Jupiter-sized planet at its outer edge. Two more asteroid belts suggest the presence of at least another two planets, orbiting within the regions of uncoalesced planetary debris between the belts (Art by Louis Irwin, adapted from original by NASA/JPL-Caltech).

exoplanets with very eccentric orbits, circling a wide variety of stars from red dwarfs to giants to multiple star systems.

During the writing of this book, nearly 500 planets outside our Solar System (exoplanets) have been detected, and that number will probably double within a year. Recent advances in exoplanet research can be exemplified by discovery of the Gliese 581 planetary system. An M-category star, also known as a red dwarf, Gliese 581 lies 20.3 light years distant from Earth and is orbited by at least four planets, ranging in estimated size from 1.9 to 13 Earth masses. These exoplanets are likely to be rocky planets with thick atmospheres ranging in size from that of Earth to Neptune.

The assumption that many exoplanets must have exomoons of their own is implicit, and some of those satellites must be of an appropriate size and composition to provide conditions suitable for life. Until a fuller repertoire of solar systems has been studied, we can't be sure how common rocky planets, water worlds, barren moons, snowball satellites, and gas giants are in the rest of the universe. We have every reason to believe, though, that all of them exist somewhere. Fortunately, our Solar System has examples of each, so that a science of cosmic biology can well begin by focusing on the planetary habitats nearest our home.

## 1.2 The Rare Earth model

In its starkest form, the Rare Earth model postulates that conditions suitable for life are restricted to the conditions that happen to be present on Earth. Those conditions, in turn, arise from a set of coincidental factors that collectively are highly unlikely to be found very often throughout the universe. Hence, Earth-like planets are a rare occurrence, and therefore life itself is a rare phenomenon. To reiterate, the focus of the book by Ward and Brownlee was on the frequency with which *complex* life is likely to arise, not on the chances for *any* form of life. We describe it in this more restrictive form, however, because the notion has given rise to related concepts like the "stellar habitable zone" and the "galactic habitable zone," which to different degrees assume that rocky planets with Earth-like properties are required for the support of life in general.

To be sure, in our own Solar System, the Earth is unique. It orbits the Sun at a convenient distance to receive plenty of sunlight but not so much that the atmosphere is boiled away. Its size is right to hold an atmosphere thick enough to protect against all but the larger meteorites, but not so dense that heat is trapped inescapably. The uniquely high percentage of oxygen in the air supports an efficient form of energy metabolism, and the ozone arising from oxygen filters out dangerous ultraviolet radiation. Above all, Earth is a planet with an average temperature of about 15°C that allows most of the water on its surface to be liquid most of the time. It has an axis inclined to the plane of its orbit around the Sun that ensures seasonal fluctuations in temperature, air and water currents, and precipitation. More frequent cycles arise from a rotation on its axis and the tidal pull of the Moon and the Sun that generate daily cycles of light and dark, tides that rise and fall, and temperatures that go up and down. The cyclic nature of the biosphere provided by these fluctuations, combined with the complex topography of the land and the global prevalence of interfaces between the water and the land, produce a rich diversity of environments. These then serve as the tapestry through which life can splinter into an ever increasing degree of complexity.

The weakness of the Rare Earth model lies, not in its requirement that environmental diversity is necessary for the evolution of biological complexity – we think that is more likely than not correct – but in the assumption that planets with a high degree of environmental fractionation are rare. They may be. We don't have enough data yet to answer that. Even if they are, and we therefore conclude that the evolution of complex life is rare, that doesn't necessarily mean that the evolution of any kind of life is rare. Therefore, the concept that the habitable zone is restricted to that region of a solar system where rocky planets with liquid water can stably exist is not very useful. If we assume that solid substrates interfacing with liquids are at a minimum necessary for the emergence of any kind of life, the galactic habitable zone, within which rocky planets can exist, makes a little more sense. But even here, the environmental requirements for life in general are not yet specified well enough to rule out many potential worlds that lie beyond the galactic habitable zone.

As a search strategy for life on other worlds, looking *first* within the stellar "habitable zone" makes sense, because life that evolves under conditions familiar to us is more likely to be recognizable by us. The galactic "habitable zone" is an interesting theoretical concept but one lacking in relevance at the level of our current technological capacity to search for alien life.

Life on Earth clearly confirms that multicellular macroorganisms require a narrower set of environmental conditions for survival than microorganisms. The history of life on Earth further shows that multicellularity and large size evolved over long periods of time under fairly stable conditions on a planet with liquid water and continental land masses. Whether that much time and those particular conditions are inherently required for the evolution of biological complexity, is unknown. Furthermore, the rarity with which the geophysical conditions peculiar to Earth are found throughout the universe cannot be assessed at this time. Nonetheless, the overall view that large, more complex organisms are relatively uncommon compared to the much higher prevalence of microorganisms on other worlds is likely correct.

## 1.3. The Life Unseen model

In *its* starkest form, the Life Unseen Model states that the life we can't see is the life that is most common and most typical in the universe. We as humans can't see it with our naked eyes because it's too small. We know it exists, though, because we *can* see it through a microscope, and we can study its biochemistry and metabolism through a variety of elegant techniques.

The unseen world of life on Earth turns out to be far more massive, widespread, variable, and adaptable than those forms of life that garner most of our attention most of the time. Prokaryotes alone, the simplest and most ancient forms of unicellular life, beneath the surface of the soil and the floor of the ocean may approach in biomass that of all plants [5]. When more complex but still microscopic organisms, like protists and yeasts, are added to the larger plants, animals, and fungi that remain sequestered out of sight in the ocean and beneath the ground, their total biomass adds to much more than all the plants and animals we can readily perceive.

It is among the smallest creatures that we find the greatest adaptations to environmental extremes. Especially among the prokaryotes, forms are known that live at temperatures close to the boiling point and below the freezing point of water, that flourish in alkaline lakes and acidic pools at the extreme ends of the pH scale in both directions, that survive crushing pressures, decades of desiccation, the vacuum of space, and levels of radiation that no human could come close to tolerating [6]. These "extremophiles" give little evidence of being bound by a habitable zone of any dimensions. To be sure, under physicochemical conditions that prevent normal chemical interactions from taking place, life has not nor is it likely to be found. Short of that limitation, however, no intrinsic restriction on living processes is evident.

The existence of the extremophiles on Earth has weakened the meaning of the stellar habitable zone in space. Since conditions considered "extreme" on Earth are common on other worlds, the existence of extremophiles in those habitats must be regarded as a distinct possibility, and maybe even a common occurrence. Thus, the whole concept of a "habitable zone" may have little meaning within the Life Unseen framework.

As noted, the extremophiles on Earth tend to be small (though not inevitably microscopic), so they may well be small wherever they exist. This means they will be unseen, at least directly. Detecting biosignatures of unseen life is a major area of astrobiology, with the technology for doing so steadily advancing. There is good reason, therefore, to expect that sequestered forms of life will be detected, to the extent that they exist, as exploration of alien worlds intensifies.

## 1.4. Strategy for the study of cosmic biology

As our own planet demonstrates, the Rare Earth model of complex life and the Life Unseen model for a biosphere of sequestered and mostly microscopic forms of life can co-exist. The relationship is not symmetrical, however. There is almost certainly no complex life without the coexistence of simpler, microscopic forms. Surely more common, however, are worlds harboring biospheres consisting entirely of microscopic ecosystems but totally lacking in larger, more complex organisms.

Assuming these two models frame the boundaries of what is possible, the form and variety of life that appears on any given planet is going to be a function of both history and circumstance. History determines what the starting point for life must have been, and the sequence of conditions through which it had to evolve to reach its present form. Circumstance dictates the nature of life that is currently sustainable. Earth, Venus, and Mars provide an illuminating example.

Earth and Venus are fraternal, if not quite identical, planetary twins. Born at the same time from the same protostellar matter, they were aggregated into rocky planets of almost the same size, spinning around the Sun in adjacent orbits. From what we know about the origin of solar systems in general, we have every reason to believe that both were bathed in warmth from the Sun and bombarded with an abundance of water and organic matter from asteroids and meteorites in their early existence. Yet today, the surface of Venus is a crushing cauldron of desiccated heat while the Earth teems with a massive and diverse biosphere. Given such a similar starting point, how could two worlds go in such different directions for the support of life? History clearly explains some of the difference. Something happened a long time ago to knock Venus upside down on its axis, so that it spins in the opposite direction from the Sun and the Earth. Whatever water and atmosphere existed at that time may well have been totally discarded. More recently, perhaps a billion years ago or less, convulsive volcanic activity resurfaced the planet and generated a crushing atmosphere of carbon dioxide and sulfuric acid that turned the planet into a runaway greenhouse oven. Current circumstance makes matters worse, in that the Sun's increasingly intense output of energy, combined with the closer orbit of Venus to the Sun, make the greenhouse effect even more severe. That nearness is also slowing the rotation of Venus, which will eventually lock the planet into holding its same face toward the Sun all the time.

Meanwhile, life flourishes on Earth as it orbits the Sun at a more comfortable distance, with just enough atmosphere to be protective and keep water liquid on the surface, but not (yet) enough greenhouse gases to trap heat inescapably. Fifty percent further from the Sun orbits Mars, the smaller sibling of Venus and Earth. Considerably colder and with a much thinner atmosphere than Earth's, Mars appears to be as lifeless as Venus on the surface. Yet it too began its voyage through time as a warmer, water world. Its smaller size and greater distance from the Sun apparently led to an insurmountable inability to hold onto its earlier, thicker atmosphere, and to the eventual loss of its large surface oceans. Thus, we see the contrast between three planets, born at the same time within the same region of the Solar System, driven by history and circumstance to three different endpoints in their ability to support life. Surprisingly, perhaps, there may still be some forms of life on Mars, and possibly even Venus, but the life there is clearly going to be different from the life on Earth. The purpose of this book is to explain why that might be, and how alternative life histories could be playing out in other parts of our Solar System and the rest of the universe.

Our strategy for visualizing a cosmic biology will be to consider the starting point of a planet's history and its geophysical transformations to the present day, generating a narrative of planetary history and circumstance that provides the fabric within which the evolutionary life on that world could have unfolded.

Using fundamental principles of evolutionary and ecological theory, we will then postulate a set of ecosystems that plausibly could exist on those worlds today. First, we need to define life in a fairly formal way (to be sure we all have the same thing in mind), then review the basic facts of chemistry and physics that underlie the science of biology. Those are the topics of the chapter to follow.

## 1.5 Chapter summary

Alien life on other worlds has been the subject of speculation since ancient times. The notion that we are not alone in the universe gained credence with our growing understanding that the Earth is not alone among the planets of our Solar System, that solar systems both like and unlike ours are abundant in our galaxy, and that the total number of galaxies is almost uncountably large.

The origin of our universe appears to have been finite, creating the space and time that we inhabit. While details of the prevailing model for the genesis of the cosmos are debatable and beyond the scope of this book, the basic mechanisms for stellar and planetary formation are generally agreed upon, and nicely explain the structure of our Solar System with its four inner rocky planets, four outer "gas giant" planets, and one binary dwarf planetary system. Recent discoveries, however, challenge the notion that the structure of our Solar System is necessarily typical.

The Earth is unique among all the planets we know in that it holds liquid water on its surface. This results from a combination of circumstances related to its mass, distance from the Sun, and Solar System companions. It also appears to have been stable for a long period, giving life ample time in which to evolve and diversify into a great variety of complex macro organisms as well as a prolific biomass of microorganisms. This combination of conditions enabling complex, multicellular forms of life to evolve has been popularized as the Rare Earth Hypothesis of Peter Ward and Donald Brownlee, which they summarize as the "... paradox that life may be nearly everywhere but complex life almost nowhere." While life on Earth clearly confirms that conditions suitable for complex macro organisms are more specific and narrow than the wide variety of environments to which microorganisms can adapt, it isn't clear that Earth-like conditions are the only ones in which complex life can flourish. Nor is it evident yet that Earth-like conditions are that rare. Nonetheless, the Rare Earth model is a useful starting point for exploring the possibilities of life throughout the universe, as long as the prevalence of habitats suitable for a vast array of unseen microbial life is borne in mind.

Venus, Earth, and Mars illustrate the fact that history and circumstance dictate the probability that life will arise, evolve, and diversify on any given world. Born as rocky planets at the same time, circling the same central star, these three planets have arrived at the present time with radically different prospects for harboring life. Using established principles of evolutionary and ecological theory, we will explore in this book how these, as well as the other planets and satellites of our Solar System, and even possible worlds beyond, could harbor life, and what that life conceivably could be like.

## 1.6 References and further reading

1   Grinspoon, D. H. 2003. Lonely Planets: The Natural Philosophy of Alien Life. New York: HarperCollins.
2   Schulze-Makuch D. and Irwin L. N. 2006. Exotic forms of life in the universe. *Naturwissenschaften* **93**:155–72.
3   Ward P. D. and Brownlee D. 2000. *Rare Earth: Why Complex Life Is Uncommon in the Universe*. Springer-Verlag, New York.
4   Dinwiddie, R. 2008.The beginning and end of the universe. In: P. Frances (ed). *Universe*. New York: DK Publishing; pp. 44–55.
5   Whitman W. B., Coleman D. C. and Wiebe W. J. 1998. Prokaryotes: The unseen majority. *Proc. Natl. Acad. Sci. U.S.A.* **95**: 6578–6583.
6   Rothschild L. J. and Mancinelli R. L. 2001. Life in extreme environments. *Nature* **409**: 1092–1101.

www.cosmology.berkeley.edu
A general resource on the origin of the universe.

http://www.umich.edu/~gs265/bigbang.htm
A clear and accessible discussion of the Big Bang theory.

http://exoplanet.eu/catalog-all.php
An up to date catalogue of exoplanets.

www.astrobiology.com/adastra/extremophiles.html
A useful overview of the relevance of extremophiles to astrobiology.

# 2

# Life, Chemistry, Action!

*The definition, composition, and activation of living systems*

Born from clumps of stardust spinning around a central sun, the planets of our Solar System, like most planets in the universe, began their voyage through time with a rich mixture of chemistry and energy on board. From such a cauldron, life emerged on our planet Earth early, and diversified into the great variety of living organisms that cover the surface, fill the air, and saturate the soil and oceans of our globe. Organic evolution (or "evolution" for simplicity hereafter) is the term for the historical process by which this diversification occurred. This book is about how that process could have happened on a number of different worlds both in our Solar System and beyond. In this chapter we will define life, to make sure we all have the same thing in mind, then lay out the basic requirements for life in terms of chemistry, energy, and function.

## 2.1 The challenge of defining "life"

While we all have an intuitive understanding of what we mean by "life," the term is not really that easy to define in ways that distinguish truly living organisms from other phenomena in the natural and fabricated world that share a lot of life's characteristics. For starters, the form that life has assumed on Earth is so diverse that no description of size, shape, or appearance is generic enough to cover examples of every living thing. We are left, instead, with a set of characteristics, like growth, ingestion, respiration, excretion, autonomous action, and multiplication. The problem is that for every characteristic we can think of, some example pops into mind that has *that* property, yet clearly is not alive. Crystals and volcanoes grow. Automobiles ingest gasoline and excrete water and carbon dioxide. Forest fires and hurricanes transform energy into action that appears to be autonomous and self-perpetuating until their life spans are exhausted. Clouds can multiply and our publisher has made a number of exact copies of this book. Yet no one will argue that clouds, crystals, cars, or the office copy machine are alive.

We might be tempted to define life the way Potter Stewart, a justice of the Supreme Court of the United States, described pornography in 1964, by simply stating that "... I know it when I see it." The problem is that even on Earth we don't always recognize life when we see it (Can we tell whether a leafless tree in winter is dead or alive?) On other worlds, where habitats and planetary history

L.N. Irwin and D. Schulze-Makuch, *Cosmic Biology: How Life Could Evolve on Other Worlds*,
Springer Praxis Books, DOI 10.1007/978-1-4419-1647-1_2,
© Springer Science+Business Media, LLC 2011

may be very different from those of Earth, the form that life would take and the functions that it would readily reveal, may be totally alien to our own experience. So we do need a formal definition that captures the essential features of living organisms, while distinguishing them from things with analogous features that are not alive in the biological sense.

### 2.1.1 Life as a duality of process and entity

To begin with, life is more about process than it is about substance. To be sure, most of the complex contents of living organisms are unique to the living world, but ultimately they derive from the same atoms and molecules that make up the non-living universe. For a period of time after death, the composition of a once-living organism is little different from when it was alive, yet the organism no longer has the attribute of "life." Just as wind is more than air, life is more than biochemistry.

Life is a lot about complexity [4]. The simplest bacterium is an incredibly complicated metabolic processor, warehouse, informational archive, and agent of action. Imagine how much more complex an organism like a redwood tree or your human body is. Of course, a lot of non-living things are complex – skyscrapers, computers, and cars to name a few. But the complexity of living things exceeds those examples many fold in two respects. First, the complexity is miniaturized: per unit volume, a living cell is far more complicated than a computer circuit board. Second, the organization is self-maintaining. If the car breaks down, or a window in the building is broken, it takes human effort (an external agent) to repair the damage, whereas a living cell manages to maintain its unique and exquisite complexity on its own.

Our most intuitive notion about life is its capacity for intrinsic action. A dead leaf blows about in the wind, but cannot move itself. When it was a living part of a tree, it grew, rotated to face the sunlight, opened its pores to let in carbon dioxide and let water out, changed colors in the fall, received nutrients from other parts of the tree, and manufactured food to keep the whole enterprise fit and growing. Even the simplest cells respond to their environments at least in simple ways, approaching the favorable and shrinking from the dangerous. Our intuitive appreciation of the importance of intrinsic action can lead us astray, however, if not further qualified. A car can move on its own, given the command to do so. A computer screen can display a lengthy sequence of complicated motion pictures and sounds if programmed appropriately. Yet no one would claim that either the car or computer is alive. Therefore, our definition of life is going to have to be able to distinguish between the tree, the car, and the computer.

Finally, life is, above all, capable of self-reproduction. Note that the "self" part of the term is critical. Living organisms, by themselves, can assemble from the raw nutrients they take in from the environment, a near-exact replica of themselves. Again, we need to make a distinction between similar phenomena that occur both in nature and in the fabricated world. Crystals placed in an appropriate mineral solution will grow by replicating their crystalline structure.

A cloud can split into two clouds, just as cells multiply by mitosis. A photocopier can replicate exactly the information on this page. And a particularly eerie analogy to biological replication is the spread of computer viruses – the very name of which connotes its close resemblance to a true biological process. Yet all of these examples can be distinguished from the type of reproduction by which living organisms perpetuate themselves. The cloud and the crystal are unbounded and capable of indefinite expansion. Documents can be replicated exactly when placed in the machine by an outside agent who has punched the proper buttons, but they can't spontaneously copy themselves. And the computer virus is just a programming code without a material existence, so the only thing that multiplies is information about a virtual reality, and that happens only when an outside operator turns on the computer and types on the keyboard. Speaking of reality, while focusing on the dynamic *processes* of life, we can't lose sight of the fact that life consists of tangible entities that exist as individuals in time and space. A living organism is indeed a *thing*, albeit a thing with dynamic properties. Even when a living being has ceased to be alive, its physical remains give undeniable evidence of something that once was a living organism, as in the case of fossils. At the other extreme, before it springs to life, a spore or seed is not very dynamic, but possesses the potential to develop into a discrete living organism. So a fossil or a spore, while not being technically alive in the moment, give material evidence of life that once was or will be.

To be alive, then, is to have a dual nature. A living organism is both a thing and a process. Defining "life" is therefore a semantic problem, in the sense that a noun has to be defined by other nouns. At the same time, "being alive" is a state that can be defined only by an appropriate set of verbs. Since this book is about how discrete types of organisms could evolve on other worlds, we can't dodge the fact that we will be imagining the existence of actual "beings." Those beings, however, must have the special properties that characterize the living state.

## 2.1.2 Defining a living organism

The cell is the minimal unit of life that has all the properties of being alive. Much if not most of the biomass on Earth is unicellular; and, as we shall see, on many if not most other worlds, the majority of living things may well be at least microscopic, if not unicellular. We use the term "organism" in this book to include both unicellular and multicellular forms of life. All the organisms that exist on Earth, and those that will compose our speculative populations on other worlds, have three features that define them as living entities [5]:

   ***First, they consist of discreet beings enclosed by physical boundaries that separate their highly organized interiors from the surrounding environment.*** The boundary helps keep the complex interior of the organism from equilibrating, or spreading out randomly into its disorganized surroundings. A hallmark of living things is that they are far more complex than their surroundings (Box 2.1). The boundary also preserves the constant form and shape of the living entity, distinguishing it from a cloud or crystal or ocean wave. In our search for life on other worlds, then, we will be envisioning real entities with regular forms and clear boundaries that separate them from their surroundings.

Because the boundary of a living entity is a physical barrier, it also acts as a gatekeeper for what gets into and out of the interior. Pores of discreet sizes, and transporter molecules with affinities for specific compounds like necessary amino acids and sugars, ensure that the contents of the interior differ from the exterior, both in composition and concentration. That further ensures a distinction between the living interior and the non-living exterior of the organism. Also, receptors at the boundary surface detect specific molecules from the environment that are meaningful to the organism, from food to toxins to the odors of mates and predators. This gives the surface a biological uniqueness that helps determine the actions and reactions typical of that particular species.

---

### BOX 2.1. Entropy and Organization

A hallmark of living organisms is their high degree of organization [4]. Physicists use the term "entropy" to formally describe organization, though the term actually denotes a measure of *disorder*, or the opposite of organization. A collection of carbon and hydrogen atoms scattered in a random fashion, as they would spontaneously be arranged if poured out of a box onto a table, is totally disorganized, hence at the highest level of entropy (Figure 2.1a).

Arranging the atoms in a precise order produces the more highly organized structure of a hydrocarbon molecule (Figure 2.1b). Getting each atom into the right position requires both an input of energy and information about where each atom goes. The molecule has less entropy than the random collection of atoms, because its degree of *dis*order is less.

Living organisms are highly ordered states (Figure 2.1c), requiring the precise arrangement and non-random distribution of a huge number of molecules. A tremendous amount of information and energy are needed to assemble and maintain such highly ordered, low entropy states of matter.

Organization and information are directly related. The larger and more complex a structure is, the more information is needed to specify the precise organization of the structure. It follows that, once assembled, a large complex structure "contains" more information than a small simple one.

The Second Law of Thermodynamics states that, left to themselves, ordered states will proceed spontaneously toward a state of disorder, i.e. entropy will increase. The precisely packed molecules of a sugar cube dissolve into a random distribution in our cup of coffee, but sugar in solution never assembles spontaneously into a sugar cube. A consequence of the Second Law is that two systems at different levels of organization will spontaneously equilibrate when brought together into a single system. They will reach an equilibrium in which the two original states are indistinguishable. The sugar cube in contact with the coffee loses its high level of organization, as the distribution of sugar molecules equilibrates throughout the volume of the coffee cup.

Another consequence of the Second Law is that the order of a system can be increased (its entropy can be decreased) by putting energy into it. Atoms can be bonded together into molecules, and molecules can be assembled into cells and organisms, provided energy and information are made available to do so. Dead organisms decay because energy is no longer being mobilized to maintain their high degree of organization.

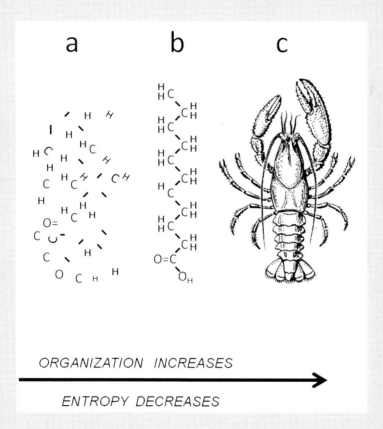

**Figure 2.1** Entropy and organization from the atomic to organismic level. (a) Randomly scattered atoms have no organization and a high degree of entropy. (b) Molecules are more highly ordered, with atoms arranged in a specific location, requiring both information and energy to place them in the correct positions. (c) A living organism is very highly organized, with billions of atoms arranged in a highly specific structure, requiring a vast amount of information and sustained energy input.

 ***Second, living organisms consume energy from their environments to maintain their high level of internal organization and carry out intrinsic activity.*** Staying organized takes work. Maintaining the highly ordered state of a living system, from replacing degraded molecules, to repairing damaged structures, to keeping ion concentrations on different sides of a membrane at disequilibrium, requires a constant transfer of energy (the meaning of "work," for our purposes). Likewise, any action, from growth to movement to emitting signals requires the expenditure of energy. A living entity, then, is an energy consumer and emitter, of necessity and by definition. To be sure, cars, hurricanes, and fires transform energy and act on their environments. Unlike the reaction inside the car's engine, though, which is fairly straightforward and explosive, or the fury of a hurricane or forest fire, which is an uncontrolled degradation of huge amounts of potential energy, the reactions inside a living cell involve many sequential steps and intermediate chemical compounds, resulting in the very gradual and controlled release of energy. Collectively, we refer to the complicated interplay of all of these chemical reactions as *metabolism.*

The metabolic transformation of energy, unlike that in the non-living examples above, is highly compartmentalized, localized, and miniaturized. The metabolic reactions inside a bacterium, which is way too small to be seen by the naked eye, are far more intricate and diverse than the large-scale explosive events inside an internal combustion engine, or the simple if expansive violence of a hurricane, or the vigor but chemical simplicity of light and heat generated by fire. In addition, the metabolism of living organisms is self-regulating. If more energy is being produced than the organism needs at a given point in time, feedback mechanisms slow down its production. If the ionic content of an intracellular compartment gets out of balance, more ions can be pumped in or out until proper balance is restored. And all of these feedback and regulatory mechanisms are *intrinsic capabilities* of the living system itself. The car is *not* self-regulating, because it requires an external agent (the driver) to control its speed and direction; nor are the hurricane and forest fire in any way self-controlling.

 ***Third, living organisms reproduce themselves by a process of nearly exact replication that is totally autonomous.*** The first critical point about this part of the definition is that living offspring are not just approximate replicates of their progenitors, nor do they merely resemble them in a range of similar shapes and sizes. Rather, the products of biological reproduction are virtually identical in general form and closely matched in size (at maturity) to the organisms that give rise to them. In this regard, they differ from the growth of mineral crystals, which add to their mass indefinitely, or from tornados, which spawn new ones but not necessarily of the same size and scope. The second critical point is that a living entity is able to assemble a replica of itself from the raw materials or nutrients it consumes, using only the form of energy available to it. For life on Earth, this process is internally directed by an information store embedded in the DNA of every living cell that itself is replicated and passed to each successive offspring. We have no way of knowing whether DNA or anything like it plays a similar role in living organisms on other worlds, but it seems self-evident that some form of information storage and readout is required for the near-exact replication that characterizes biological reproduction.

The need for reproduction to be *autonomous* to qualify as a living process is obvious, since non-living replicas produced by human hands and machines are an obvious fact of everyday life. Cars are replicated on assembly lines, a new house is built from the same blueprint as an older one, and words and pictures on paper can be photocopied without end. All these cases involve the reconstruction of a preexisting entity or the transmission of information, based on instructions for a precise sequence of activities that will give rise to products indistinguishable from their sister products. But in each case, outside agents and external information are required to do the constructing or copying. If and when a machine that meets the other requirements of a living entity (i.e. a robot) can create a copy of itself *and* imbue that copy with the ability to make a copy of *itself* in turn, then the robot arguably will have to be considered to be alive.

## 2.2 Matter gone wild: the special chemistry of life

In their evocative attempt to capture the essence of life, Lynn Margulis and Dorian Sagan [1] described it as "... planetary exuberance ... a whirling nexus of growing, fusing, and dying ... matter gone wild..." By comparison with the relative simplicity of sand and stones and air, life can certainly be regarded as exuberant. And it undeniably consists of matter in a very special state – not *wild*, exactly (since matter without order cannot be alive) – but matter that is self-actuating, self-organizing, and self-perpetuating.

For all its remarkable capabilities, a living organism on Earth is composed of a very short list of elements (Figure 2.2). Carbon, oxygen, nitrogen, and hydrogen account for over 96% of its biomass. Expanding the list to calcium, phosphorous, sodium, potassium, chlorine, and silicon would provide all that an inventive chemist would need, save for a few trace elements, to construct a living creature. Yet no one has done so. In an equally evocative testimony to the mystery of life, Loren Eiseley [2] pointed out,

> *The ingredients are known; they are to be had on any drug-store shelf. You can take them yourself and pour them and wait hopefully for the resulting slime to crawl. It will not ... Carbon, nitrogen, hydrogen and oxygen you have mixed, and the same dead chemicals they remain.*

For this reason, humans have assumed since their earliest understanding of the physical world that *mere* matter and energy cannot explain the living state – there must be an immaterial component that provides a vital essence that brings the chemicals to life. Among scientists, that notion began to lose power in the 19th century, as different functions of living organisms, bit by bit, began to be explained as a special case of chemistry – behaving, to be sure, in an intricate and complicated manner, but within the rules for normal interactions among ions and atoms confined to aqueous solutions in non-living test tubes. The idea of life as matter imbued with an immaterial vital force gradually gave way to the alternative notion of life as a very special case of known (or at least knowable), confined, chemical systems.

What, then, is special about the chemistry of life? Recall, first, that no mention of chemistry is made in our definition of living organisms. This was deliberate, in part to avoid any bias that our knowledge of life on Earth would impose on our consideration of living systems on other worlds. And that bias would be severe, because life on our home planet is remarkably uniform at its core. From bacterium to baboon, from hyacinth to human, the essence of life at the compositional level can be described as "carbon-based polymeric and ionic chemistry in solution."

### 2.2.1 The elemental composition of living things

The four most abundant components of living systems on Earth – hydrogen, carbon, nitrogen, and oxygen – are among the eight smallest atoms. An additional group of larger atoms, along with magnesium and calcium, make up the vast majority of the mass of living organisms that we know of (Figure 2.2).

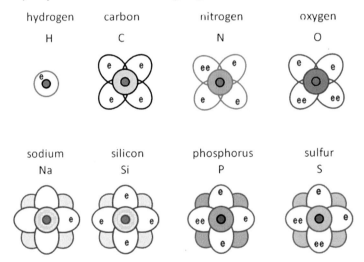

**Figure 2.2** Schematic representation of the major chemical components of living systems. Each atom consists of a nucleus (innermost sphere) of protons and neutrons with a cloud of surrounding electrons. The negatively-charged electrons occupy quantum spaces, or orbitals, around the positively-charged nucleus in complex geometric configurations shown here schematically to approximate the space they occupy. The first electron orbital (shown as a concentric circle around the nucleus) holds one or two electrons. The second orbital consists of four suborbitals (shown as diagonal ellipses), each of which can hold two electrons, for a total of eight. The third orbital is like the second, holding up to eight paired electrons in four suborbitals (shown as a cross of vertical and horizontal ellipses). Outer electrons are the most important for forming chemical bonds, and are shown as individual particles. Inner orbitals filled to their maximum capacity of electrons are shaded. The chemical symbol for each element is shown beneath its name.

The four most common constituents of life also comprise four of the five most abundant elements in the universe (helium excepted).

Atoms consist of positively charged protons and neutrons packed into a central nucleus, and negatively charged electrons equal to the number of protons, held at a distance from the nucleus. Electrons distribute themselves into distinct compartments, or energy levels, called orbitals.

The first orbital of electrons holds a maximum of two, which are symmetrically arranged around the nucleus. The simplest atom, hydrogen, has only a single electron. For atoms with at least three electrons, a second orbital tier is needed, and its geometry is more complicated. It consists of four suborbitals, each of which can hold two electrons. However, electrons are distributed as far apart as possible, so when the total number of electrons in the second tier is less than eight, the unpaired single electrons spread out into different suborbitals. The first two of carbon's six electrons lie in the first tier, leaving four to distribute themselves as far apart as possible in each of the four suborbitals of the second tier. These four outermost electrons account for carbon's unique chemistry.

### BOX 2.2. Chemical Bonds

Chemical bonds are formed when suborbitals of the outermost tier of electrons of adjacent atoms overlap in order to form a shared electron pair (Figure 2.3). Carbon's four single electrons in each of its four maximally-spaced suborbitals give carbon the ability to form bonds at four maximally and symmetrically spaced angles. Thus, the simplest organic molecule is methane, formed from a single carbon atom with four hydrogen atoms each donating an electron to complete the electron pair in each of carbon's four outermost suborbitals. Oxygen, with only two unpaired electrons, binds two hydrogen atoms to fill out electron pairs in all of its outer suborbitals, to form water.

When two atoms share electrons without letting go, like two kids pulling on the same toy, the bonds are said to be *covalent*. Often, one kid is stronger than the other, and pulls the toy closer to himself, but the toy is still shared, if unequally, as long as the weaker child holds on. In the water molecule, oxygen holds on to the electrons more strongly than hydrogen. But some atoms, like chlorine, are strong enough to pull electrons completely away from atoms that hold them very weakly, like sodium, leaving sodium with its lost electron positively charged, and chlorine with its extra electron negatively charged. Atoms with more protons than electrons are positive ions, while those with more electrons than protons are negative ions. Chemical bonds formed by attraction between ions of opposite charge are said to be *ionic*

The electron affinity of an atom, or strength with which it holds on to its electrons, is measured by that atom's *electronegativity.* Elements like sodium with one outer electron and calcium with two hold their electrons much more loosely (have weaker electronegativity) than elements that are closer to filling all 8 spaces in their outermost tier. Thus, oxygen with 6 and chlorine with 7 outer electrons have a strong tendency to pick up the last electron or two that will complete their quota of 8, giving them a stronger electronegativity. Carbon has 4 of its 8 outer orbital spaces filled, so holds on to its electrons with an intermediate degree of strength. This means that it can form bonds fairly easily, by sharing its electrons part way, but those bonds are also fairly stable because carbon won't give up its electrons entirely.

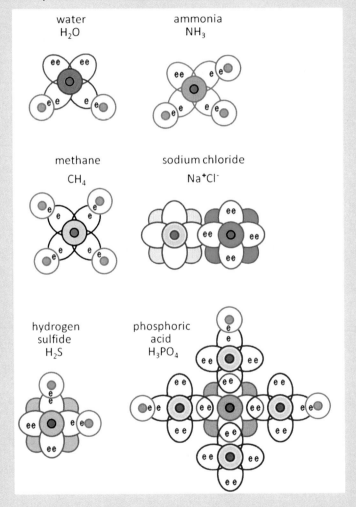

**Figure 2.3** Schematic representation of molecules formed by the overlap of the outermost electrons from contiguous atoms. Stable molecules are formed when every atom of the molecule has completely filled its quota of electrons (two for the first and only tier of electrons in hydrogen; eight for the second or third tiers of the other atoms).

### 2.2.2 Biomolecules

Molecules are combinations of atoms, formed when two or more atoms are joined together by a chemical bond (Box 2.2). Biomolecules are those molecules found in living organisms – often, but not always unique to living systems. Since living systems are highly organized and complex, containing a lot of information, the molecules that make them up tend to be large and complicated. Before discussing the different classes of biomolecules, we need to consider one of their most critical characteristics – their polarity (Figure 2.4).

Carbon and hydrogen have very similar electronegativities, so they share electrons when bonded together almost equally. Since the electrons in such a molecule are distributed symmetrically around the whole molecule, no part of it becomes more or less negatively charged than any other. This type of molecule, therefore, is *nonpolar.* Methane is a good example of a nonpolar molecule because four evenly spaced H nuclei distributed symmetrically around a central C nucleus all share their electrons equally, leaving no part of the molecule more negatively or positively charged than any other part.

Water, by contrast, is a *polar* molecule, not only because the highly electronegative O hogs the electrons it gets from the two weakly electronegative H atoms, but also because the suborbitals on the opposite side of the O nucleus are already filled, leaving no opportunity for positively charged protons from H to neutralize the negative electron cloud on that side of O. Water is thus a lopsided molecule, with two positive bulges on the front and a big negative bulge on the back.

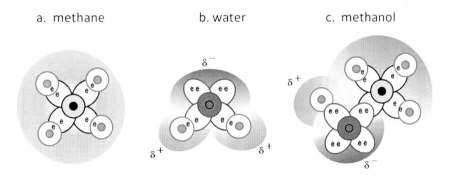

a. methane    b. water    c. methanol

**Figure 2.4** Polarity. Methane is non-polar because carbon and hydrogen form covalent bonds, holding on to their respective electrons with almost equal strength, and because the electrons are distributed symmetrically around the molecule. All the bonds in water and methanol are covalent as well, but O pulls electrons from C and H, and has electrons in filled suborbitals on one side that are not countered by neutralizing protons from H. This gives one part of the molecule a more negative charge (red), leaving the other part of the molecule more positive (blue). More complicated molecules such as methanol have a combination of polar and non-polar regions. The delta ($\delta$) before the + or − means that the charges are only partial, rather than fully positive or negative.

Carbon can also form polar molecules, when bound to elements with a high electronegativity like oxygen or nitrogen, which suck carbon's electrons closer to themselves. Methanol is an example of this (Figure 2.4), because the O draws electrons from both H and C, making its end of the molecule negative relative to the H attached to it, while electrons shared equally between C and H leave the $CH_3$ end of the molecule relatively nonpolar.

Carbon is the third most abundant element in the universe, and the fourth most abundant element on Earth. In addition to its availability, the following properties make it a good building block for living systems:

1. Carbon forms chemical bonds with up to four other atoms at a time, giving rise to a large number of different molecular possibilities. Carbon's partners might all be the same, as in the simplest organic molecule, methane, with four hydrogens bound to one carbon (Figure 2.3). Or, carbon might form bonds with four different combinations of atoms for a more complicated structure like the amino acid alanine (Figure 2.5b).

2. Carbon shares its electrons with other atoms more or less equally, forming bonds said to be *covalent*. Carbon's moderate degree of *electronegativity* means that it can form bonds readily with atoms that are either more or less electronegative than itself (see Box 2.2). This means that it can react with other compounds fairly easily, by releasing its electrons part way or pulling the other atom's electrons closer to itself, but those bonds are also fairly stable because carbon won't give up its electrons entirely nor take over completely the electrons from another atom. That makes carbon the centerpiece of a vast array of organic molecules that are stable enough to remain intact for long periods, but flexible enough to undergo transformations when necessary.

3. Another of carbon's great advantages again relates to its modest degree of electronegativity, making it able to both donate and accept electrons – what chemists refer to as oxidation and reduction, respectively – with relative ease. This makes carbon a great medium for the exchange of small packets of energy. More about that in the next section, though.

4. Carbon-based compounds are well suited for interacting with water, the most readily available liquid on Earth, and perhaps throughout the universe. Water is a polar molecule, because oxygen hogs the electrons it gets from the two hydrogen atoms, making the oxygen end of water more negative than the side left with the positively-charge protons from hydrogen (Figure 2.4). When a polar molecule like ethanol is added to a polar solvent like water, the relatively positive and negative parts of both molecules attract one another, and the ethanol dissolves readily in the water. Polar molecules like ethanol are said to be *hydrophilic* ("water loving") for that reason. Biomolecules like carbohydrates, amino acids, and nucleic acids are readily soluble and therefore easy to disperse in water. But carbon also forms nonpolar molecules when bound to itself or hydrogen, in that no part of the molecule gets more negative or positive than any other.

When a nonpolar molecule like the long hydrocarbon chains in vegetable oil are added to water, the nonpolar carbon chains repel the polar water molecules but fit comfortably next to one another. The oil forms droplets that do *not* dissolve in water, and are therefore said to be *hydrophobic* ("water fearing"). This is a useful property for components of living systems that form barriers within aqueous solutions (Figure 2.8). In summary, the many carbon compounds that are hydrophilic can dissolve readily in water and engage in all the intricate dances of life, while those carbon compounds that are hydrophobic form solid barriers that resist the passage of water and other molecules through them, providing the enclosures and boundaries required by our definition of living systems.

At this point, we need to consider whether silicon, carbon's larger sibling in the same column of the periodic table, has the power and versatility to support a different biochemistry of its own. To be sure, silicon-based biology is a staple of science fiction, and the prevalence of silicon in living organisms such as plants and marine diatoms gives some credibility to the possibility of silicon-based life. Imbued like carbon with four electrons in orbital spaces for eight, silicon shares a lot of chemical characteristics with carbon. The short answer, though, is that silicon could provide a suitable backbone and intermediate element for biomolecules only under vastly different conditions from those on Earth. Such conditions, of course, might be out there, so we will briefly consider the possibilities.

Silanes are molecules of silicone and hydrogen. The simplest is silane, $SiH_4$, the analog of methane, $CH_4$, in the carbon-based world. Silanes in principle could form long –Si-Si-Si – chains just like polymers of carbon, but the Si-H and Si-Si bonds are weaker than the comparable bonds with carbon, so silicon-based polymers are less stable than those based on carbon. In fact, silanes break apart into $SiH_4$ gas molecules at such a low temperature, that any habitat where polymeric silanes could remain stable would have to be far below the freezing point of water.

Silicates are molecules based on units of $Si-O_4^{-4}$. The simplest example that is stable at Earth-like temperatures is silicon tetrahydroxide, $Si(OH)_4$, the fundamental building block of rocks and many minerals. Its actual structure is usually that of an extended polymeric crystal that forms rock-solid minerals at temperatures up to that of the inside of a molten volcano. This is because the Si-O bond is much stronger than the C-O bond, and therefore requires very high temperatures to make possible any type of interactive chemistry. In an oxygen-rich environment, silicon would soak up oxygen much quicker than carbon, rapidly converting silicon to its inert mineral form.

Silicones are molecular hybrids of silicon, oxygen, carbon, and hydrogen. As you would expect from the repeating–O-Si-O– units that form the silicone backbone, these molecules are also very stable at the lower temperatures that most compounds require to stay in the liquid form. Organosilicones have been synthesized in the lab to a large size and high degree of complexity, but their

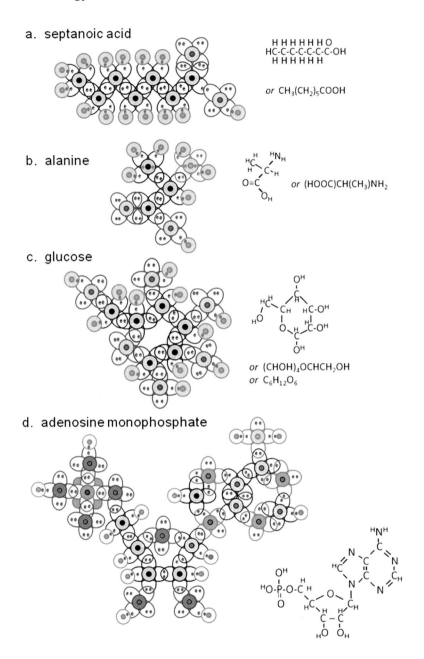

**Figure 2.5** Major categories of biomolecules. (a) A major component of lipids are fatty acids, such as septanoic acid. (b) Alanine is one of about 20 amino acids that make up proteins. (c) Glucose is the major energy fuel for most living organisms on Earth. (d) Nucleotides such as AMP are important for their role in energy metabolism and also as components of hereditary macromolecules like DNA and RNA.

occurrence in the natural world is much more limited in variety, and restricted to stable structures rather than interactive metabolism.

In summary, a world in which silicon would provide a better backbone for life than carbon would be either extremely cold, with no oxygen, and a stable liquid other than water (for silanes), or a world so warm that the only liquid would likely be molten minerals (for silicates). In either case, the chance for complicated chemical reactions comparable to carbon-based metabolism would appear to be much less likely because of the temperature extremes. Finally, the empirical evidence argues against silicon-based life. Though silicon is much more abundant than carbon on most of the rocky planets, including our own, no forms of life based exclusively on silicon have been found. While many complicated organic compounds, including amino acids (protein building blocks) have been found in meteorites brought to Earth from outer space, they have not been found to contain any complex polymers of silicon. On the other hand, silicon *is* found as a trace element in many organisms in which they provide structural rigidity, like the stems of plants, the bones of animals, and the skeletons of diatoms and echinoderms like sand dollars. These examples hint at an earlier, more prominent role that silicon may have played at the dawn of life, as we shall discuss later.

### 2.2.3 Macromolecules
Macromolecules are biomolecules assembled into huge aggregates. Their role in living processes depends on their unique composition, which translates into either a unique 3-dimensional structure (proteins) or a specific linear arrangement of variable subunits (nucleic acids).

Amino acids are the monomers ("single units") for the polymer ("many units") protein (Figure 2.6), and nucleotides are the monomers for the nucleic acid polymers, DNA and RNA (Figure 2.7). Since the monomers come in different varieties (20 amino acids and 4–5 nucleotides), and since the monomers can theoretically be arranged in any order, the number of unique configurations possible for any macromolecule is vast. The larger the molecule, the greater the number of possible shapes (what chemists call "configurations"). As we've already discussed, size equates roughly with information, so the molecules of life are largely macromolecules, in order to carry the great amounts of information on which living systems depend.

About 20 different amino acids make up the proteins known to be present in living organisms on Earth. Each differs in the chemical makeup of the variable side group (*R*) attached to the fourth bonding site on the central carbon atom. The peptide shown in Figure 2.6b consists of the amino acids alanine ($R = CH_3$), glycine ($R = H$), serine ($R = CH_2OH$) and phenylalanine ($R = CH_2C_6H_6$). Because the side groups differ in size and chemical properties, their sequential arrangement causes the polypeptide chain to assume a unique spatial arrangement. This, in turn, gives rise to a specific 3-dimensional structure capable of recognizing distinct configurations of *ligands*, or molecules that attach to them, such as the specific substrates for a particular chemical reaction.

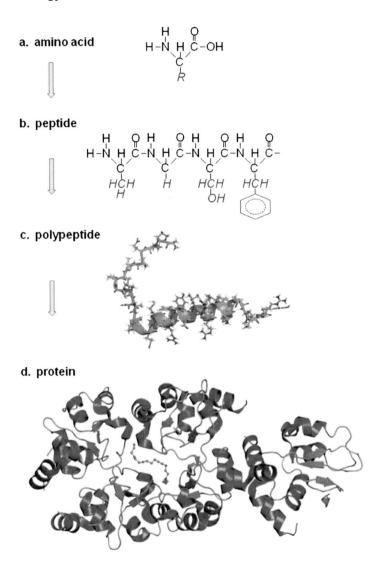

**Figure 2.6** Proteins. Amino Acids are the monomeric units that make up proteins. (a) All amino acids are configured in the same way, with a central carbon atom bonded to an amine ($NH_2$), a carboxylic acid (COOH), a hydrogen (H), and a variable group (R) which gives each amino acid a unique shape. (b) Peptides are formed by linking the amine N of one amino acid and the carboxylic C of another. The resulting peptide bonds link adjacent amino acids together. (c) Polypeptides consist of strands of amino acids linked together in a chain. (d) Proteins are very large polypeptides. The unique composition and sequence of their amino acids gives each protein a unique structure, enabling it to interact only with specific configurations of other molecules, such as the enzyme pictured above with a pocket that fits the particular substrate (small linear molecule in green) (Protein drawing courtesy European Bioinformatics Institute, www.ebi.ac.uk).

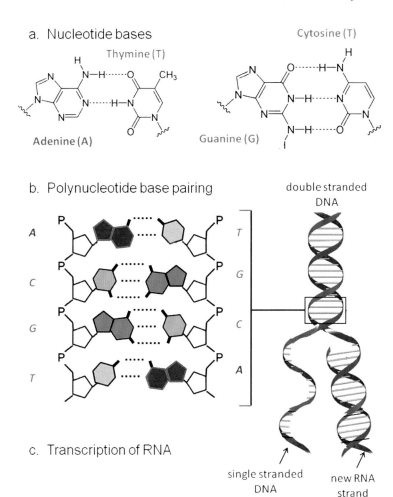

a. Nucleotide bases

Cytosine (T)

Thymine (T)

Adenine (A)

Guanine (G)

b. Polynucleotide base pairing

double stranded
DNA

A          T

C          G

G          C

T          A

c. Transcription of RNA

single stranded
DNA

new RNA
strand

**Figure 2.7** Nucleic acids. The monomeric units for nucleic acids are nucleotides, which consist of a 5-carbon sugar, a phosphate group, and one of several N-containing ring structures called bases attached to the sugar. (a) In DNA the sugar is deoxyribose and the bases are either adenine [A], cytosine [C], guanine [G], or thymine [T], shown for simplicity without the ribose to which they are attached. Weak bonds can form between A and T, and between G and C. (b) Nucleotides are strung together by linking the phosphate group from one nucleotide to the sugar group of the adjacent one, generating a long chain with the bases sticking out to the side. When held together according to the base pairing rules [A:T or G:C], two complementary chains can form a long double-stranded sequence of nucleotides held together like the two sides of a zipper. (c) RNA is normally a single-stranded molecule (containing ribose instead of deoxyribose, and uracil [U] instead of thymine). Its base-pairing rules are similar to those for DNA [A:U and G:C], so that a particular sequence of bases on one strand of DNA can induce the formation of a complementary sequence of bases in a single-strand of RNA, as shown in the lower right.

Nucleic acids are composed of nucleotides strung together in long linear sequences to make up either deoxyribonucleic acid (DNA) or ribonucleic acid (RNA). The main feature of both molecules is that each provides a long sequence of four different bases, theoretically arranged in any order, so that the total number $N$ of different possible combinations is huge: $N = 4^n$, where n is the total number of nucleotides in the chain. Since a DNA molecule can be hundreds of thousands of nucleotides long, the amount of information carried by DNA is vast enough to carry the genetic instructions for the entire developmental program and functional repertoire of a living organism.

A second feature of great importance for nucleic acids is that the base pairing rules, which dictate which of the four bases can hook up with any given base, provide a mechanism for copying information, since the base sequence of one strand can serve as the template for another.

## 2.3 The advantages of liquids for life

The mantra of the search for life on other worlds by all the space-faring nations at the present time is "follow the water." This is because all life as we know it is based on polymeric and ionic chemistry that takes place in aqueous solutions. We don't know that the liquid has to be water. Though water is abundant in the universe and has some compelling advantages that we'll discuss below, we will also consider other possibilities in later chapters. We begin with a general argument for the overwhelming advantages that liquids provide over solids or gasses as the preferable state of matter for hosting the dynamic properties of living systems.

### 2.3.1 General properties of liquids

The general properties of liquids that make them most suitable for living systems include the following:

1. Liquids keep concentrated but do not constrain the movement of the large number of reactants needed for the complicated chemical interactions necessary for metabolism. Molecules in a gas would be too dispersed for frequent metabolic interactions, while the solid state would render reactants too stationary. The chemistry of life is a social endeavor, requiring a critical but mobile population density of participants.

2. Liquids promote chemical interactions through their influence on the shape that biomolecules and macromolecules assume. Protein catalysts (enzymes), for instance, have to have a particular configuration in order to bring two reactants together so they can form a bond, or create the strain in a bond that needs to be broken to release energy for another purpose. Polar solvents like water promote the unfolding of polar macromolecules, thus exposing reactive sites, while keeping non-polar structures like membranes sequestered and stable. In a gaseous state, surrounding chemical influences

would be too dilute to influence molecular configurations, while the solid state would render molecular shapes too inflexible.

3. Liquids dissolve substances so that individual molecules can break away from their aggregated forms, move about freely, and readily come in contact. In the solid state, reactants would stay compacted and inaccessible to interactions with one another. A special case of dissolution in polar solvents is the ability of salts (molecules held together by ionic bonds) to separate into charged particles. The concentration and movement of small, charged ions plays a critical role in the cellular physiology of life as we know it. Most biomolecules and macromolecules are largely polar, so having a polar solvent to dissolve them into individual units is a necessity. While non-polar solvents are not a factor for living systems on Earth, under other circumstances they could critically affect the behavior of a different set of reactants, providing the general benefits of a liquid medium for a rather different type of chemistry.

4. Liquids maintain a constant volume, enabling concentration-dependent phenomena (which are critical in chemistry) to be influenced simply by increasing or decreasing the number of reactants in the system. Volume changes would alter the concentration of all constituents, significantly muting the capacity for controlling the concentration (hence impact) of individual reactants. Keeping volumes constant also allows living organisms to maintain a fixed size, making it easier to adapt to other variable aspects of their environment.

5. Liquids exist at intermediate temperatures between the solid and gaseous state. Chemical reactions are greatly affected by temperature. Having both an upper and lower limit to the temperature within which a biochemical system can operate reduces temperature as a variable and makes it possible for metabolic interactions to be controlled primarily through changes in the concentration of specific constituents. In fact, the metabolic adjustment of living processes to a particular "preferred" temperature is so pervasive that it is one of the most consistent observations in biology. Organisms adapt through evolution to carry out metabolic interactions optimally at the prevailing temperature of their habitat. Since liquids provide thermal buffers by absorbing heat before they evaporate and losing heat before they freeze, biochemical systems can operate at narrower thermal ranges than their external habitats typically display. And the exceptions prove the rule. Many microorganisms, some plants, and a few animals that survive seasons of extreme temperatures (most notably the cold of winter, but for some desert amphibians and reptiles, the heat of summer) shut down most of their physiological functions until the environmental temperature returns to a range within which more efficient metabolism and dynamic actions can take place. Seeds and spores similarly remain in a quiescent state until the temperature is right for them to resume effective metabolic activity. In fact, as we shall see in later chapters, retreat to a spore-like vegetative state may be one way that living organisms

have developed on other worlds where conditions optimal for living processes recur only sporadically.

## 2.3.2 The special properties of water

Water may be the most abundant molecule in the universe, after the diatomic gases $H_2$, $N_2$, and $O_2$. It is particularly favorable as a medium for complex chemistry on Earth because the average temperature on our planet for most of its history has stayed within the range at which water is in the liquid state.

Water is a polar molecule because oxygen pulls electrons toward its nucleus more strongly than hydrogen holds on to them (Figure 2.4). Because of this polarity, individual molecules adhere weakly to one another through hydrogen bonding (the partially positive H end of one molecule attracted to the partially negative O of another). This makes it harder for water to break free into independent gas molecules, despite its small size (formula weight of only 18), meaning it can absorb more heat than most liquids before evaporating. Thus water can buffer, or absorb and then release, substantial amounts of heat energy while remaining itself in the liquid state. On the one hand, this stabilizes the environment for biochemical interactions within the liquid, while on the other providing a greater thermal range within which biochemical systems have to operate.

The weak adhesion of water molecules to one another gives water a fairly high surface tension, shown by its tendency to form beads on non-polar surfaces like glass. This may have been particularly important for the origin of life, where concentration in water droplets may have been an important requirement. However, the small size of the molecule means that its viscosity (ease with which molecules slide past one another) is reasonably low, making it easier for water to flow through channels such as membrane pores or blood vessels, and enabling diffusion of solutes to occur over necessary distances in a reasonable amount of time.

Perhaps the greatest advantage of water's polarity is its ability to dissolve ionizable compounds – molecules that are charged because of the uneven distribution of electrons. Salts like sodium chloride ionize completely in water, meaning that the $Na^+$ and $Cl^-$ ions separate entirely and act as independent charged particles. Many living processes depend on the presence or movement of such ions. Biomolecules like carbohydrates and amino acids (Figure 2.5), and macromolecules like nucleic acids and proteins are all polarized, at least in part, so they all dissolve readily in water. Some biomolecules, like fatty acids and phospholipids, are made up mostly of non-polar hydrocarbon chains, but they carry organic groups like carboxylic acids and phosphates at the end of the chain, which give one end of the molecule polar characteristics. This enables the non-polar portions to aggregate together, providing a barrier to the passage of water, other polar molecules, and all macromolecules, while the structure as a whole can interface comfortably with the water solvent because of the polar groups on its surface. This is the basic chemical nature of *membranes*, which form the barriers and boundaries for cells in an aqueous environment.

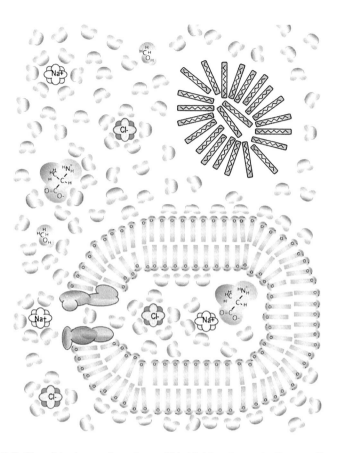

**Figure 2.8** Chemistry in a polar solvent. This highly schematic diagram illustrates several important features of the special properties of water as a solvent and the substances (solutes) that are dissolved or dispersed within it. Water is a polar molecule (Figure 2.4). In the liquid state, individual water molecules move freely about one another, but show some weak interactions between their partially positive and negative poles. These partial charges also help disperse molecules like sodium chloride, which dissolve into positively-charged sodium ($Na^+$) ions attracted to the negative O end of water, and negatively-charged chloride ($Cl^-$) ions attracted to the positive H side of the solvent. Other polar molecules, such as ethanol and alanine (Figure 2.5b) likewise dissolve readily in the polar solvent, because their negative and positive regions become surrounded by the opposite charges of water. Non-polar molecules like hydrocarbons (Figure 2.5a) shown figuratively as gray rods in the upper right corner are repelled by the polar water, but readily aggregate among themselves to form insoluble droplets. Otherwise insoluble hydrocarbon chains can be made soluble by adding charged units, such as a negative phosphate group, to one end of the molecule. The polar heads can then interact with the polar solvent, while the non-polar tails attract one another. Known biological membranes make use of this property by forming bilayers of phospholipids, with their polar heads in contact with the polar water solvent, while their non-polar tails provide a barrier for the free movement of larger molecules and polar solutes of any size (© Louis Irwin).

Complexity can be added to such structures by the insertion of tube-like proteins through the plane of the membrane. The top, bottom, and inner lining of the tube are made of polar amino acids, while the middle, outer portion of the tubular protein is made of non-polar amino acids that form a stable fit with the non-polar regions of the membrane core. Such tubes can act as channels through which water and other molecules that normally could not pass through the non-polar membrane core can move in and out of the compartment enclosed by the membrane.

Finally, water is an important reactant in biochemistry itself. The splitting off of $H^+$ ions from water by sunlight is the first step in photosynthesis. Water is removed from amino acids, carbohydrates, and nucleic acids in the process of forming macromolecules, while it is added to macromolecules in the process of breaking them down. Water is also a major end product of energy metabolism.

## 2.4 The need for and sources of energy for living systems

Inherent in the definition of life is the fact that it both consumes and expends energy – to maintain its highly ordered state, to persist in disequilibrium with its environment, to reproduce and grow, and (for some forms) to move about [3]. The universe is teeming with energy, so you wouldn't think that harvesting energy for living processes is such a problem. The question is, what energy and from where?

From our terracentric vantage point on the third rocky planet from the Sun, 150 million kilometers (93 million miles) from the star that bathes us in warmth and light, a nearly infinite supply of energy seems to be available. And indeed, light is a very effective source of energy for those forms that have evolved the ability to make use of it. The smallest quantum of light, the photon, is powerful enough in principle to provide the energy for adding a terminal phosphate group to six molecules of ADP, forming six molecules of ATP (Figure 2.9) that can be used for all sorts of metabolic purposes. The process by which this occurs is far from simple, but over the course of three billion years of evolution, plants and some microorganisms on Earth have perfected a highly efficient photosynthetic mechanism for capturing the energy from sunlight and turning it into carbohydrates and other metabolic intermediates that end up ultimately as ATP.

But what of those worlds too far away from a sun for its light to be strong enough, or for all the habitats on other worlds as well as our own too far beneath the surface for any type of radiation, even if available, to penetrate? In fact, what of our own planet in its earliest days before the sophisticated apparatus of photosynthesis had evolved? What energy besides sunlight is available to power biospheres other than our own, as well as the portions of our own that light can't reach?

### 2.4.1 Oxidation-reduction chemistry
Since life is fundamentally a chemical process, it should come as no surprise that chemistry itself is a major source of energy. Of the nine potential sources of

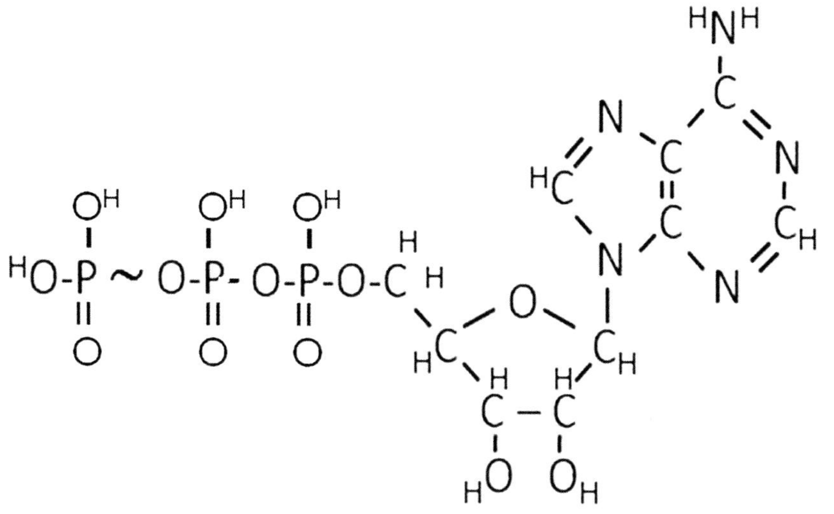

**Figure 2.9** Adenosine triphosphate (ATP). With highly electronegative P and O atoms bound to one another very tightly, a lot of energy is released when the bonds are broken, especially when the third phosphate group is broken away from the rest of the molecule. To indicate its high energy content, the last P-O bond is shown as a wavy line.

energy listed in Table 2.1 that living systems theoretically could harvest, five are from reactions that occur in the abiotic (non-living) environment. Of note is the fact that all are oxidation-reduction reactions.

When electrons are taken away from an atom, its valence (tendency to form bonds) becomes more positive. Oxygen is a powerful electron attractor, so when it takes electrons away from an atom like C or H, we say that the atom losing the electrons has been "oxidized." The valence of O itself is thereby made more negative, or reduced. For every atom that is oxidized, another is reduced. When H reacts with O to make $H_2O$, hydrogen is oxidized and oxygen is reduced (Table 2.1, second source). When sulfur (S) is converted from its solid elemental state to part of the sulfuric acid ion ($HSO_4^-$), it is oxidized (Table 2.1, third source). In this case, iron (Fe) is the oxidizing agent, because its valence is reduced from III to II. Oxidation-reduction reactions release energy basically because molecules are more stable when electrons go from being shared to being held more securely by one atomic nucleus or another. As oxidizing agents pull electrons more strongly toward their nuclei and reducing agents let go of them, energy is released.

The oxidation of H by O to make water is very energetic, but the amount of free H in the environment is too small to make this particular reaction biologically useful. As Table 2.1 reveals, however, several other well-known reactions produce energy yields that are very competitive with sunlight, and examples of each are known to be used in certain organisms. In view of the complicated molecular machinery required for photosynthesis, its evolution must have taken a long time. During those early stages of life on Earth,

**Table 2.1** Theoretical energy yields from different sources, and the equivalent number of ATP molecules that each could generate.

| Source | unit | eV unit | kcal mole | ATP equivalents |
|---|---|---|---|---|
| ADP + $P_i$ → ATP | 1 molecule | − 0.3 | − 7.0 | 1 |
| $2H_2 + O_2$ → $2H_2O$ | 1 reaction | 4.92 | 113.4 | 16.2 |
| $S + 6Fe(III) + 4H_2O$ → $HSO_4^- + 6Fe(II) + 7H^+$ | 1 reaction | 2.6 | 59.9 | 8.6 |
| sunlight | 1 photon | 2.0 | | 6.6 |
| $4Fe(II) + O_2 + 4H^+$ → $4Fe(III) + 2H_2O$ | 1 reaction | 2.0 | 46.1 | 6.6 |
| $H_2 + 2Fe(III)$ → $2H^+ + 2Fe(II)$ | 1 reaction | 1.54 | 35.5 | 5.1 |
| $4H_2 + CO_2$ → $CH_4 + 2H_2O$ | 1 reaction | 1.40 | 32.3 | 4.6 |
| ionic diffusion down 100-fold concentration gradient | 1 ion | 0.12 | | 0.40 |
| kinetic energy of water striking 1 $\mu m^2$ at 0.1 m/s | 1 $\mu m^2$ | 0.07 | | 0.22 |
| osmotic pressure down 0.7 osmole gradient | 1 molecule | 0.006 | | 0.02 |

*Notes:* The synthesis of one mole of ATP from its precursors requires about 7 kcal, or 0.3 eV per molecule. Potential energy sources are listed in descending order of their theoretical yields. Actual yields vary according to temperature and other variables, so values should be regarded as relative comparisons only.
eV = electron volts; kcal = kilocalories

oxidation-reduction chemistry is most likely to have been the creative force that turned inanimate matter into a living substance. Or it could have been heat.

## 2.4.2 Thermal energy
Heat is generated by atomic fusion in stars and radioactive decay in planetary cores. It thus is common and readily available throughout the universe. Since heat is a manifestation of molecular motion, it lends itself well, in principle, to direct absorption by chemical reactions, yielding products with a higher total energy content than their precursors. Not only that, but the potential yield of energy from heat is very high. A water vacuole making up 10% of a typical microbe would yield enough energy to form 8 million ATP molecules for every degree of cooling, if every bit of the energy could be captured for that purpose. The problem is that heat (molecular motion) is a highly disordered form of energy to begin with, and its dissipation results in even greater disorder. Thus, relatively little of the energy available from heat can be used to do work, because most of it goes into increasing the disorder of the system. (In thermodynamic terms, free energy for doing work is small in comparison to the energy given up to increase entropy). The bottom line is that thermal energy is very inefficient. Furthermore, unless the influx of heat continues, the thermal gradients on which heat flow depends will rapidly be degraded.

Highly inefficient forms of energy might still be usable if the source is abundant and persistent. The heat that flows continually from a thermal vent might be such an example. The early stages of a planet's history, when heat from the process that formed it provides a steady, intensive flow of energy from its interior, might be another case. Under the special circumstances where, for a time at least, the supply of thermal energy is effectively inexhaustible, it could well be competitive with more efficient but weaker sources of energy. Where heat is less intense, or thermal gradients cannot be sustained, it probably loses out to the more efficient and reliable, if weaker, forms of energy. That may well be why today on Earth no forms of life that rely on thermal energy are known. But at the dawn of planetary history, before a process for converting sunlight or taming energetic chemical reactions had evolved, heat may have played an important role in bringing the world to life.

### 2.4.3 Kinetic energy
Any mass in motion has kinetic energy. The wind in our face or the current in the river that makes it harder for us to swim upstream are examples of kinetic energy that are commonplace in most environments. There is no reason in principle for this form of energy not to be usable by living organisms. In fact, our ability to feel the touch of a feather on our skin, to hear a sound, or to sense that our head is spinning are all made possible because our nervous systems have evolved mechanisms to detect kinetic energy. In an aquatic environment – probably the most common habitat for life in the universe – the motion of water due to gravity flow downhill or convection currents generated by temperature differences provide a ready source of energy. A current of 10 cm/sec, the speed of a slowly moving stream, could deliver enough kinetic energy to power the synthesis of one ATP molecule per second for every 5 square micrometers of surface that the water current strikes (Table 2.1, second source from bottom). By comparison with sunlight and most oxidation-reduction reactions, this is a rather weak energy yield, but on a world where neither light nor an appropriate set of reactants are available, kinetic energy would clearly provide an alternative.

### 2.4.4 Ionic diffusion
Whenever two solutions containing a diffusible substance are brought together at different concentrations, the substance moves from its area of higher concentration to regions where it is less concentrated. Thus, the sugar that was compacted into a cube moves, or diffuses, into the larger volume of coffee where the sugar initially is less concentrated. One of the major ways in which living systems maintain disequilibrium from their environments is by containing substances that are either more or less concentrated than those substances are in the environment. Thus, a molecule of anything that is more concentrated outside a cell tends to flow into the cell spontaneously if allowed to. Or, substances held inside the cell but not outside will diffuse outwardly if passage is made possible. This tendency for diffusion to occur is proportional to the difference in concentration, or concentration gradient, between the two sides of the membrane that separates one compartment from another. So just by finding itself surrounded by a solution of something more concentrated in its environment than within its interior, a living cell has access to a potential

source of energy. If it can find a way to capture the energy of the moving particles as they diffuse into the cell, perhaps through channels adapted specifically to couple that movement to an energy-absorbing chemical reaction, a cell can convert this form of energy to biologically useful purposes.

When the diffusing particles happen to be ions, they provide the added advantage of carrying electrical charges that generate attractive or repulsive forces that can do work like other forms of energy. A battery works in just this way, by separating charges that can do work when the charges flow. All known biological membranes have the capacity to generate ionic gradients through several mechanisms. Because they are selectively permeable, allowing some ions in and keeping other ions out, biological membranes can harvest the energy of diffusion that the resulting concentration gradients generate. This, in fact, is precisely the way that the phosphorylation of ATP happens: the diffusion of hydrogen ions ($H^+$) down their concentration gradients is coupled to an enzyme that adds the terminal phosphate to ADP. Calculations show that a concentration differential of 100 fold on either side of a membrane – a gradient commonly found in living organisms – can generate enough energy for the synthesis of one new ATP molecule for every three ions that diffuse through the enzyme channel (Table 2.1, third source from bottom). Though weaker than most oxidation-reduction reactions, this still provides a substantial source of energy wherever living organisms surrounded by salt solutions are found. And organisms living in high salt environments, like some microorganisms in hypersaline lakes, have access in principle to much more energy than the example above – enough to be very competitive with sunlight and oxidation-reduction chemistry.

### 2.4.5 Osmosis

A wilted plant regains its stiff, healthy structure when water is poured into the soil around it. Every woman knows that salt retention causes tissue swelling and a bloated feeling during the latter part of her menstrual cycle. Both phenomena are due to osmosis, a special case of diffusion in which water is the diffusing substance. When compartments enclosed in membranes permeable to water are immersed in a dilute solution, water will move from the solution into the contents enclosed by the membrane. This is because water moves spontaneously from a dilute solution into a more concentrated one – the process we call osmosis. Living cells typically have a large number of dissolved metabolites and ions, collectively called solutes, enclosed by membranes which keep the solutes inside but allow water to enter. Osmosis is thermodynamically favorable because it reduces the difference in solute concentrations on the inside and outside, thereby increasing the entropy of the system as a whole and releasing energy. As molecules of water flow spontaneously through channels into the cell, molecular machinery attached to the channels could couple the energy released to a chemical reaction that would add the energy to a biomolecule like ATP.

The amount of energy that can be harvested from osmosis depends on how steep the concentration gradient is between the inside and outside of the cell. We have calculated that for a gradient of 0.7 osmoles, something like the difference between the inside of a fresh water fish and the stream in which it lives, each molecule moving by osmosis into a cell of the fish could release about six thousandths of an eV (Table 2.1, last line). This is pretty weak energy, but enough to power the phosphorylation of one molecule of ATP for roughly every 45 water molecules that diffuse in. No such mechanism is known for any organism on Earth, probably because all the other potential sources in Table 2.1 yield more energy. But on a world where no other energy source is available, osmosis could be the engine for life.

### 2.4.6 Other sources of energy

Table 2.1 is just a partial listing of the energy that is readily available in the universe. Magnetic energy is also abundant throughout the cosmos, but we have previously shown that it is very ineffective under nearly all environmental scenarios, except for those in which a planet orbits a very powerful magnetic source, such as a magnctar – a neutron star with an extremely strong magnetic field [5]. Radioactive decay, ionizing radiation, tectonic stress, gravity, and pressure are present in every star and on the planetary bodies around them. Could these be possible sources of energy for life as well? Though none can be ruled out, all have apparent problems that make them less likely than the sources listed in Table 2.1.

For energy to be biologically useful, it has to be reliable and available in the right amount to be captured efficiently by chemical reactions that create biomolecules. The decay of radioactive elements can emit very energetic particles, but radioisotopes are not uniformly distributed, and their energetic emissions are too sporadic to be predictable. Ionizing radiation is probably emitted by every star, but at energy levels too high to be captured effectively by biochemical systems. On the other hand, magnetic fields which are likewise likely to be present around most stars and many planets, are too weak at molecular dimensions to be competitive with other readily available energy sources. Tectonic stresses could be present on any solid planetary body, but only certain minerals can tap this energy source, and a mechanism for converting it into a biologically useful form is hard to envision. Gravity and pressure are both pervasive and powerful sources of energy in the universe, but at molecular dimensions their gradients are so small that the energy they provide is orders of magnitude less than other equally available sources of energy.

We should note that we are talking only about the direct conversion of energy from these exotic sources into biologically useful forms. Their indirect effects may be very important for supporting life in some circumstances. Radioactive decay in a planet's core generates heat that rises by convection to the surface where it can be emitted at thermal vents or in hot springs, and can power the tectonic movements that ultimately emit sulfur and other oxidizable chemicals from volcanoes. As we shall see in later chapters, ionizing radiation could be the means for releasing oxygen from water ice at the surface of frozen bodies, providing an oxidizing agent for energy-yielding chemical reactions. Tidal excursions due to gravity, and convective currents due to differential heating from thermal vents are two examples of kinetic energy – a biologically useful form set in motion by primary sources too erratic, powerful, or disorganized themselves to power biochemical processes directly.

In conclusion, the universe is awash in energy. Light and oxidation-reduction chemistry are particularly well suited for the small scale energy transactions needed to maintain the organization and power the activities of living systems. They are by no means the only potential sources of energy, however. Where they are not available, alternatives would have to serve. In speculating on the course that evolution may have taken on other worlds, the energy available on that world needs to be a central consideration.

## 2.5 Chapter summary

We begin with a definition of life – a challenging task since our concept has to be generic enough to apply wherever life might be found in the universe, yet restrictive enough to exclude those phenomena in nature or the fabricated world that share some of the characteristics of living things. Adding to our challenge is the dual nature of a living organism, as both an *entity* and a *process*.

We propose three necessary and sufficient criteria for defining a living organism. First, it consists of a highly organized system separated by physical boundaries from its more disordered surroundings. Second, it maintains its high level of organization and performs work by transforming energy from its environment in a self-regulating manner. Third, it reproduces itself autonomously by assembling raw materials from its environment into near-exact replicas according to information perpetuated through an indefinite succession of generations.

Both the highly ordered structure and the dynamic properties of living organisms are manifested as polymeric and ionic chemistry in solution. Polymers are large molecules, which in living things on Earth are based on chains of carbon atoms, bonded to themselves or to oxygen, phosphorus, or nitrogen through covalent bonds strong enough to be durable yet flexible enough to undergo rearrangement. Carbon is well suited to be the backbone for biological polymers on Earth because of the prevailing temperature and the availability of liquid water, its natural partner, on our globe. Silicon is chemically similar and could possibly serve as a polymeric backbone, but only under very different environmental conditions from those prevailing on Earth.

Biomolecules consist of organic (carbon-based) polymers that transfer energy, build structure, carry out dynamic activity, and coordinate the complex interplay of chemical reactions that we refer to collectively as metabolism. Biomolecules also constitute the monomeric, or single, units for construction of much larger macromolecules. Life on Earth relies on three major classes of macromolecules: carbohydrates, proteins and nucleic acids. Carbohydrates store energy in long chains of sugar, and (with the smaller molecules of lipids) also contribute to many cellular structures. Proteins provide the building blocks for cellular structures and provide shape-specific surfaces for catalyzing and regulating specific chemical reactions. The unique shape of any given protein is determined by the sequence of its monomeric units, amino acids, which come in about 20 different varieties. Nucleic acids encode information in the form of a long linear sequence of four different bases that determine the order in which amino acids will be lined up in a protein. They constitute the hereditary material that is passed from one generation to the next. Whether carbohydrates, proteins, or nucleic acids as we know them exist in living systems elsewhere in the universe will remain an unanswered question until we can analyze samples of alien life. It's reasonable to assume, however, given the need to store energy and the high information content of living systems, that macromolecules of some sort must play analogous roles wherever life exists.

There are compelling reasons to believe that living processes are restricted to a liquid environment. These include the fact that liquids maintain critical concentrations without restricting mobility, dissolve most substances so they can act as independent reactants, promote chemical interactions through their influence on the chemical state and shape of molecules, maintain a constant volume, and provide a filter and buffer that insulates the living system from greater fluctuations in the external environment. For life on Earth, water is an optimal solvent, since it dissolves the major macromolecules listed above, but leaves another class of biopolymers, the lipids, insoluble and therefore capable of forming the core of biological boundaries, or membranes. Under different physical conditions, a solvent other than water in principle could serve as the medium for a form of biochemistry unknown on Earth.

Life depends on a constant flow of energy, which it takes in from the environment in order to maintain its own highly ordered (low entropy) state, repair itself, react to and act upon its environment, grow, and reproduce. While the universe is filled with energy, light and oxidation-reduction chemical reactions are particularly well suited for delivering energy in a form and in quantities that living systems can use. Other forms of energy are widely available, however, and could serve as sources for biogenesis in the absence of light or appropriate chemistry. These include heat, ionic and osmotic gradients, and various forms of kinetic energy.

The prospect of life on other worlds is a function of circumstance. By that we mean, first and foremost, the physical setting, including the sources of energy and chemical resources that are available. But life also has a history, which means that starting conditions and trajectories over time will play a large role in the way life ends up. That is the topic to which we now turn.

## 2.6 References and further reading

1   Margulis L. and Sagan D. 1995. *What Is Life?* New York: Simon & Schuster.
2   Eiseley L. 1946. *The Immense Journey*. New York: Random House.
3   Morowitz, H. J. 1968. *Energy Flow in Biology*. New York: Academic Press.

4   Kauffman, S. A. 1995. *At Home in the Universe: The Search for Laws of Self-Organization and Complexity.* Oxford: Oxford University Press.
5   Schulze-Makuch D. and Irwin L. N. (2008) *Life in the Universe: Expectations and Constraints.* Springer-Verlag, 2nd ed.

http://www.ibiblio.org/jstrout/uploading/potter_life.html
An informative, critical essay on the difficulties of defining life.

http://www.chem1.com/acad/webtext/virtualtextbook.html
An excellent on-line reference for general chemistry, including the concept of entropy.

http://www.squidoo.com/central-dogma
Entry to a variety of links with excellent visual representations of the structure and function of nucleic acids and proteins.

# 3

# Life's Fundamentals

## Origin, evolution, and the great pyramids

Now that we've talked about the source of solar systems, the nature of life, the relevance of chemistry, and the importance of energy, we need to think about how life can arise in an otherwise non-living world, then give rise to the interdependent web of organisms that constitute ecosystems. And we need to do so in a fairly general way, so our assumptions can apply throughout the great range of habitats that are likely to be host to living organisms, wherever they are found in the universe.

No one knows how life arose on Earth, much less anywhere else. In fact, we're much less certain now than we once were that life did in fact originate on Earth, as opposed to somewhere else prior to its arrival on Earth by transport on meteorites or other objects from space. This interplanetary spread of life, or *panspermia*, may be a general process for the seeding of life among neighboring planets [1]. What we do know is that life had to start somewhere. Earth is as good a place as any (and more likely than many) for serving as the cradle for *a* form of life, so we'll begin with what we can infer about the origin of life on our home planet, then generalize from the one case we know to a broader range of possibilities.

## 3.1 Beginnings

While we don't know how life began, anywhere, we're far from devoid of ideas [2–6]. In fact, we know quite enough about biology and chemistry to sketch out a plausible set of stages through which incipient life must have passed, at least on Earth [6]. They include the following:

### 3.1.1 A nine-step program for the origin of life on Earth
1. Simple organic compounds, like amino acids, fatty acids, and sugars, accumulated in the primordial waters of the Earth. They may have been generated from simpler molecules like carbon dioxide, ammonia, formaldehyde, and hydrocyanic acid in the early eon's of the Earth's history when energy from lightning, volcanic convulsions, and infalling meteorites provided abundant sources of energy (Table 3.1a). Or they may have come from space, where they are known to exist both from the evidence of radioastronomy and from chemical analysis of meteorites (Table 3.1b).

L.N. Irwin and D. Schulze-Makuch, *Cosmic Biology: How Life Could Evolve on Other Worlds*,
Springer Praxis Books, DOI 10.1007/978-1-4419-1647-1_3,
© Springer Science+Business Media, LLC 2011

**Table 3.1.**  Prebiotic availability of biomolecules

| a. Products of Miller's "sparking" experiments | | b. Found in Murchison Meteorite, 1969 | |
|---|---|---|---|
| *Amino acids* | *Others* | *Amino acids* | *Others* |
| glycine | glycolic acid | glycine | aminoisobutyric acid |
| alanine | lactic acid | alanine | amino-*n*-butyric acid |
| glutamic acid | fatty acids (C2-C5) | glutamic acid | acetic acid |
| aspartic acid | formic acid | aspartic acid | propionic acid |
| non-protein amino acids | | isovaline | |

2. Lacking enzymes or preexisting life to break them down, the growing number of organic molecules would have accumulated to a higher concentration than that at which they are found in nature today. Under appropriate circumstances, which could have included drying out, heating, or catalysis by inorganic substrates, condensation reactions could have occurred which joined together simple organic monomers in different combinations of larger, more complex *nitroglycopeptides,* or *NGPs* (Figure 3.1). The composition of the various NGPs would have varied almost randomly at first; but over time, certain chemical configurations would have become more common because their synthesis was easier or they happened to be more stable.

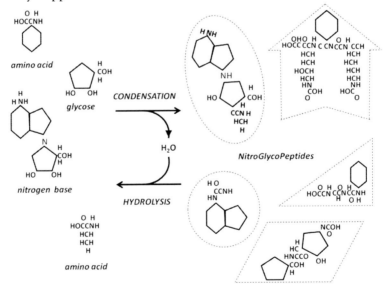

**Figure 3.1** Nitroglycopeptides (NGPs). Under appropriate conditions, condensation reactions among different combinations of nitrogen bases, sugars, and amino acids could form a variety of more complex molecules. Hydrolytic conditions would break them down into chemical building blocks for recycling into a new set of combinations. Over time, certain NGPs would recur more often than others because their formation would be favored, or their breakdown would be less likely. Unsequestered and lacking a mechanism for self-replication, such aggregates cannot be said to constitute a living system, but they may be characterized by what could be called a prebiotic "NGP world."

3. Under recurrent cycles of desiccation and rehydration, different NGPs could have begun to form larger aggregates, or protopolymers (Figure 3.2). These too would have been highly variable, a function perhaps of local substrate conditions; but integrated over time and a global extent, certain protopolymers may have begun to recur or survive with greater frequency.

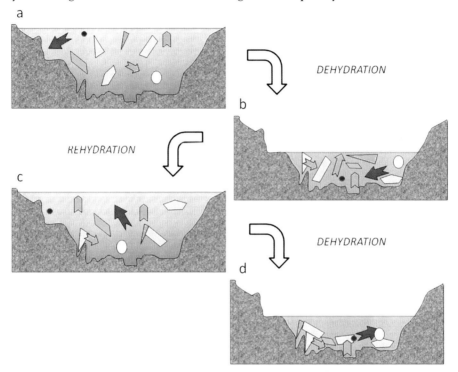

**Figure 3.2** Formation of protopolymers. In tidal or ephemeral pools, recurrent cycles of dehydration and rehydration would give rise to different protopolymers. (a) When a pool is full, a particular combination of NGPs is diluted into isolated monomers, shown as geometrical abstractions of the actual chemical structures in Figure 3.1. (b) Desiccation would concentrate the individual monomers and favor their condensation into larger molecules, or protopolymers. The physical structure of the substrate could favor the alignment, hence joining together, of specific NGPs. (c) Rehydration of the pool would favor hydrolysis of the protopolymers back toward smaller units, with many but not all protopolymers breaking apart. (d) A new round of dehydration would once again favor condensation. Since some of the starting reactants that had survived the previous desiccation cycle would already be more complex, even larger and more complicated combinations could result.

4. As aggregates of molecules like fatty acids, with their hydrophobic tails and polarized heads (Figure 2.4a in chapter 2) formed into bilayered or spherical protomembranes (Figure 3.3c,d), they would have been able to sequester ions, organic monomers, and protopolymers within protocellular structures. Once

entry into and exit from the protocells became selective, the contents inside would have begun to differ from the solution on the outside, establishing a disequilibrium between the inside of the protocell and its surroundings.

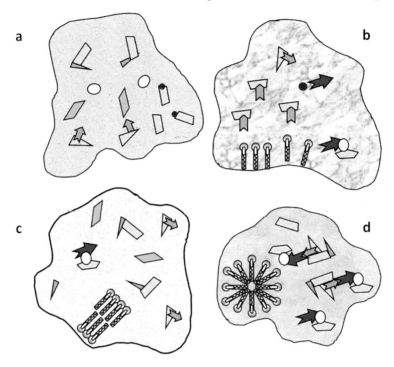

**Figure 3.3** Prebiotic evolution. Protopolymers could evolve into aggregates of increasing complexity over time, as pictured in this series of different pools, having started with a slightly different mix of constituents, at different stages of evolutionary development. Pool (a) at an early stage of evolution consists of a dilute solution of NGPs and small protopolymers. Pool (b) at a mid-early stage of evolution, having started by chance with a different mix of protopolymers, has achieved a little more complexity following a greater number of desiccation-rehydration cycles, including the presence of some long-chain hydrophobic hydrocarbons along the hydrophobic substrate. Pool (c) at a mid-late stage of evolution contains a variety of protopolymers found in other pools, with hydrophobic molecules that have begun to associate with one another. Pool (d) at a late stage of evolutionary maturation contains a mixture of more complicated protopolymers, including spherical structures stabilized internally by hydrophobic attraction and externally by hydrophilic interaction with the water solvent molecules (Drawing by Louis Irwin).

5. Thus concentrated and enclosed, the mixture of reactants and protopolymers could have begun to react in a complex web of chemical interchange (Figure 3.4). Generally speaking, the synthesis of more complicated molecules absorbs energy, while the breakup of larger molecules into smaller ones releases energy. Reactions less favorable because they required an input of energy could have

still proceeded if coupled to another reaction that released energy. Some reactions not prone to occur readily (what chemists call reactions with high energies of activation) may have been helped to proceed by catalysts, either in the inorganic substrate or from polymers now grown large enough to assume configurations that provided *active sites* for the binding of substrates.

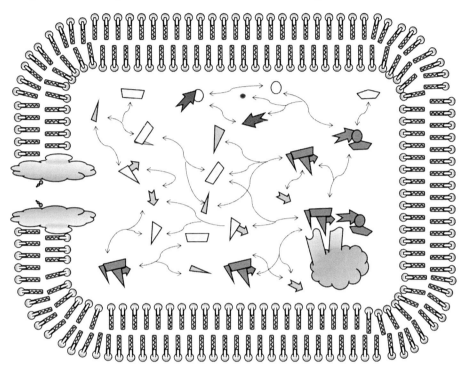

**Figure 3.4** Protometabolism. A major step toward ignition of living processes would have been enclosure of a rich mix of constituents by hydrophobic boundaries (membranes). Thus concentrated and isolated from their surroundings, they could undergo a combination of reactions with one another, alternately building up and breaking down different protopolymers. Some reactions could be aided by the presence of macromolecules whose configurations may have helped bring specific reactants together or promoted their breaking apart (complex at lower right), representing a primitive form of catalysis. Other macromoleculues inserted into the membranes (left side) could have served as channels for the passage in or out of selected constituents (Drawing by Louis Irwin).

6. Some of the protopolymers could have been strung together in such a way that they carried information in the sequence of their monomeric units. The sequences may have been largely random in the beginning, but over time, certain sequences may have become more common. Perhaps their arrangement was initially dictated by the contour or chemical features of an inorganic

substrate like clay. Maybe some sequences recurred more often because they were able to promote their own survival. Ultimately, these "informational" polymers could have acquired the ability to replicate themselves, giving rise to the first protogenetic system (Figure 3.5).

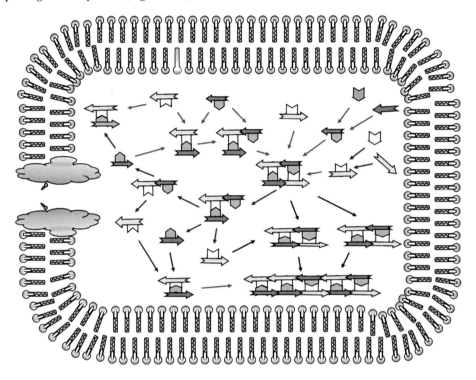

**Figure 3.5** Protoreplication. Early forms of replication would have required only a simple set of complementary molecules with directionality. In the scheme above, four building blocks (upper right) give rise to two complementary structures with opposite directionality. They could form, break apart, and reform repeatedly. A long string of constituents, which could be arranged in any order, would determine the precise order of a complementary string running in the opposite direction (lower right) (Drawing by Louis Irwin).

7. Inefficient and inexact at first, the mechanisms of replication could have progressively become more efficient and accurate. In some cases, these "informational" biopolymers may have assumed catalytic as well as genetic roles. The ability to reliably replicate polymers that carried highly specific information, both for catalytic and genetic functions, would have enabled the beginning of true metabolic regulation. Most scientists believe that the first polymer to master these functions during the evolutionary history of life on Earth was ribonucleic acid (RNA). The evolutionary stage at which this occurred is referred to as "the RNA world" [12].

8. As catalytic, structural, and genetic functions became more sophisticated, a division of labor occurred. With their huge capacity for different three-dimensional configurations, proteins assumed the primary function of catalysis and structural building blocks. With their simple structure but extensive capacity for information coding in the sequence of their monomers, nucleic acids took over the primary function of replication and hereditary transmission. As the store for genetic information, RNA became the molecule that directed the proper alignment of amino acids into proteins, the technical term for which is *translation*. As the catalyst for metabolic reactions, proteins became the metabolic partners that enabled RNA to replicate itself and provide a template for translation.

9. In the version of life that unfolded on Earth, a more stable form of nucleic acid called deoxyribonucleic acid (DNA) emerged to take over the long-term storage of genetic information. This rendered RNA an intermediate in the coding of protein structures. DNA and RNA are very similar chemically and use the same logic for storing information in a sequence of just four different monomers (Figure 2.6 in chapter 2). This made the overall readout of information to flow in a sequence from DNA to RNA to protein. The synthesis of RNA on a DNA template is known technically as *transcription.* The persistence of DNA as the ultimate form of hereditary storage for all living organisms (with some minor exceptions) ushered in the "DNA world" from which all life on Earth today is descended.

### 3.1.2 Qualifications and limitations

Of the nine stages listed above, the final two are established facts. None of the first seven is free of controversy, though most have been modeled, and a few have been demonstrated experimentally. In stage one, for example, many biomolecules have been created in the laboratory from simple precursors known to have been present on the early Earth and found in abundance on other worlds (Table 3.1a). The delivery of organic molecules, including amino acids, to the Earth on meteorites from space is also well documented (Table 3.1b).

Condensation reactions (stage 2, Figure 3.1) are difficult to carry out in water, but have been shown to be possible, under a variety of specialized experimental conditions which could occur in nature [7]. Stage 3, the next step up in complexity, is more difficult to show empirically, but has been achieved to a limited degree. This is the stage at which the inorganic substrate may play a particularly important role. The catalytic properties of minerals and the regularity of their structure makes them particularly strong candidates for a role in biasing the alignment of monomers and protopolymers, and promoting their condensation into more complex macro-molecules (Figure 3.2).

Stage 5, at which a critical mass of reactants, protopolymers, and polymers begin to interact in a self-sustaining way (Figure 3.4), has not been achieved, but has been modeled effectively [8]. Similarly, the origins of replication (stage 6) are shrouded in mystery. Here again, the inorganic substrate may play a critical role by favoring certain alignments over others. Experimental attempts to achieve self-sustaining examples of both proto-metabolism and proto-replication constitute an active area of research from which breakthroughs should emerge in the near future [9].

The limitations of the scheme above have more to do with the transitions between stages, than with the nature of the stages themselves. The transition from a primitive form of proto-replication to the sophisticated sequence of replication, transcription, and translation that life on Earth uses today is particularly difficult to envision. But just as the development of symbolic language and domestication of animals by humans presumably occurred in gradual stages that are lost to history, so the evolution of sophisticated metabolic and genetic mechanisms must have occurred in stages that have left no trace.

Another qualification of the general scheme for the origin of life outlined above is that the order of stages in some cases is hotly debated. The greatest controversy centers around whether metabolism can evolve without having a mechanism for replication in place first [10]. The difficulty of imagining a self-sustaining metabolic system is counteracted by the need to have sufficient metabolism to engineer replication. Until highly plausible models or empirical evidence settle the issue, it seems most fruitful to proceed on the assumption that proto-metabolism and proto-replication evolved to a significant if not large degree in parallel.

Another center of controversy relates to the environment in which each of these stages occurred. Figure 3.2 envisions the formation of protopolymers as occurring at water-substrate interfaces subjected to recurrent cycles of dehydration and rehydration, like tidal pools or ephemeral ponds [11]. This is clearly not the only possibility. Earth provides a number of other candidate environments, which is fortunate since the habitats in which life might arise on other worlds is bound to vary considerably throughout the universe. Among the habitats that scientists have argued as the most likely cradle for life are the following.

### 3.1.3 Alternative origin scenarios

#### 3.1.3.1 A lukewarm water origin for life
In a letter to Joseph Hooker in 1871, Charles Darwin imagined that prebiotic chemistry may have begun in a "warm little pond, with all sorts of ammonia and phosphoric salts" which light or other forms of energy could convert into a protein "ready to undergo still more complex changes." Too little was known of chemistry at the time to give detail to his imagination. Building on these concepts and early experimental findings, the Russian biochemist, Aleksandr I. Oparin, in the early decades of the 20th Century began to argue that energy from lightning or other sources, in an atmosphere rich in molecules like hydrogen, ammonia, and methane but poor in oxygen, could have produced a "primeval soup" of organic compounds in the oceans. J.B.S. Haldane, an English chemist, advanced similar ideas in the middle of the century in England. Stanley Miller, a graduate student of Harold Urey, a noted cosmochemist at the University of

Chicago who was familiar with these ideas, tested them by passing electric discharges through a mixture of methane, ammonia, water, and hydrogen. The resulting production of several organic compounds including amino acids (Table 3.1a) provided a dramatic proof-of-concept for at least the evolution of organic chemistry in the primeval oceans of the Earth [12].

The growing evidence that organic molecules have been brought to Earth by meteorites from space has lessened the pressure on chemists to describe how those earliest amino acids and like compounds formed from the simplest precursors on Earth. But, like life itself, organic compounds had to form somewhere, so the abiotic synthesis of biomolecules remains a relevant question. "Sparking" experiments like those of Miller have been repeated countless times under different conditions, including the use of gases thought to be more realistic for the atmosphere of the early Earth. Yields are usually low, and the products vary with the starting reactants, as would be expected. But organic compounds are inevitably generated. There seems little doubt that a combination of abiotic synthesis supplemented by cometary delivery of organic compounds could have seeded the early oceans with the precursors for more complex biomolecules.

Getting to the next step is a little harder, because condensation reactions (Figure 3.2) – the formation of chemical bonds by removing water (a hydrogen from one molecule and an –OH group from another) – is difficult to bring about in a water medium. For that reason, many scientists believe that special conditions, such as heating, tidal cycling, or other factors need to be invoked to explain how monomers like amino acids could link together into polymers like peptides and proteins. In the 1950s and '60s, Sidney W. Fox [13] provided the strongest evidence that high temperatures in the presence of amino acids and a mineral substrate could produce peptides capable of forming into "microspheres" that could represent an early step toward protocells.

### 3.1.3.2 A cold water origin for life

While some reactions are favored by high temperatures, others are favored by cold. Miller and his colleague, Leslie Orgel, have shown that adenine and guanine, two of the nucleotide monomers of RNA and DNA, form best in ice [3]. As water freezes, microchannels of liquid brine concentrate organic reactants more and more as the ice crystals surrounding them exclude dissolved material. This hastens chemical reactions, while lessening the destructive power of heat. A number of experiments have shown that certain polymers can form under these conditions. However, other nucleotides appear to require high temperatures for their formation, so thermal cycling would probably be required to form a complete set of today's biomolecules. Whether the same set was required at the dawn of life is another matter. The generative potential of the ice-water interface could still be operating to provide a beginning set of biomolecules wherever icy worlds are found in the universe.

*3.1.3.3 A deep, dark, and superheated origin for life*

The deepest roots in the tree of life – that is, the oldest known organisms – are bacteria that thrive at very high temperatures, for example in hot springs and at deep sea thermal vents. This has led to speculation that the cradle of life on Earth may have been at the bottom of the ocean, where energy is provided by heat or chemistry rather than light, and where the temperature is well above the boiling point of water. Günter Wächtershäuser, a German chemist, has been the foremost proponent of an origin of life under hot, high pressure conditions through oxidation-reduction reactions involving metals like nickel and iron, in conjunction with sulfur [14]. Wächtershäuser refers to his metabolism-first model as the "iron-sulfur" world, which like the "NGP world" proposed by us above, would have preceded the "RNA world" of Orgel and others. This possibility is also relevant to our quest for life on other worlds, since volcanic activity has now been found at several sites in our Solar System, including beneath at least a couple of the icy moons of Saturn and Neptune.

## 3.2 Organic evolution: the process of biological change through time

The overarching assumption of modern biology is that life has changed over time. We call this process evolution. Since this is a book about how life could evolve on other worlds, we need to have a clear understanding of what evolution *is* and how it works.

The driving force behind evolution is assumed to be natural selection acting on variable traits. In modern terminology, the focus of evolution is on *populations* of organisms. Individuals do not "evolve" during their lifetimes – a given animal doesn't shrink in size just because food becomes more scarce when climate change causes a lasting reduction in rainfall. But the average size of animals in a population may grow smaller over time under those conditions, because the individual animals that were naturally smaller to begin with have an advantage. Since they can get by with less food, they are "selected by nature" to survive in greater numbers than those naturally endowed with a larger size. To the extent that body size is a characteristic that can be inherited, the offspring of the smaller survivors will tend to be smaller as well. Over many generations, the advantage of small size will be compounded, and the average size of individuals in the population will become smaller, resulting in a permanent evolutionary change in the characteristic size of animals in that population.

Those who don't understand evolution often characterize it as a random process. There is nothing random about natural selection, however. If the process were truly random, all variants would have an equal probability of surviving, and those that did would survive by chance alone. But the opposite is true: those individuals most likely to survive are selected by nature – not at random, but by virtue of characteristics that have survival value.

That's not to say that chance plays no role in evolution. To a large degree, the *variation* present in a population of animals, microbes, or molecules is due to chance factors. At the dawn of life, the earliest biomolecules that were formed from abiotic precursors or that arrived from outer space may have constituted a fairly random collection. Much of the variation seen in populations of living organisms is assumed to result from more or less random mutations in genetic material. (Some parts of a chromosome are more prone to mutation than others, but over a large extent of the chromosome, the chance of mutation is unpredictable and therefore essentially random.) Chance also plays a role in the starting genetic material for a new population, as when a tiny fraction of a population survives a catastrophic event, or a very small number of individuals make it through a "bottleneck," like a land bridge between two continents.

Note the other essential part of the process. There must be a mechanism of inheritance for a trait to be acted upon by natural selection; that is, the trait must have a genetic basis. We will consider each of these three factors – selection, chance, and heredity – in a little more detail, in order to judge what course is plausible for evolution to take under the variety of circumstances that may occur on other worlds.

### 3.2.1 Selection

Prior to the emergence of a self-sustaining form of life, selection operates but only in a very inexact way. At the stage when organic compounds are first beginning to join together into what may become biomolecules (Figure 3.1), the combinations that join together may be haphazard and essentially random within a broad range allowed by the rules of chemistry. However, features of the local substrate, which well may be unique to a local setting, can bias the formation of bonds between particular reactants (Figure 3.2). If some mineral structure shows preferential binding to particular organic molecules, and that mineral structure recurs, then it will promote more frequent binding between the particular reactants that bind to it. Multiplied over many cycles of bond formation, breakdown, and rejoining, and perhaps repeated at multiple sites with the same substrate, a particular mix of proto-biomolecules will be produced at a higher frequency.

On substrates lacking electrically charged sites, hydrophobic molecules might tend to form (Figure 3.3b). In time these water-repelling molecules may assemble into hydrophobic aggregates (Figure 3.3c,d), or into extended membrane-like structures that would surround and isolate a mix of protobiomolecules from their aqueous environment, thus forming protocells (Figure 3.4). Any given protocell would contain a mix of molecules determined largely by chance but to some degree by the selective bias imposed by the substrates in which they formed. These various reactants, in turn, might start interacting to a variable degree, with some mixtures leading to longer-lasting self-sustaining reactions more than others. Some of these reactions would either assist their protocell's survival, or at least be more likely to resume interacting after a protocell had been destroyed, then reconstituted with the remnants of the reactants from the previous protocellular generation. Well before complex chemical systems have come "alive," therefore, a crude form of natural selection would be at work.

Once truly living systems emerge, meaning enclosed cell-like structures with self-sustaining metabolism *and* a mechanism for hereditary transmission, natural selection can gain control of the trajectory that that particular form of life will take through time. Not all natural selection operates in the same way, however, and understanding the different types of selection is essential to predicting what will be plausible in different environments, and therefore on different worlds.

Assume a broadly distributed animal species occupying the coastal range of a continent. As long periods of time pass without any changes in the environment, animals at either extreme of size will be slightly less well adapted than those of intermediate size – the larger ones require more food than is readily available, and the small ones have too hard a time staying warm in the winter. Some intermediate size is the optimal compromise between being large enough to stay warm but not large enough to require too much food. Natural selection will tend to promote survival of individuals closest to the optimal size, which therefore will become the mean for the population as the range of variation narrows (Fig 3.6a). This is an example of *stabilizing selection* – natural selection acting to *resist* change, except for a narrowing of the variation around a stabilized mean.

Now assume that as further time passes, geological forces gradually cause the uplifting of a mountain range parallel to the coast (imagine the coastal range of California, if you like). Those individuals of our sample species that find themselves living at higher altitudes now need to be larger to retain body heat in the colder climate at higher elevations. Assuming that the food supply is sufficient to support larger body size, those individuals on the larger end of the frequency distribution for size will be favored over the smaller individuals, including those at what used to be the optimal size for the species. The selective pressure toward greater body size at higher altitudes will favor a shift of the mean value upward, though without necessarily changing the degree of variation in body size (Fig 3.6b). This is an example of *directional selection* – natural selection acting to *promote* change in the direction mandated by the nature and direction of change in the environment.

The lowland populations now face different circumstances, depending on which side of the range they find themselves. The newly uplifted mountain range creates a rain shadow on the side away from the coast, while promoting increased rainfall on the coastal side. Individuals at the same elevation on the two sides of the range, confront different environmental conditions, with the coastal populations living in a cooler environment with greater rainfall (hence more food), while their relatives on the other side of the range find themselves in a warmer, drier climate (hence less food). What was once a single species exposed to a uniform climate has now been split into two populations subjected to two very different environments – one favoring large body size on the costal side, the other favoring small body size on the desert side of the range. The result will be evolution in opposite directions, with two different mean values emerging from one (Figure 3.6c). This illustrates *disruptive selection* – natural selection acting to split an existing biological group into two descendant groups with different biological characteristics. If and when the two populations become reproductively

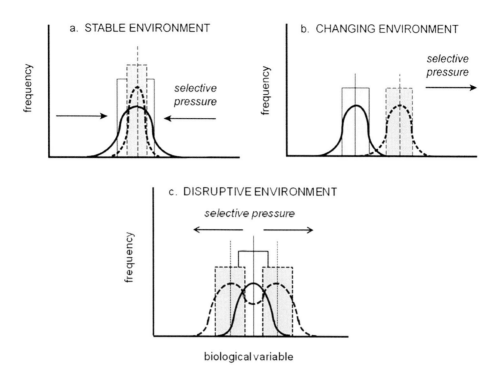

**Figure 3.6** Natural selection under different types of selective pressure. The number of individuals in a population (frequency) is plotted against a quantifiable trait, such as a physical or physiological characteristic that favors survival. The population mean is shown by the vertical line. Bars with solid lines represent the standard deviation of the bell-shaped variation in the original population, and dashed lines the standard deviation in the descendant populations. (a) Stabilizing selection occurs over time in constant environments, because those individuals with traits closer to the mean value will survive in greater numbers, resulting in a reduction in variance but no change in the mean. (b) Directional selection results from environmental change in a particular direction, giving an adaptive advantage to those individuals with traits on one side of the bell-shaped curve, causing the mean value for the trait to shift in the favored direction. (c) Disruptive selection results when a previously homogenous environment is split into new environments changing in different directions, subjecting populations found in each of the new environments to selective pressures in opposite directions, replacing the single bell-shaped curve with two new ones, each with a new mean optimized for their new environment. This disruptive selection is often the primary mechanism for the evolution of new species (Drawing by Louis Irwin).

isolated, they will evolve into two different species. Disruptional selection is thought to be one of the precursors to the origin of new species.

Keeping these different forms of natural selection in mind will be very important for assessing the plausibility of how life could evolve on other worlds, some of which have provided homogenous stable environments for billions of years, while others are dynamic and ever-changing in the challenges they present to whatever life may exist there.

### 3.2.2 Chance

Chance plays a role in the composition of a given star, in the nature, size, and location of planets in a solar system, and in the chemical makeup of every planetary body. Before life can begin, then, chance has already narrowed the possibilities for the molecules that make it up, the solvent from which it emerges, and the energy source that will drive it.

Within the range of possibilities mandated by the physical environment, chance continues to affect the direction that life will take. There are many more amino acids theoretically possible and known to exist in the universe than are found in living forms on Earth. We know this in part because we find them in meteorites – amino acids like aminoisobutyric acid that are not naturally occurring on our planet (Table. 3.1b). We don't know how large a role natural selection has played in reducing the number of possible amino acids to the 20 we now find, but the starting collection was surely determined by a great deal of randomness in the organic chemicals that fell to Earth or got synthesized under early Earth conditions. And the 20 that life on Earth has ended up with may have depended on which ones were first incorporated, by chance, into self-sustaining metabolic pathways.

Chance surely plays a part in the composition of the precursors for life wherever it arises [15]. To reiterate, whatever the protocellular aggregates are that compete for survival prior to the natural selection of one or more types of true cells that launch the evolutionary trajectory for life on a given planet, their initial composition may well be determined largely by chance (Figure 3.3). That composition may be biased by the substrate or other aspects of the environment, as suggested in Figure 3.2, so the chance distributions are not totally random. Note that the nature of the substrate or other factors themselves may be the products of chance as well. Because of chance, every molecule, protocell, or organism that constitutes a living system shows variation, and this is the raw material upon which natural selection acts to first give birth to life.

Once formed, living systems continue to flourish in an opportunistic world, where variation precipitated by chance is molded by natural selection into forms of life best fit for the environments in which they thrive. After natural selection has designated the general nature of the living system, the tendency *in the short run* is for stabilizing selection to reduce variability. Chance continues to generate variation in living organisms by several mechanisms over the long run, however. The most important are the following:

*Mutation*
Whatever the chemical basis is for storing information, it is subject to mutations which are largely random events. For life on Earth, that means that the sequence of nucleotides in DNA is subject to alteration by radiation, chemicals, or other physical factors. If one base along a strand of DNA that codes for a specific gene is

destroyed or replaced by another one, the genetic message for the gene is altered. This means, in turn, that the amino acid sequence, and hence the structure, of the protein that the gene codes for, will be altered. The change most often is detrimental (since natural selection presumably has already produced a well adapted structure), but occasionally it can be beneficial, and will be favored by natural selection in the struggle for survival. Since most mutations are detrimental, most are not perpetuated. Some mutations are neutral, altering a part of the DNA that doesn't code for a gene, or affecting protein structure in such a minor way that natural selection has no effect. And some mutations that initially are neutral, become favored or disfavored when the environment changes, raising or lowering their value under the altered conditions.

*Sexual Reproduction*

At its most general level of definition, sexual reproduction is simply exchange of genetic material. The onset of this process in the history of life was a monumental step forward in the ability of living organisms to maintain a continual mixing within their gene pool (all the genes of an interbreeding population collectively). With each genetically unique organism donating (male) or accepting (female) a unique sample of the gene pool toward the makeup of their offspring, the product of the union is guaranteed to be genetically unique as well. This process inevitably gives rise to some genetic variants that are more or less favorable, so natural selection operates anew on each generation. Biologists debate whether the increased variability produced by natural selection is the main reason for its persistence, but there is no debate that increased variability, hence an increased repertoire of gene combinations, is produced by chance for natural selection to act upon.

*Genetic Bottlenecks*

In those circumstances where a very small segment of the gene pool forms the foundation for subsequent generations, the "founding" pool of genes may differ just by chance from the composition of the original gene pool from which it was derived. This shifts the starting point for the subsequent evolutionary history of that particular group of organisms. This is known as passing through a genetic bottleneck. Examples include the establishment of a new population on an island by a small number of immigrants from the mainland, the movement of a small number of migrants through a temporary or difficult passage from one habitat to another, or the survival by a small number of individuals of a natural catastrophe that wipes out most of the parental gene pool. The populations that pass through the bottleneck typically give rise to descendants that have much less variation than the larger, original population. Some anthropologists believe that modern humans are descended from survivors of several genetic bottle-necks, ranging from 60,000 to 2 million years ago, when the entire human species may have dropped to a few thousand, or fewer, individuals. Other examples are thought to be cheetahs, the human population of Finland, and Darwin's finches in the Galápagos Islands. Humans in general have relatively little genetic variation, consistent with the bottleneck hypothesis. Cheetahs and the Finns are known for their genetic homogeneity, and the differences among the various species of finches on each island in the Galápagos reflect the "founding" effect of the first colonizers.

*Genetic drift*
In any large population, a local segment may consist of a collection of genes that differs from the average composition of the population as a whole. This chance deviation could provide fertile raw material for the natural selection of new biologically useful characteristics. Of course it could also doom the local group to a higher than average mortality. The point is that natural selection can act on local variations in genetic composition that vary by chance.

### 3.2.3 Heredity
Heredity is the passage of information about how an organism builds and maintains itself from one generation to the next. For life on Earth, genetic information is stored in chemical form – mostly as a sequence of nitrogen bases along a strand of DNA (Figure 2.7 in chapter 2), a huge macromolecule that would extend for miles if stretched out in a line. Extensively folded, however, it can be tucked into the microscopic nucleus of a single cell. When the cell divides, a complete replicated copy of each DNA molecule passes to the descendant cell. As long as parents have offspring, genetic information lives on. This feature, in which the repository of hereditary information survives independently of the organism through which it passes, is a signature feature of living systems.

The DNA that constitutes life on Earth uses four different bases arranged in groups of three (Figure 2.7 in chapter 2), for a total of 64 different possible combinations ($4^3=64$) of *codons*. This is more than enough to code for the 20 amino acids found most commonly in living organisms on Earth. There is circumstantial evidence that earlier genetic codes may have used only two bases ("letters") for each codon ("word"). This would still have provided enough coding for 16 amino acids ($4^2=16$). An even simpler code using just two variable molecules can be envisioned (Figure 3.5). Depending on the length of each "word," such a code would provide enough information for 2, 4, 8, or 16 different possibilities, based on groupings of 1 ($2^1=2$), 2 ($2^2=4$), 3 ($2^3=8$), or 4 ($2^4=16$) different bases.

If this is starting to sound like an essay on information theory, that's really the point. Biology is about information as much as it's about energy, structure, and function. Whatever alien life we find, wherever we find it, there will need to be some mechanism for information storage and transmission that is found within, but survives independently of, that form of life. A linear array of chemical monomers, like the four nitrogen bases in DNA, serves as a simple, elegant solution to this problem. Whether a linear array of molecular components will be found to be a universal mechanism for genetic storage, whether they will be similar to or the same as the ones in DNA, and whether the number of variable forms of those components will be less or greater than three –these questions are among the most intriguing of all in the quest for life on other worlds.

## 3.3 Ecosystems: from populations to pyramids

Have you ever wondered what those big-headed slender aliens with huge slit eyes in B-grade science fiction movies have for dinner? Are any of them ever portrayed as farmers? Do they have other species for pets? If they are green, are they photosynthetic? Except for a few *avant-garde* episodes of *Star Trek* or *The Twilight Zone*, the feeding habits of aliens have seldom been a source of popular concern. This, however, is one of the most critical bits of information needed in assessing the scientific plausibility of any alien form of life.

### 3.3.1 Food webs
The fact is that organisms evolve as part of an ecosystem that includes many forms of life, all interacting in a complicated way. What happens to one group can have a profound effect on another, so the existence of any form of life presupposes an array of others with which it interacts.

Consider the simplified marine ecosystem shown in Figure 3.7. Sharks and swordfish top the food chain. They both eat larger fish, like perch and tuna, and sharks prey on seals, which also eat perch. Tuna and perch feed on smaller fish, like sardines, which derive their food from even smaller invertebrates. The source of food for the smaller invertebrates is ultimately a large biomass of microorganisms and a few plants occupying the upper layers of the ocean where light penetrates enough to drive photosynthesis. Scavengers like lobsters, clams, and flounders live on the ocean floor, where they harvest the dead remains from all parts of the food chain.

Organisms that derive their energy directly from the environment, as opposed to indirectly by consuming other organisms, are known as *autotrophs* (loosely translated from the Greek as "self feeding"). All producers are autotrophs. Those organisms that derive their energy from other living, or formerly living, organisms are known as *heterotrophs* ("other feeding"). Among the heterotrophs, nutrients flow from one level of the food chain to the next – *primary consumers* feeding on producers, *secondary consumers* feeding on primary consumers, and *tertiary consumers* feeding on both primary and secondary consumers. Another type of consumer, the *detrivore,* recycles nutrients by consuming food from dead organisms. Not illustrated in Figure 3.7 but commonly found in terrestrial ecosystems are heterotrophs that are both primary and higher order consumers. Humans, for instance, eat plants as primary consumers, beef as secondary consumers, and tuna as tertiary consumers.

### 3.3.2 Trophic structures
Whatever life is found on another world, unless it consists of autotrophs alone, is going to be part of a community of organisms, whose individual members

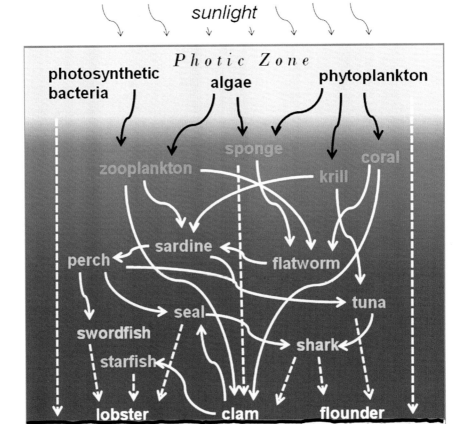

**Figure 3.7** Food web in a marine ecosystem. Arrows show the direction of nutrient flow. Photosynthetic producers (black) convert sunlight and simple molecules into organic compounds. They are eaten by primary consumers (orange) which serve as food for secondary and tertiary consumers (yellow). All eventually die, their remains falling to the ocean bottom (dashed lines) to serve as food for a variety of scavengers (white). Most consumers except for detrivores actually live in the photic zone, though they are shown beneath it here to provide a clearer diagram (Drawing by Louis Irwin).

interact extensively. The flow of energy through a community, like the food web illustrated in Figure 3.7, can be simplified into a formal diagram of the total amount of living organisms (biomass) at each level of the food chain, or each trophic level (Figure 3.8).

Energy enters the ecosystem from a primary source, such as sunlight. Phytoplankton and other photosynthetic microorganisms (producers) in the

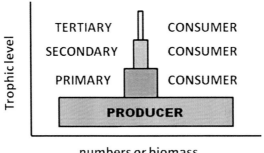

**Figure 3.8** Community structure pyramid. The number of individuals or total biomass is indicated by the width of the polygon (horizontal axis) for each trophic level (vertical axis). Detrivores are mostly microorganisms too numerous to be shown on this scale.

photic zone convert elementary molecules like $CO_2$ and $NH_3$ into organic molecules using light energy. Because they manufacture their own "food" from non-living resources, producers are known as autotrophs. Producers can also be designated according to the source of energy they draw upon. Thus photosynthetic producers can also be called *phototrophs*.

Producers are eaten by primary consumers, like zooplankton. The organic compounds made by the producers serve as the source of energy and structural building blocks for all subsequent consumers. Secondary consumers, like small fish, feed on the primary consumers. Tertiary consumers, such as large fish, feed in turn on the secondary consumers. In contrast to autotrophic producers, consumers are known as heterotrophs.

When organisms at any level of the food chain die, their remains become detritus which is consumed ultimately by detrivores, some of which can serve as a source of food as well.

Because energy conversion can never be 100% efficient from one level to the next (see Box 2.1 in chapter 2), the total amount of energy available for biosynthesis and maintenance decreases at each higher trophic level. This means that the total biomass of any given trophic level is less than that of the level below it. Since larger organisms are generally required for consuming smaller organisms (except for detrivores), the average size of individuals increases at each trophic level, hence the total number at that level goes down (number = biomass/size). Thus, if either total number of individuals *or* total biomass is plotted as a function of trophic level, the result is a pyramid, with biomass highest for producers and lowest for the top consumers in the food chain.

The structure of such communities can be estimated if a few critical facts are known. If the total amount of available energy, and the efficiency with which that energy is converted into biomass, are known, the total biomass supportable by that energy can be calculated. Furthermore, if the efficiency of energy conversion from one trophic level to the next is known, the total biomass for each trophic level can be calculated. And from biomass, the number of organisms can be calculated if their size is known, or their size can be calculated if their total number is known. This means that, in principle, the number and size of organisms at different trophic levels for a given amount of energy input can be calculated, and the overall structure of the community described in quantitative terms. We have done this for putative life on Europa, for instance [16].

The significance of this for astrobiology, and for the purpose of this book, is that reasonable speculations can be advanced about the nature and structure of ecosystems on other worlds, provided that plausible facts about the environment are available at the outset. Most often, unfortunately, the crucial facts are not yet known with sufficient detail to speculate with confidence about any alien ecosystem. But if boundary conditions can at least be placed on some of the underlying assumptions, such as the amount of energy available for support of the ecosystem, then plausible boundaries for what an ecosystem *could* look like in that environment, on that alien world, can be outlined. That is an exercise we will undertake in the chapters to follow on how life could evolve on other worlds, after reviewing how life *did* evolve on Earth.

## 3.4 Chapter summary

Informed speculation about how life can originate from non-living matter, and an understanding of how evolution works and ecosystems are structured, is necessary to provide a framework for imagining how life could evolve on other worlds.

A number of different ways, in a number of different environments, have been suggested for the origin of life on Earth. Almost all proposed schemes take place in water, since Earth (and many other bodies in our Solar System) were water rich early in their history; and most of the proposed schemes have a number of generalizations in common. All assume that organic compounds were made available either from abiotic synthesis or by delivery from comets or meteorites, as precursors. Energy consuming reactions are presumed to have bound different precursors together on occasion. Usually, these proto-biomolecules would have quickly come apart, but under occasional favorable conditions, some may have survived to another cycle of synthesis, perhaps leading to a more complex mixture. The early aggregates would likely have included molecules with significant hydrophobic properties, capable of aggregating apart from their hydrophilic surroundings, and occasionally enclosing a portion of the organic-rich contents of the hydrophilic environment within proto-membranous barriers. These could have constituted early protocells. As protocells became more common and more complexly constituted, some may have harbored a set of biochemical interactions that became self-sustaining. Some of these interactions could have included self-directed replication of information capable of holding and transmitting hereditary information. Once the metabolic and replicative processes became consistent enough to reproduce new protocells of the same kind from pre-existing ones, a living process by definition would have been underway.

There is much debate about the details, necessary environments, and sequence for the scheme outlined above. There is not much disagreement among scientists, however, that life could arise from non-living precursors under appropriate conditions, in principle.

One of the conceptual hurdles that a mechanistic model for the origin of life has to overcome is the apparent "directedness" of that emergence – its transformation from disorganized and chaotic (high entropy) matter through a progressive sequence of increasing complexity leading to highly organized (low entropy) self-maintaining and self-perpetuating entities. Natural selection is the mechanism that provides that apparent direction to the trajectory of living systems.

Natural selection, acting upon the inherent variation in biological populations, is now understood to be the major factor that drives evolution. Before the evolution of definable species even begins, however, natural selection works to shape the early precursors of life. Through non-random topographies in the substrate, chance aggregation, and other random factors, some combinations of interacting components turn out to be uniquely favorable for survival. Natural selection screens these early aggregates of organic matter, leading to the perpetuation of those protocells most likely to survive long enough to replicate themselves in whole or in part.

Once a population of true cells can reliably replicate itself, true evolution proceeds under the mutual influence of natural selection acting on inherent genetic variation, introduced by mechanisms such as mutation, sexual reproduction, and genetic drift. Environmental conditions drive natural selection in different ways. Homogeneous environments that remain stable for long periods of time promote *stabilizing selection*, which resists evolutionary change and diminishes variation. *Directional selection* occurs in changing environments as selective pressures drive biological change in a particular direction. When a population becomes split by environmental changes in opposite directions, *disruptive selection* is the result, often culminating in the origin of news species.

While the view propounded above is focused on a single population of organisms, in reality life evolves as a community of organisms, tied together by intricate food webs that distribute the flow of energy through the community. Biologically useful energy is brought into a biotic community from the abiotic environment by autotrophic organisms that constitute the *producers* for all the organic precursors the community will use. *Primary consumers* incorporate these organic nutrients into the food web, while *secondary* and *tertiary consumers* move it up the food chain. Ultimately, *detrivores* recycle organic material from the remnants of dead organisms. Because none of the consumers or detrivores manufactures their own food, they are known as heterotrophs.

Each of these steps in the hierarchical food chain captures less than the total energy available from the trophic level below it, resulting in a pyramidal structure of biomass and numbers, whereby biomass and number of organisms progressively goes down as trophic level rises. If quantitative values for the energy available to the system, and certain other parameters are known, the overall structure of the ecosystem, including number and size of organisms at each trophic level, can be estimated. This enables speculation about life on other worlds to be a scientific rather than purely imaginative exercise, even if the details are highly uncertain.

## 3.5 References and further reading

1  Melosh H.J. 2003. Exchange of meteorites (and life?) between stellar systems. *Astrobiology* **3**: 207–15.
2  Miller S.L., Lazcano A. 1996. The origin and early evolution of life: prebiotic chemistry, the pre-RNA world, and time. *Cell* **85**: 793–799.
3  Miller S.L., Orgel L.E. 1974. *The Origins of Life on the Earth*. New York: Prentice-Hall.
4  Orgel L.E. 2003. Some consequences of the RNA world hypothesis. *Orig. Life Evol. Biosphere* **33**: 211–218.
5  Popa R. 2004. *Between Necessity and Probability: Searching for the Definition and Origin of Life*. Berlin: Springer-Verlag.
6  Schulze-Makuch D. and Irwin L. N. 2008. *Life in the Universe: Expectations and Constraints*. Berlin: Springer-Verlag, 2nd ed.
7  Mason, S. 1992. *Chemical Evolution: Origin of the Elements, Molecules, and Living Systems*. New York: Oxford Univ. Press.
8  Luisi P.L., Ferri F., Stano P. 2006. Approaches to semi-synthetic minimal cells: a review. *Naturwissenschaften* **93**: 1–13.
9  Luisi P.L., Varela F.J. 1987. Self-replicating micelles: A chemical version of a minimal autopoietic system. *Origins Life Evol. Biosphere* **19**: 633–643.
10  Andras P. and Andras C. 2005. The origins of life – the 'protein interaction world' hypothesis: protein interactions were the first form of self-reproducing life and nucteic acids evolved later as memory molecules. *Medical Hypotheses* **64**: 678–88.
11  Lathe R. 2004. Fast tidal cycling and the origin of life. *Icarus* **168**: 18–22.
12  Miller S.L. 1953. A production of amino acids under possible primitive earth conditions. *Science* **117**: 528–529.
13  Fox S.W. and Dose K. 1977. *Molecular Evolution and the Origin of Life*. New York: Marcel Dekker.
14  Wächtershäuser G. 1988. Before enzymes and templates: a theory of surface metabolism. *Microb. Rev.* **52**: 452–584.
15  Hughes R.A., Robertson M.P., Ellington A.D., et al. 2004. The importance of prebiotic chemistry in the RNA World. *Current Opinion in Chemical Biology* **8**: 629–33.
16  Irwin, L.N. and Schulze-Makuch, D. 2003. Strategy for modeling putative multilevel ecosystems on Europa. *Astrobiology* **2**: 813–821.

http://www.talkorigins.org/faqs/abioprob/originoflife.html
Good overview of ideas about and experimental approaches to the abiogenic
origin of life, with many relevant links.

http://www.globalchange.umich.edu/globalchange1/current/lectures/selection/
selection.html
Excellent summary of natural selection, including its different varieties.

http://www.indiana.edu/~ensiweb/lessons/ns.cum.l.html
A clever demonstration of the non-random nature of natural selection using
playing cards, suitable for classroom use.

http://evolution.berkeley.edu/evosite/evohome.html
Excellent primer on evolution, designed especially for teachers.

http://scienceaid.co.uk/biology/ecology/food.html
Brief, clear illustration of energy flow and ecological pyramids.

# 4 Fire and Water

## *Life on a geologically complex, water-rich world with an oxidizing atmosphere like Earth*

Suppose there were two intelligent beings – let's call them scientists – sitting on a planet orbiting Epsilon Eridani, a star just over 10 light years from Earth. They are writing a book about how life could evolve on other worlds. Their technology is a little more advanced than ours, but not by much. For example, they've developed powerful remote detection technologies that can gather a good bit of information on solar systems up to 12 light years away (while we can barely detect large planets in such a system, and only indirectly at that). Like us, though, they haven't yet achieved the capacity for exploration by robotic probes of a world that distance from their home planet, so they don't have any close-up high-resolution images of the worlds that far away.

They are aware of our Solar System, and think it has some interesting possibilities for harboring life. From the size, luminosity, and temperature of our Sun, they can tell it's in the middle of its life cycle, somewhere around 5 billion years old, perhaps. That means the eight planets they can make out that orbit our Sun are probably about that age as well. With their high-powered telescopes, they can see that four of our planets are quite large, especially the biggest one nearest the Sun. But they can also make out that there are four much smaller planets orbiting inside the giants. Using conventional spectrophotometry, they can tell that the giants probably consist mostly of gasses like nitrogen, hydrogen, and helium. The smaller planets with the inner orbits have higher densities, a lot more silicon, and a lot less in the way of volatile gasses. That makes them smaller, rocky planets.

Of the four rocky planets, the third one out from the Sun appears to be special in two respects. First, it has an unusually strong spectral signature for water. Second, it has a surprising abundance of oxygen. We don't know what those alien authors would call that third planet from the Sun, but we call it Earth.

## 4.1 Nature of Earth

Knowing the value of water as a cradle for life, and the unexpectedly high content of oxygen suggesting a biogenic source, our authors would have focused on the Earth from their earliest discovery of it. Observation over the years would

L.N. Irwin and D. Schulze-Makuch, *Cosmic Biology: How Life Could Evolve on Other Worlds*,
Springer Praxis Books, DOI 10.1007/978-1-4419-1647-1_4,
© Springer Science+Business Media, LLC 2011

have given them a lot of information about this planet, so inconspicuous that far away but obviously special in several respects.

### 4.1.1 Atmosphere

At the very limit of the resolution of their telescopes, they could tell that the Earth is covered by a thin layer of gases, consisting of five times more nitrogen than oxygen, with a smattering of carbon dioxide and an amount of methane that is higher than expected as well. Together, these gases must make up an atmosphere of moderate density. That would tell them two things: Earth has weather, and its surface has some protection from radiation and incoming debris from space.

### 4.1.2 Building blocks

At first, our alien scientists would have been impressed with the high content of silicon and sulfur – thousands of times higher than their abundance in the Sun. While much if not most of these two elements would probably be tied up in the rocky mantle and crust, they couldn't help noting that these elements can also, in theory, be building blocks for life. The favored constituent of living systems, they would have known for reasons we discussed in chapter 3, is carbon; but carbon, they discovered, makes up less that 0.1% of the Earth's mass. Still, this makes it three times more abundant on Earth than in the Sun. So carbon, silicon, and sulfur, all three could in principle serve as building blocks for life on Earth.

### 4.1.3 Energy

At 150 million kilometers from a star the magnitude of the Sun, Earth would clearly be bathed in sunlight, a very favorable source of energy for living systems. That high oxygen content would provide the prospect of oxidative chemistry as well. Assuming there were plenty of reduced compounds, like $H_2$, $H_2S$, iron sulfide, methane, or other organic substances, the possibility of extracting energy from oxidation-reduction chemistry would be a second, enticing possibility for the energetic support of life. Finally, the presence of an atmosphere and large bodies of liquid water would mean the possibility that kinetic energy, as well as osmotic and ionic gradients in the water, could be harvested by living systems.

### 4.1.4 Temperature

Their powerful infrared spectrophotometers would tell them that Earth has an average global temperature of about $15°$ C ($59°$ F). This, plus the fact that Earth has an atmosphere and gravity strong enough to hold on to molecules as small as $H_2O$, would mean that much if not most of the water is liquid, with some of it as water vapor in the atmosphere. Once they knew the prevailing temperature on Earth, they would have ruled out silicon and sulfur as building blocks for life, because there are no stable polymers of those elements flexible enough to serve as a basis for biochemistry at that temperature.

**Figure 4.1** Third rocky planet from the Sun. This view of Earth acquired by the Galileo spacecraft in 1990 from a distance of 2.4 million km (1.5 million miles) reveals the planet's essential features: a water-rich world with continental masses, an atmosphere, and weather (NASA/JPL).

### 4.1.5 Topography

From a distance of 10 light years, our alien authors' view of Earth would probably not be as clear as the one in Figure 4.1, but if they could barely make out some masses of land, they would be able to deduce that the surface is not uniform – that it consists of boundaries between land and water. From this they would be able to add the reasonable speculation that there might well be various depths to the surrounding oceans, that the masses of land might hold pools of water, and that the land itself might vary in elevation. Indeed, their information on the size and density of Earth might lead them further to assume that it has a metal-rich core likely generating heat internally from radioactive decay. This is a circumstance that could lead to plate tectonics, which would suggest changes in the topography over time, as well as a heterogeneous surface, ranging from mountains, to lowlands, to deserts.

### 4.1.6 Cycles

If by chance they could detect the period of the Earth's rotation (its daylength), the angle of its tilt from the vertical (its obliquity), and its orbital period (length of one year), they would know that it has alternating seasons of warmth and cold that differ in the northern and southern hemispheres. They would also know how long a day lasts at any spot on Earth, and how long the seasons last in each hemisphere.

### 4.1.7 Conditions for life on Earth

It's no wonder that scientists on that planet in the Epsilon Eridani solar system would be very interested in Earth. With only the information outlined above, they would be able to draw up the following list of conditions favorable for finding life, and a rich variety of it at that, on Earth:

1. A large amount of liquid water spread over the planet, held in by gravity and an atmosphere, but able to cycle between solid (frozen), liquid, and vapor phases, would allow life-sustaining water to be broadly and variably distributed.
2. A global temperature averaging 15° C would keep most of the water liquid, but would allow for water vapor to form and fill the atmosphere, then return to the surface through rain and perhaps snow at the colder latitudes. This would mean pools of water, forming ponds and lakes on the land, as well as rivers running from higher to lower elevations. This would be "fresh" water, since dissolved salts would be left behind when water evaporates. Since most of the water would evaporate from the oceans, they would become saltier over time as minerals from the crust dissolved but were left behind when the water evaporated.
3. With water as a solvent, carbon-based chemistry would be well suited to form the polymers necessary to give rise to the molecular variability and complexity of a living system.
4. The abundance of sunlight would provide a constant, inexhaustible source of energy well suited for supporting biological processes.
5. Internal heating and plate tectonics would indicate a geologically active planet. This would enable the cycling of minerals from the interior to the surface and back to the interior, aiding oxidation-reduction cycles, and resculpturing the planet on an ongoing basis – forming new continents and waterways, raising up mountain ranges and sinking others beneath the sea. Such change over time would be a driving force behind the ongoing evolution of whatever life had emerged.
6. The presence of an atmosphere, especially one rich in ozone derived from oxygen, would filter out a lot of the ultraviolet radiation that would tend to break down many organic compounds. It would also protect the surface from bombardment by small to medium sized meteorites. Finally, it would mean that weather in the form of wind and precipitation would be changing the surface continually.

7. Topographic variation, ranging from shallow shorelines, to deeper seas, to deep oceans, and from low to high elevations on land, coupled with weather patterns and ocean currents that would create areas ranging from very wet to very dry, would mean a lot of habitat fragmentation, a situation conducive for diversification.

8. Seasonal cycles, which would vary at different latitudes, would provide a basis for biological cycles and adaptations that would promote further diversity.

9. An age of close to 5 billion years would mean a long time for life to take hold and evolve.

### 4.1.8 Facts consistent with the existence of life on Earth

The fact that all the conditions listed above would be present does not mean that life would have to exist on Earth. Our cosmic colleagues, after all, are good enough scientists to be appropriately skeptical. However, there are two facts that probably would persuade them that Earth very well may harbor life.

**1. The large amount of oxygen and the presence of methane would most likely mean a biogenic origin.** The persistence of oxygen and methane in Earth's atmosphere would be significant in two ways. First, neither would be that concentrated in the first place, and certainly wouldn't stay that concentrated, if they weren't being continuously generated. Oxygen is highly reactive with almost anything capable of being reduced, and methane in the presence of oxygen is not stable for very long. Thus, a biogenic origin for both oxygen and methane are much more likely than any exotic geochemical cycling that could theoretically replenish the atmosphere with both compounds. Second, the coexistence of oxygen and methane would indicate that Earth has the capacity for robust oxidation-reduction cycles, which provide energy for biological systems just as efficiently as sunlight.

**2. Liquid water, at least two abundant forms of efficient energy, and probably organic compounds have been present over a long period of time.** The presence of liquid water, abundant sunlight, and the capacity for oxidation-reduction cycling would be readily evident. The presence of organic compounds would be less certain. Our remote scientists could undoubtedly pick up spectral signatures that could indicate organic molecules, but the blurring effects of the atmosphere, the spectral coincidence of signals from inorganic compounds, and the interference from other molecules, including organics in space, over such a distance would make detection of organic compounds on Earth problematical. Nonetheless, let's assume that the high density of plant cover over all the equatorial land masses, with their concentration of complex organic molecules like chlorophyll and cellulose, would be detectable. Even if our remote observers would not be able to identify these compounds, they would likely recognize them as something different, and therefore possibly indigenous to our Solar System. Knowing the approximate age of the Earth, they would assume that the persistence of organic compounds in such an oxidizing environment must mean a biogenic origin and continual replenishment of them as well.

### 4.1.9 Possible assumptions about the nature of Earth's biosphere

Once our alien authors decided that the probability of life on Earth is high, they would be tempted to make a number of predictions about the characteristics of the biosphere on our planet, including the following:

- Life on Earth is probably diverse because of the
  - large degree of habitat fragmentation and environmental variation on the planet, and
  - the long span of time available for evolution to occur.
- At least three broad categories of life are probably found on Earth, occupying subterranean, aquatic, and terrestrial habitats, respectively. The atmosphere itself may constitute a fourth habitat.
- Each habitat category probably supports multiple trophic levels: two or more consumers as well as producers and decomposers.
- Given the abundance of energy and only moderate gravitational constraints, the size of individual organisms at the top of the food chain could be large. They would likely be largest in the oceans, because of the buoyancy of water, and smallest in the atmosphere because of the thinness of air.
- Given the relatively warm prevailing temperatures and the availability of oxygen for supporting the metabolic efficiency of oxidative metabolism, the upper level consumers might also be quite active.

### 4.2 A model for the history of life on Earth

Having assembled the facts and inferences listed above, our alien scientists would now be prepared to outline a plausible scenario for the evolution of life on Earth:

### 4.2.1 Origin of life on Earth

From their vantage point, the Earth would appear so much more favorable for life than any other planet in our Solar System, our alien authors would surely make the simplifying assumption that life most likely arose *de novo* on Earth. All the conditions (water, carbon, energy) were present at an early stage, and interfaces between water and mineral substrates, as along a shoreline, in ephemeral inland pools, or on the ocean bottom, would have provided potential launching sites for life.

Their assumptions about how fast this would have happened would likely depend on their own experience. If life had arisen early in the history of their planet, they would probably assume it could have emerged early on Earth as well. Had that not been the case, they would at least know that a window longer than 4 billion years was available for life to evolve on Earth [1].

How fast life can emerge in principle, and how rapidly it can undergo major transitions (from unicellular to multicellular, for example), is one of the greatest unknown, and therefore limiting, factors in any speculation about the nature of life on other worlds [2].

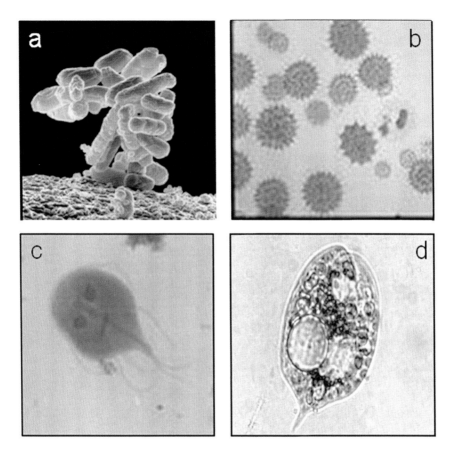

**Figure 4.2** Ancestral forms of life on Earth. Microorganisms that represent a plausible evolutionary sequence in the early history of life on Earth. (a) Simple chemoautotrophic cells with undifferentiated cellular interiors. (b) Photoautotrophs following the evolution of molecular machinery for converting the energy of sunlight into high-energy chemical bonds. (c) Motile unicellular microbes showing signs of increasing complexity, such as this binucleated ancestral protozoan, *Giardia*. (d) Larger eukaryotic microbe, with more complex subcellular organelles, such as this flagellate.

## 4.2.2 Early stages of life on Earth

Our authors would assume that the earliest forms of life were simple cells devoid of complex internal structures. They likely would further assume that those earliest forms of life were chemoautotrophs (Figure 4.2a). While they would know that sunlight has always been readily available on Earth, they would also realize that harvesting energy from straightforward oxidation-reduction of inorganic substrates would be simpler than coupling sunlight to the generation of biologically useful energy [3]. That would come about in time, however, leading to simple unicellular photoautotrophs (Figure 4.2b), once the molecular machinery for harvesting sunlight had evolved.

At this point, two producers would be present. Soon, more complicated primary consumers, in the form of motile, unicellular organisms would be assumed to appear (Figure 4.2c). Larger microbes with subcellular differentiation for metabolic specializations (Figure 4.2d) might have represented the next evolutionary step.

### 4.2.3 The dilemma of oxygen and the earliest forms of life

Earth's high oxygen content would present a genuine puzzle to our alien scientists, who could only guess how and when it came about. There is no comparable atmosphere on any other planetary body known to us at the present time, nor is there any known geophysical process that generates and *maintains* such a high level of atmospheric oxygen. The possibility that some exotic geochemical cycling or unusual atmospheric chemistry could generate such an elevated, persistent level of oxygen cannot be excluded. Until such an example is found, however, or a geochemical system capable of generating it is convincingly modeled, it seems safer to assume a biogenic origin for oxygen; and that, in turn, would suggest that the composition of the atmosphere has changed considerably over Earth's planetary history.

A question with which our distant observers must have grappled would have been what life was like during the early stages of Earth's history, when the atmosphere was deficient in oxygen? Life clearly must first have arisen in the absence of oxygen (anaerobically). Knowing the role that oxygen could play in energy-yielding oxidation-reduction cycles, our scientists may have concluded (correctly, as we know), that life on Earth was microscopic and low in energy consumption for a long time, until sufficient oxygen had built up to support more energetic forms of life.

### 4.2.4 Transition to multicellularity

Now our alien authors would be in a quandary. Should they assume that Earth became filled over evolutionary time with single celled forms of life, saturating the air and oceans and covering all the land with infinitely variable and increasingly sophisticated organisms, but never exceeding the size of a single cell? Or should they decide it would be more likely that, sooner or later, some of the single cells would begin to aggregate into collections of cells, giving rise to larger organisms capable of processing more energy more efficiently, and carrying out a greater variety of functions?

Our temptation is to assume that if they are intelligent enough to get this far, they must be (like us) relatively large, complicated, multicellular forms of life. Based on their own apparent history, therefore, they would almost surely assume that life on Earth would sooner or later have made the transition to multicellular forms. On the other hand, our assumption that intelligence presupposes multicellularity may be biased by our own nature. In principle, there seems to be no law of chemistry, physics, or information theory that would preclude a complex cell of some sort from being intelligent (see chapter 12).

There is a more compelling argument for the eventual evolution of multicellular forms of life, however. Once producers have become readily available, the evolution of consumers would seem inevitable, because deriving energy from organic nutrients is much simpler than converting energy from the abiotic environment into organic form. That being the case, the size of consumers would have been larger than the organisms being consumed (at least in the beginning when the easiest form of consumption was simply engulfing the producers serving as food). Extrapolating this trend to higher level consumers would mean an ever increasing size of consumer organisms. Single cells, however, cannot become very large because of the diffusion distances from the cell periphery to its center, and because of the drastically diminishing surface to volume ratio as a cell grows larger. The obvious solution to the restraint on the size of single-celled organisms is to allow them to grow larger by making them multicellular.

So we will suppose that our alien scientists would probably have postulated that multicellular life would eventually have arisen on Earth. The aggregation of single cells into a sponge-like organism specialized for filter feeding (Figure 4.3a) would be a reasonable assumption. Intriguing examples of possible transitional forms are the calcite microbialites, found in British Columbia and Turkey (Figure 4.3b). These stationary structures are produced by colonial mixtures of prokaryotic and eukaryotic cells living together but carrying out no obvious coordinated function, though some benefit must be obtained from their association.

We should note that had our alien colleagues decided that, well before multicellular life made its appearance on Earth, the planet would in fact have been saturated with unicellular microorganisms, they would not have been

**Figure 4.3** Multicellularity. (a) Sponges are the most ancestral form of multicellular eukaryotic life surviving on Earth. (b) Microbialites, a loosely organized colony of prokaryotes and eukaryotes, may be relics of an even earlier transitional form, as shown here in Pavilion Lake, British Columbia, Canada (Photographs by Jessie Cohen, National Zoological Park (a), and Donnie Reid (b)).

wrong. Indeed, every cubic centimeter of the upper layers of the ocean, fresh water, and land, holds millions of single celled organisms. Even the air to a considerable altitude is filled with floating microbes. While multicellularity has its advantages, the elegant simplicity of the single cell has kept it the most abundant form of life on Earth [4]. The same probably holds true wherever life exists throughout the universe.

In time, other multicellular configurations of greater complexity clearly carrying out coordinated functions would have begun to litter the floor of shallow bodies of water (Figure 4.4). Among the organisms that might reasonably be expected would be upright flat-faced structures designed to capture sunlight, and a host of consumers, such as stalk-like filter feeders akin to modern sponges, bottom grazers like modern snails that would slide along the bottom engulfing surface-dwelling microbes, and filter feeders of various other shapes. Nor would it be unreasonable to suppose the existence of organisms with tentacles, for harvesting kinetic energy from the ambient water currents.

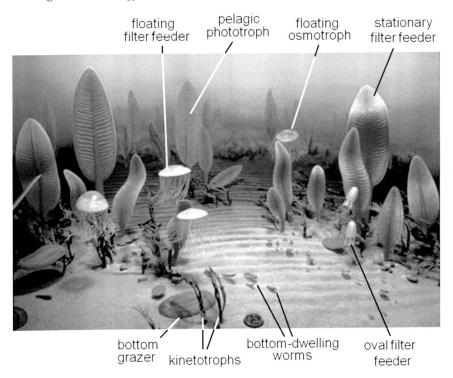

**Figure 4.4** Ancestral sea floor. This artist's conception of what a sea floor may have looked like on Earth 600 Mya shows a variety of increasingly complex multicellular organisms. Having no way of knowing any details of these organisms, our alien scientific colleagues would only be able to speculate on the generic forms of life that might be present (Image from the National Museum of Natural History, courtesy of the Smithsonian Institution).

Eventually, some consumers would be expected to develop an appetite for other consumers. These secondary consumers would be found in every habitat where primary consumers were found to be feeding on producers. The simplest strategy for secondary consumers would simply be to rely on the mobility of primary consumers to come to them. Filter feeders like sponges, floaters like jellyfish that entangle whatever prey swims into their tentacles, or clams that slowly pump nutrient-filled water currents through their shell-enclosed digestive systems could all be logically anticipated. Others, like frogs, crocodiles, and carnivorous plants would be sedentary most of the time, waiting to ambush their passing prey. Some would be very active, in order to chase and catch their mobile food – sharks in the ocean and hawks in the air, for example. The need for speed and strength in bringing down large, fast prey, in fact, would be a strong selective force in the evolution of animals like wolves and lions.

The evolution of active organisms, and possibly of large organisms in general, probably depended on transformation of Earth's atmosphere from oxygen-poor to oxygen-rich air. The rise of atmospheric oxygen presented both a challenge and opportunity for the support of life. On the one hand, oxygen is such a strong reactant that highly reduced compounds, such as organic molecules with all their carbons and hydrogens, would be in danger of being oxidized – literally "burned up" – whenever they came in contact with oxygen. In that sense, oxygen is very toxic to living systems. On the other hand, such reactions release a lot of energy (Table 2.1 in chapter 2); so the potential benefit to life of such a pervasive ability to oxidize organic compounds is also clear, assuming that processes could have evolved to carry out the reactions in a controlled fashion. This could be accomplished through a series of incremental steps releasing small amounts of energy at a time. If these small packets of energy could be captured in the high-energy bonds of other organic molecules like ATP (Figure 2.7 in chapter 2), the challenge of creating and storing biologically useful energy would be met.

In time, we know that photosynthesis and oxidative metabolism came to be linked together, in a biogenic cycle that both *generates* and *consumes* oxygen (Figure 4.5). In plants, $CO_2$ from the atmosphere serves as a carbon source for the reductive synthesis of glucose – a process that consumes energy from ATP. H is derived from $H_2O$, which is split by sunlight, releasing $O_2$ into the environment and energy which ultimately ends up in ATP. In animals, glucose is oxidized by $O_2$ to $CO_2$ and $H_2O$, with the energy released being captured in the high-energy bonds of ATP.

We have no way of knowing how well our alien colleagues would have been able to deduce the planetary and atmospheric history of Earth. If they would have at least figured out that our atmosphere started out with little or no oxygen, then gradually accumulated a great deal of it, they well might be able to guess the broad outlines of the history of life on Earth.

## 4.2.5 A brief descriptive history of life on Earth

What actually happened on Earth was that life emerged close to four billion years ago, and persisted in microbial form until about a billion years ago. Over this

**Figure 4.5** Cyclic production and consumption of glucose ($C_6H_{12}O_6$), $O_2$, $CO_2$, and $H_2O$. Plants synthesize glucose from $CO_2$ and generate $O_2$, while animals oxidize glucose with $O_2$ to produce $CO_2$. (Elephant photograph by Ann Batdorf, National Zoological Park).

period, photosynthesis evolved and began to change the atmosphere, consuming $CO_2$ and generating $O_2$. And life, while remaining unicellular, took on more complex forms, as simple cells merged into a symbiotic union of more complicated cells, with various compartments and specializations, including an increasingly complex system of heredity involving genetic exchange. It took a long time for enough oxygen to build up to the point where multicellular organisms of significant size and activity could be supported energetically.

That point occurred about 540 million years ago (Mya), with a sudden proliferation of mostly larger, more active animal phyla – an event so sudden and dramatic it has been termed the "Cambrian Explosion" [5]. Figure 4.4 is an artist's conception of what the shallow sea floor may have looked like a little prior to that, when a lot of evolutionary experimentation was going on with all sorts of body forms, functions, and life styles – most of which did not survive the filter of natural selection.

Our alien scientists would have no way of knowing, from contemporary remote observations of Earth, whether or when an event like the Cambrian Explosion would have occurred. But they probably would have inferred that life on Earth would sooner or later generate some larger, more active organisms. Unless they were extremely insightful, or very lucky, it seems unlikely that they would have been able to guess what those shallow seas of the Earth would have looked like as life emerged from its Precambrian experimentation. From the fossil record we know that mollusks like the nautiloids, and arthropods – animals with jointed appendages and exoskeletons – became prevalent in the early seas and oceans (Figure 4.6a). The development of hard, outer armor would have been a reasonable conjecture for organisms increasingly dependent on one another for food. It seems most unlikely that an animal as objectively bizarre as an octopus (Figure 4.6b) would have been anticipated by alien scientists, much less that it would have evolved from a much simpler mollusk into the dominant and most intelligent invertebrate on Earth. By the end of the Cambrian, the trend toward increased size and specialization for capturing prey was well underway as depicted in Figure 4.6.

*4.2.5.1 Transition to larger, active organisms*
Water is a buoyant medium, especially when salty, so size in a marine habitat is restricted more by the availability of food and other ecological considerations than by gravity. Aware of this, our alien authors may well have predicted that a succession of giant predators would come to dominate the seas. As larger, more mobile animals (Figure 4.6) evolved to feed on sessile producers and the slower, smaller primary consumers like those pictured in Figure 4.5, they in turn would have spurred the evolution of ever larger and more aggressive predators.

The evolution of jaws was a major transition among animals with an internal skeleton (vertebrates). They first appeared in the mid-Silurian, a little over 400 Mya, among armored fish known as Placoderms (Figure 4.7a). The largest was about 10 meters long. Throughout the Devonian (410–360 Mya) they ruled the seas, but became extinct at the catastrophic Devonian-Carboniferous boundary, when the Earth's climate turned colder, glaciers covered the continents, and even marine ecosystems were affected.

The demise of the Placoderms, however, was followed by larger monsters of the deep, such as the Ichthyosaurs (Figure 4.7b) of the Jurassic, who died out by the start of the Cretaceous 345 Mya, to be replaced by even larger Plesiosaurs (Figure 4.7c) who survived to the end of that period, 65 Mya.

Ichthyosaurs and Plesiosaurs were both reptiles that evolved from terrestrial ancestors. When the Plesiosaurs died out with the dinosaurs at the end of the Cretaceous, they were eventually replaced by whales (Figure 4.7d), the largest animals in the history of life on Earth, and also descendants of ancestors living on land. Therefore, at an earlier point, some forms of life had obviously moved from the sea to the land, marking another major transition in the evolution of life on Earth.

*4.2.5.2 Transition from water to air*
From the beginning, unless our alien authors were living on an oxygen-rich world like ours, they would have been concerned about the potentially poisonous effect of oxygen on living systems. The toxicity of oxygen is a less severe problem in water, because the solubility of oxygen in water is low. They would realize that whatever process was generating oxygen, would take a long time to build up the content of oxygen in the air. As it increased in the air,

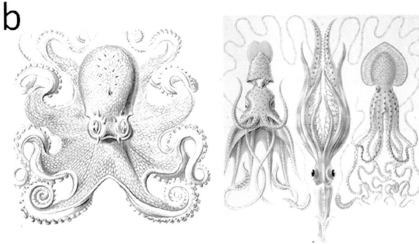

**Figure 4.6** Descendents of the Cambrian Explosion. (a) This artist's conception of the diversity of forms that survived the Precambrian experiment 500 Mya includes a variety of plant-like sessile forms, hard-shelled animals similar to modern clams, and Nautiloids, a branch of mollusks with a head and tentacles protruding from a long conical shell. (b) Early mollusks evolved into a variety of cephalopods – active marine predators with multiple arms, fine sensory and manipulative skills, and complex nervous systems (Image from the National Museum of Natural History, courtesy of the Smithsonian Institution (a) and drawings by Ernst Haeckel (b)).

**Figure 4.7** A succession of large animals have dominated the food chain in the oceans at different times. Pictured in order of their evolutionary appearance are (a) Placoderms, (b) Icthyosaurs, (c) Plesiosaurs, and (d) whales (Art by NobuTamura (a) and Heinrich Harder (b,c)).

however, it would have gradually become more concentrated in the ocean. With a gradual enough rise in the oxygen content of the water, organisms could adapt over time by converting the toxic vice of oxygen into an energy-yielding virtue (Figure 4.5). Once aquatic organisms had mastered the ability to harness oxidative metabolism for energy-yielding purposes, then gaining access to even more oxygen would have been value added. With the concentration of oxygen much higher in the air than in water, life on land would have become highly favored by natural selection. Thus, invasion of the land was bound to happen.

The first to leave their watery home of origin would have been simple producers, perhaps unicellular organisms like photosynthetic algae or chemo-lithotrophs able to oxidize mineral substrates for energy, both using atmospheric carbon dioxide as a carbon source. With sunlight even brighter outside the water than beneath its surface, multicellular photosynthetic forms of life – what we call plants – would probably have carpeted the surface once they solved the problem of avoiding desiccation. A succession of the simpler plants, beginning with low-lying forms like mosses, then with increasingly upright varieties like liverworts and horsetails (Figure 4.8), would have been a reasonable supposition for the composition of the early coastal lands and incipient forests of the Earth.

Once producers were well established on land, consumers would be sure to follow. The fossil record does in fact show that primitive plants and some

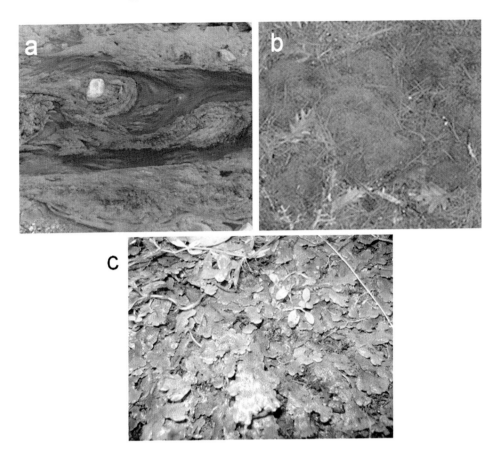

**Figure 4.8** Early evolution of plants on land. (a) Thin films of algae would probably have been the first phototrophs to invade the land. (b) Low-lying phototrophs lacking true leaves and stems like these mosses were probably the first upright multicellular plants. (c) Liverworts represent a more complex stage of plant evolution (Photographs by Johann Dréo (a), djpmapleferryman (b), and Aleksandr (c)).

arthropods came on shore during the Silurian, sometime after 435 Mya. By the dawn of the Carboniferous, 345 Mya, global rainfall was increasing, forests were covering the landscape, and amphibians, the first terrestrial vertebrates, had evolved from lungfish the capacity to leave the water entirely, except to reproduce. By the end of this period, the Earth's most successful group of animals in terms of both variety and abundance – the insects – had made their appearance. The first reptiles, evolving from a line of amphibians, and the first true seed plants, the conifers, had begun to populate the forests with the largest and longest-living organisms of any type in the history of life on Earth, by 140 Mya.

**Figure 4.9** Life covers the land. (a) Heavy global rainfall during the Carboniferous 330 Mya supported the growth of dense forests of primitive stemmed plants. Huge insects were also part of the terrestrial biota. (b) Ferns became prominent in the Triassic 240 Mya, as the earliest dinosaurs appeared. (c) By the Cretaceous 90 Mya, a rich variety of plant life was dominating the landscape. (d) During the Quaternary, starting 65 Mya, deciduous forests and flowering plants became dominant over much of the planet, with mammals replacing reptiles as the largest terrestrial animals (© Karen Carr, Karen Carr Studio, Inc., and the Indiana State Museum Foundation, with permission).

### 4.2.5.3 Flight, Fur, and Flowers

Something terrible happened to life on Earth 251 Mya, marking the boundary between the Permian and Triassic (P-T) periods. Termed the "Great Dying," it was the single greatest crisis in the history of life on our planet, an event that wiped out over 80% of all animals and much of the continental flora. The causes are unclear. Evidence of widespread volcanism, dramatically reduced rainfall, a nearby supernova, and collision with an asteroid have all been put forward as explanations. Whatever happened, it was a major crisis that eventually turned into opportunity for some new and revolutionary innovations.

**Figure 4.10** Major innovations during the Mesozoic. While the dinosaurs ruled the Earth for over 100 million years, a few significant changes were occurring in the underbrush. (a) Small homeothermic mammals would appear, maintaining high internal body temperatures during the chilly nighttime to which they were restricted by larger predatory reptiles. (b) Birds would evolve from dinosaurs. (c) Flowering plants would join the naked-seeded conifers as prominent terrestrial plants, supporting increased biodiversity of pollinators like flying insects and birds (Art by NobuTamura (a,b) and photograph courtesy National Park Service (c)).

The largest land animals at the time of the crisis, the reptilian Sauropods, were extinguished. This opened niches for new forms of reptiles, one of which would lead to the dinosaurs, who would rule the land for 160 million years. Then, one or more species of dinosaurs would see their scales transformed into feathers over evolutionary time, becoming the birds that have now filled the air for more than a hundred million years (Figure 4.10b).

An even earlier inconspicuous branch of reptiles surviving the P-T Transition would lead to mammals (Figure 4.10a). These animals, with their ability to maintain a constant elevated body temperature with the help of fur for insulation, would eventually spread into every corner of the globe, though that expansion would await a later catastrophe that would clear away the dinosaurs. Finally, another innovation that followed the widespread demise of the older form of plant life was the evolutionary emergence of flowering plants (Figure 4.10c), spurred by the recovery of insects as agents for cross-pollination.

### 4.2.5.4. Transition to the Modern World

If you can call something that's 65 million years old modern, then the transition to the modern world came about – again, suddenly and catastrophically – 65 million years ago. All the evidence points to the impact of an asteroid the size of Manhattan (off what is today the Yucatan Peninsula in the Gulf of Mexico) as the precipitating if not the causal event. An impactor of that size would have sparked a widespread conflagration, wiping out vast amounts of plant life at the bottom of the food chain, and darkening the sky with smoke over a wide enough area and long enough interval to cool the planet and cut its plant production perhaps below a sustainable level for the huge herbivorous consumers on which the carnivores depended for food. Whether the doomsday scenario was as simple as that or somewhat more complicated – perhaps spurred by the onset of more

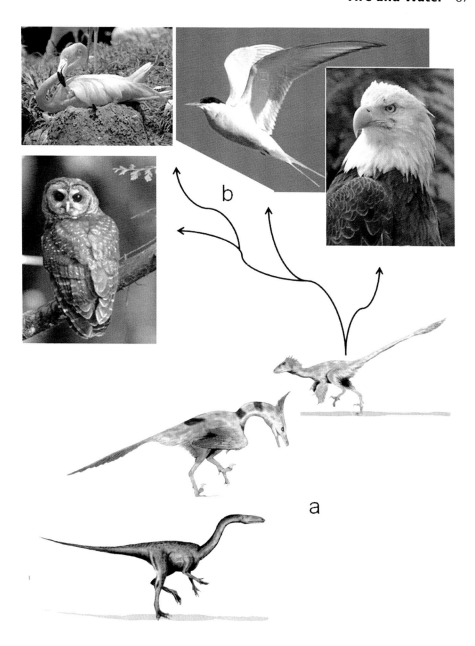

**Figure 4.11** Evolution of birds. (a) Ancestral birds evolved from dinosaurs. (b) A great variety of birds evolved as soon as the dinosaurs disappeared. Like the mammals that preceded them, their high metabolic rates and ability to maintain a constant, elevated body temperature enabled them to occupy every terrestrial niche (Art by Nobu Tamura (a) and photographs courtesy US Fish & Wildlife Service, and Smithsonian Institution (b)).

volcanic and tectonic agitation – the fact remains that the land was transformed from a warm, relatively flat and dry habitat dominated by reptiles, to a churning surface of recurrently colliding continental plates, subject to generally cooler, wetter climates, with mammals ascending to a dominant position on land and new forms taking over in the seas. That world, starting 65 Mya, is the world we live in today; so in that sense the modern world is about that old.

Some of the dinosaurs disappeared virtually overnight, while others had perished already or were in decline when the Chicxulub impactor crashed into the planet. In the seas, many forms of life disappeared as well. But the incredible variety of life that a long period of evolution on a geologically differentiated and climatically variable planet had brought about was ready and able with pre-adapted forms to occupy the vacated niches.

In the oceans, modern bony fishes (teleosts) radiated into a greater variety of species than all the other vertebrates combined, while the cephalopods with reduced shells, like octopi and squids, rose to the peak of invertebrate size, mobility, and intelligence.

On land, the feathered reptiles whose progression from bipedal walking to running to flight had been hampered by the abundance of large, aggressive terrestrial predators, underwent a radiation comparable to that of the teleosts, filling the air with birds in greater variety than any other terrestrial vertebrate (Figure 4.11).

The disappearance of the large dinosaurs left the surface of the land open for occupation by a different type of vertebrate. We can't say that mammals were new, because their ancestors are traceable to the Carboniferous, 300 Mya. They persisted through the Mesozoic, confined to a nocturnal lifestyle by their diurnal reptilian competitors. By elevating their body temperatures and giving birth to live young, they were able to live and reproduce at night while their predators slept. As soon as the dinosaurs were gone, the mammals were able to radiate into nearly every corner and climate of the globe, including the oceans where whales would grow to be the largest animals ever (Fig 4.12).

The period since the disappearance of the dinosaurs has seen a lot of vacillation in climate and precipitation, with frequent oscillations between warmer wetter, and cooler drier climates. During the late Miocene, about 10 Mya, one of those cooler and more arid periods occurred, diminishing forests and providing selective pressures for survival on prairies and plains. Those were the conditions under which arboreal primates, already adapted for life in the trees with excellent binocular color vision and grasping appendages, would leave the forests and walk out onto the open savannahs, where social behavior, bipedal locomotion, and prolonged juvenile development, among other adaptations, would give rise to anthropoids, leading eventually to *Homo sapiens* [6].

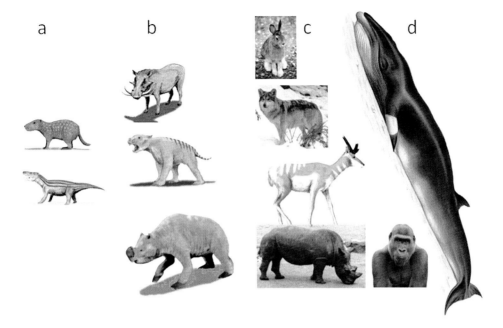

a        b        c        d

**Figure 4.12** Evolution of mammals. (a) Ancestral mammals evolved from reptilian precursors 300 Mya. The earliest forms were mostly small, carnivorous animals something like modern day insectivores. (b) When the dinosaurs disappeared 65 Mya, mammals entered into a period of experimentation that produced a number of odd and now extinct forms. (c) With the cooler climates and greater topological fragmentation of the land, mammals radiated into all conceivable niches. (d) Even the oceans became populated by mammals, including the largest of all modern animals (Art by Nobu Tamura (a-c) and Marie-Firmin Bocourt (d)).

## 4.3 A deduced biosphere for Earth

Our alien observers would know nothing of the rich and dramatic history of life on Earth described above. If their planet were anything like Earth, with a geological history anywhere comparable, nothing of the foregoing narrative would surprise them, as they too would probably have garnered evidence on their own planet for cyclic changes in climate, geological upheavals, occasional catastrophes, and the associated biological revolutions and transitions that accompany such changes. Not knowing the specifics or sequence of the changes on Earth, however, the details of Earth's biosphere would be a complete mystery.

Still, they would not be totally clueless about the possibilities. Once they concluded that life is highly likely to have arisen on Earth, they would see in the planet's topological variability and rich supply of energy the probability of a lot of biodiversity. This would suggest to them complex food webs and a hierarchy of trophic levels. At the very least, they should be able to postulate in broad

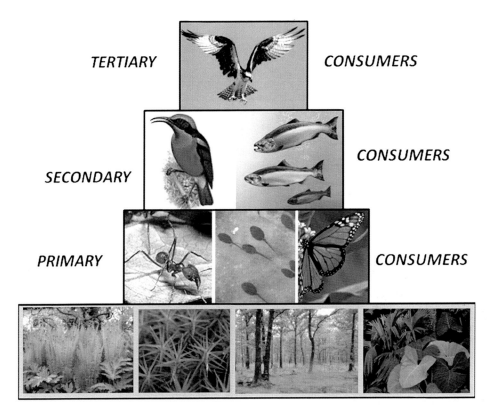

**Figure 4.13** Ecolological structure for the biosphere on Earth. Producers convert energy from the environment into biologically usable energy. Producers on the surface of the Earth are nearly all phototrophs, so photosynthesis is used to fix carbon into larger molecules containing energy upon which all consumers depend. Primary consumers are organisms that consume producers. Herbivores like ants, tadpoles, and butterflies all live off plant material. Secondary consumers eat primary consumers, while higher order consumers top the food chain by consuming lower level consumers, and in some cases (like humans) even plant material from the producer level. The pyramidal structure represents the fact that the total biomass at any trophic level is less than that of the level below it (Photographs courtesy US Forest Service (osprey), Robert W. Hines (fish) and E. B. Forsythe (butterfly) - US Fish & Wildlife, Jessie Cohen, National Zoological Park (ant and water plant), Paul Nelson-US Forest Service (woodland), and others courtesy of www.imageafter.com).

outline the basic ecosystems that must exist on Earth. With a little more deductive reasoning, they would probably be able to subdivide those ecosystems into several plausible biotic communities, leading them to a minimal list of generic organisms that would likely be found on Earth (Table 4.1).

### 4.3.1 Trophic levels of life on Earth
Intelligent observers on another world, knowing the general physical and

**Table 4.1** Summary of putative organisms on Earth

This lists the minimal set of organisms that an alien observer, knowing the general features of the planet but lacking detailed information on its biota, could expect to have evolved.

| TROPHIC LEVEL | ENERGY SOURCE | GENERIC ORGANISM | ACTUAL EXAMPLES |
|---|---|---|---|
| Producers | light | algae | *Euglena* |
| | light | fresh water plant[1] | water lily |
| | light | marine plant | brown algae |
| | light | terrestrial plant | clover |
| | oxidation of sulfur | sulfur chemotroph | *Paracoccus* |
| | reduction of iron | benthic iron lithotroph | *Deferribacter* |
| | reduction of $CO_2$ | methanogens | *Methanococcus* |
| | water current flow[2] | pelagic reeds | shoal grass |
| | air flow[2] | terrestrial reed | marsh grass |
| Primary Consumers | lithotrophic producers | subterranean | earthworm |
| | lithotrophic producers | lithotrophic feeders | tubeworm |
| | marine phytoplankton | marine zooplankton | larval jellyfish |
| | fresh water plant | fresh water herbivore | tadpole |
| | marine plant | marine herbivore | sea hare (*Aplysia*) |
| | terrestrial plant | terrestrial grazing herbivore | deer |
| | terrestrial plant | terrestrial seed eater | squirrel |
| | terrestrial plant | terrestrial leaf eater | ant |
| | terrestrial plant | flying seed eater | crow |
| Secondary Consumers | subterranean | subterranean carnivore | roundworm |
| | lithotrophic feeders | marine filter feeders | clam |
| | marine lithotrophic feeders | very small marine carnivores | marine flatworm |
| | marine zooplankton | small marine carnivore | sea anemone |
| | fresh water herbivore | small fresh water carnivores | salamander |
| | terrestrial leaf eater | small terrestrial carnivore | lizard |
| | terrestrial seed eater | flying carnivore | hawk |
| | terrestrial grazing herbivore | large terrestrial predator | lion |
| Tertiary Consumers | subterranean carnivore | surface carnivore | mole |
| | marine filter feeders | marine grazers | starfish |
| | very small marine carnivores | small carnivores | perch |
| | small fresh water carnivores | land-based aquatic carnivores | water snake |
| | small marine carnivore | large marine predator | shark |
| | terrestrial carnivore | large terrestrial predator | crocodile |
| Decomposers | benthic detritus | benthic microscavengers | marine bacteria |
| | benthic detritus | benthic megascavengers | lobster |
| | pelagic detritus | pelagic detrivores | flounder |
| | subsurface detritus | subsurface microbes | protozoa (*Amoeba*) |
| | subsurface detritus | subsurface fungi [2] | slime mold |
| | surface plant detritus | surface microbes | yeast |
| | surface animal detritus | surface fungi | Mushroom |

[1] any multicellular phototroph
[2] a reasonable inference by an alien, though incorrect; actual energy source is light
[3] any multicellular organism that ingests externally digested food by absorption

chemical features of Earth, should be able to anticipate the basic ecological structure of life on our planet. Their logic could go something like this.

*Producers*
The abundance of sunlight and its efficiency in creating high-energy chemical bonds in biomolecules would suggest that the surface of Earth should be covered with phototrophic producers. These could be expected to range from microscopic single cells to huge multicellular organisms, limited in size only by gravity and their ability to remain stable in whatever physical forces (wind, water currents) act upon them.

However, since harvesting sunlight in a biologically useful way requires fairly sophisticated biochemical machinery, simpler forms of energy probably drove the earliest producers. This would most likely have been oxidation-reduction reactions involving elements that would have been available on the early Earth, like sulfur, iron, hydrogen, and carbon dioxide. The oxidation of sulfur in water, or the reduction of iron or carbon dioxide by hydrogen, yield enough energy to support the metabolism of microbes today. It would be reasonable to assume that the first producers were something like today's chemotrophs, some of which have obviously survived to the present and persist in habitats where light can't reach.

*Primary Consumers*
Wherever producers thrive, primary consumers would be supported. This would include at least the photic zone of all the oceans (the upper levels where light penetrates), and the surface of all the land masses where water is sufficient to support life.

Where the producers are microscopic, the primary consumers would likely be small as well. Photosynthetic algae, for example, are consumed by protists like amebae in fresh water streams, and phytoplankton are eaten by zooplankton in the ocean. Where the producers are large, like trees on land, or abundant, like grass growing over a vast expanse, they would support large consumers like herbivores. The largest land animals of every age have in fact been herbivores.

As a general rule, primary consumers would be more mobile than producers, since the availability of their food would vary. Producers would grow only where energy and water allowed them to; then depletion by consumers would drive the consumers elsewhere to find undepleted supplies of more food.

*Secondary Consumers*
At this level of biological organization, life would be getting pretty interesting. The variety of primary consumers would mean an even greater variety of secondary consumers. Some would still be small, like the flatworms that consume the herbivorous protists in a mountain stream. Others would be bigger, like spiders that prey on smaller insects. Some would be as large as sharks and tigers.

Just as the primary consumers themselves might have become specialized to feed on certain producers, or on different producers under different circumstances, so would the secondary consumers have to specialize to keep up. An example would be a primary consumer like a rodent that evolved a nocturnal lifestyle to evade a diurnal predator like a snake, only to become the target of selective pressure toward the evolution of a nocturnal predator like a fox. In like fashion, adaptations for wet or dry habitats, cold or warm climates, fresh or salty water, great subsurface pressure or thin air at high altitudes, and other forms of habitat fragmentation that would characterize a heterogeneous planet like Earth would generate a great variety of consumers.

### Tertiary Consumers

Consumption at the tertiary trophic level is complicated. A clear-cut tertiary consumer would be the shark that eats a fish (secondary consumer) that subsists on zooplankton (primary consumer) that fed on phytoplankton (producer). But most food webs are not that simple (see Figure 3.7 in chapter 3). An eagle is a quaternary consumer when it catches a snake (tertiary consumer) that ate a lizard (secondary consumer) that lived on leaf-eating ants (primary consumers). The same eagle is a secondary consumer, though, when it preys on a seed-eating rodent (primary consumer). Humans are primary consumers when they eat vegetables, secondary consumers when they eat beef, and at least tertiary consumers when they eat any kind of fish.

### Multi-level consumers: parasites

An exception to the generalization that consumers are equally or more mobile than what they consume is seen in the evolution of forms of life that permanently attach themselves to their food source – what we would call parasites. These organisms don't need to be mobile at all as long as they are contained by their hosts. Enmeshed continually in their food at all trophic levels, in fact, they have also lost their ability to live independently. We can only conjecture whether alien scientists without a knowledge of specific forms of life on Earth could imagine something like viruses. That would almost surely depend on what they would know about the nature of their own molecular biology.

### Decomposers

Until the remains of dead organisms are totally broken down, they have some nutrient value. This provides food for a whole category of consumers, the decomposers (also known as detrivores). All dead material, whether from producer or the highest level of consumer, contributes energy to this trophic level. On Earth, the fungi, an entire kingdom of organisms, serve almost solely as decomposers. But detrivores also range from bacteria to worms to crayfish to buzzards; they are found in all taxonomic categories, in all habitats, feeding on organic remains from all trophic levels.

These examples show that trophic levels are an abstraction lacking fixed dimensions. The key point remains that within any given food chain, total biomass (or technically, total productivity) is highest at the bottom and lowest at the top. In other words, there will always be more biomass at the producer than at the consumer levels, and there will be more primary than secondary consumers, more secondary than tertiary consumers, etc. This enables the prediction of how any biosphere will be structured, in principle.

### 4.3.2 Ecosystems on Earth

Knowing of Earth's differentiated oceans and land masses, and a climate conducive to liquid water with an evaporative water cycle, alien scientists could anticipate at least three distinct ecosystems on the planet: a marine ecosystem in the planet's stable oceans, a terrestrial ecosystem, and a fresh water ecosystem on land where rainfall would collect into streams, rivers, and lakes.

#### 4.3.2.1 Marine ecosystems

The marine ecosystems would likely be subdivided into a surface and pelagic (continental shelf seafloor) environment on one hand, and a benthic (deep ocean seafloor) environment on the other. The surface environment would be based on phototrophic producers, and would likely be assumed to be widespread, since the surface of the ocean ought to be highly productive because of the vast area available to abundant sunlight. The benthic environment could be widespread as well, but alien scientists could not know that for sure without knowing how productive chemoautotrophy would be on the ocean floor. We know the reality to be that the benthic ecosystem (exclusive of the subsurface) consists of a much lower biomass than that of the photic zone.

#### 4.3.2.2 Terrestrial ecosystems

The land would likewise be subdivided into surface and subsurface environments. At the surface, phototrophs would surely be the dominant producers. Where water and soil nutrients were abundant, the phototrophs would be expected to be large and abundant. Such areas would be expected to be covered – perhaps almost saturated – with plants. The plants, in turn, would support a complex food web of consumers, all of which would generate ample detritus for an extensive community of decomposers. The subsurface would be less heterogeneous, with generally smaller individuals because of the confined spaces. At the upper levels of the subsurface, the food web would still be based largely on surface plants and photosynthetic microbes, with a large component of detrivores. At deeper levels, the producers would more likely be chemoautotrophs. The magnitude of the subsurface ecosystems (both terrestrial and marine) would be difficult for alien observers to guess. We know that in fact the biomass of subsurface life on Earth may equal that of life on and above the surface [4].

#### 4.3.2.3 Fresh Water Ecosystems

Aquatic life in fresh water would depend primarily on phototrophs as producers. Given the intimate contact between rivers, ponds, and lakes with the land, an interdependent interface would exist between the terrestrial and aquatic ecosystems. For example, terrestrial plants on the banks of streams could provide food for primary consumers in the water, while terrestrial predators could seek food often among the water-dwelling consumers.

### 4.3.3 Biotic Communities on Earth

Each of the major ecosystems on Earth could be expected to be fragmented into a number of different communities based on climate alone. That Earth has climatic variations would be deduced by its orbital tilt and (to a lesser degree) the eccentricity of its orbit, subjecting its northern and southern hemispheres to different degrees of heating in a seasonal cycle. Differential heating, with the presence of an atmosphere and an active water cycle, would mean winds, rain, and global variations in temperature. The obvious biotic communities would include the following.

#### 4.3.3.1 Marine Tropics

Within the equatorial latitudes, the more even heating of the surface would lead to constant, relatively warm ocean temperatures. Such areas on Earth would be predicted to have a high productivity (support a large total biomass), even if the details of habitats like coral reefs were unknown. If the interface between land and sea is complex and subject to tidal rhythms (as would be inferred by the presence of the Moon), the associated niche fragmentation would lead to predictions of considerable biodiversity.

#### 4.3.3.2 Terrestrial tropics

Any land masses straddling the equatorial latitudes would likewise be expected to have a warm, relatively constant temperature. This would probably mean high humidity and therefore probably high rainfall, suggesting dense plant growth. Subjected to a moderate degree of gravitational force and an atmosphere not dense enough to support destructively high winds, plants in the tropical zone could grow quite large and high. This, in turn, would provide an expanded opportunity for niche fragmentation, meaning a large number of different forms at every trophic level. Life in the terrestrial tropics, in other words, would be expected to be more diverse than it is anywhere else on the planet (as indeed it is).

#### 4.3.3.3 Marine temperate zones

Cooler waters north and south of the marine tropics would support animals adapted to cooler and more variable temperatures. This might select for larger body size. The producers, however, would still be largely photosynthetic microbes and phytoplankton. The tropical and temperate zones of the oceans might also be expected to differ in direction and extent of ocean currents, and in rainfall; but these details would be hard to detect from a very remote distance. Alien observers would probably be able only to speculate that the tropical and temperate zones of the oceans would differ in the forms of life that dominate each, without being able to specify much about what those forms would be like.

*4.3.3.4 Terrestrial temperate zones*
Seasonal fluctuations in temperature which would characterize the large land masses north and south of the equatorial latitudes would probably mean variations in rainfall, and that would determine to a large degree what those biotic communities would consist of. That, in turn, would depend to a large degree on (a) prevailing wind patterns, and (b) topography. The reality is that the temperate zones of Earth are home to great deserts, highly productive but treeless grasslands, and dense forests. Each of these habitats would have its own characteristic populations of microbes, fungi, plants, and animals. With increasing distance toward the poles, selective pressure for adaptation to cold would become more severe, surely giving rise to quite a different mix of organisms than in the tropical regions. But absent the details of wind currents, rainfall, mountain ranges, continental masses, and other physical characteristics, as well as ignorance of the history of life that gave rise to the contemporary biota, the biological details would be impossible to predict other than in a most general way.

*4.3.3.5 Fresh water communities*
Having an active water cycle would mean that Earth must have streams, rivers, ponds, and lakes of fresh water scattered across the continental masses. Producers such as algae and plants along the margins of the waterways would support a full repertoire of fresh water aquatic organisms, interacting to a degree with the terrestrial biota with which they would be in contact. The animals in such a community should resemble marine organisms to which they would likely be related evolutionarily.

*4.3.3.6 Polar life zones*
Earth has experienced geological ages in which climate at the poles has been relatively mild, as well as frigid. Unless our alien scientists had a good idea of the temperature at the poles, it would be difficult for them to speculate on the biotic communities on either land or in the oceans near the north and south poles. If they knew it to be very cold, they would strongly suspect that biotic diversity is limited to a few, relatively large consumers, living off the productivity of the seas. It's questionable whether they would imagine anything like penguins or polar bears, but those are prototypical of what they could reasonably expect.

## 4.4 Characteristics of biota on Earth

Having established that life on Earth is likely to be highly diverse, the biological characteristics of those organisms would have to be extremely varied. Examples from some of the major physiological systems will suffice to make the point.

### 4.4.1 Metabolism
Even the simplest producers would have many complex metabolic pathways. The chemoautotrophs would have a way of extracting energy from their abiotic

chemical substrates and coupling the energy released to a biologically useful storage form. The phototrophs would have even more complicated pathways, since coupling the capture of photons to the generation of high-energy chemical bonds is not a trivial feat. In addition to the pathways for energy release and capture, a whole set of biosynthetic pathways would be needed for growth and maintenance of the simplest cells.

At the consumer level, metabolism would undergo a shift in emphasis from biosynthesis to regulation. Producers synthesize most of the biomolecules that consumers need. Many of those nutrients have to be remodeled and rearranged, but biosynthetic pathways leading to the basic building blocks might be skipped if the building blocks themselves (such as certain amino acids or fatty acids) can be obtained directly from the producers. With the increased cellular specialization that comes with multicellular architecture, however, regulation of which metabolic pathways are turned on when and where becomes the primary role of metabolism. Natural selection would mold pathways accordingly, in a manner specific for the needs of each organism.

Because some forms of life would have to cope with environmental extremes, some types of metabolism might emerge to handle those circumstances specifically. For example, homeothermy, or the ability to maintain a constant internal body temperature, is a great advantage to animals living in cold climates, because it enables them to stay active (in search of prey or shelter) and resist freezing. This requires the consumption of a great deal of energy, some of which is geared exclusively toward generating heat. In very dry habitats, metabolic adjustments need to be made for water conservation. Organisms at high altitudes have to deal with lower levels of oxygen. These and countless other examples illustrate the great variety of metabolic specializations that could be predicted for life on Earth.

### 4.4.2 Reproduction

Our alien authors' speculations about reproduction would surely be based on their own experience, since our premise is that they would have no detailed knowledge of reproductive strategies on Earth. At a minimum, they would recognize the need for some way of reliably copying genetic material and passing it from parent to offspring. One of the great uncertainties in astrobiology is whether anything like the encoding of genetic information as it happens on Earth operates in forms of life with an origin and evolutionary history independent from that on Earth. The only safe conclusion is that there must be some coding system, it must be capable of replication with high reliability, and it must be preserved from one generation to the next.

Biologists on Earth have long argued the utility of sexual reproduction. It is certainly widespread, from the simplest bacteria to the most complex multicellular organisms. But exceptions to it are widespread as well, so natural selection seems to have made a judgment on a case by case basis as to whether sexual reproduction is a net gain over the safer, less complicated alternative of asexual reproduction.

It is tempting to assume that our alien scientists would look at the heterogeneity of environments and selective pressures that must be constantly at work on Earth, and conclude that some mechanism for genetic innovation – whether it be a robust rate of mutation, or sexual reproduction, or cellular mechanism akin to chromosomal crossover – would be helpful in keeping up a steady stream of new, raw genetic material for natural selection to act upon. In a highly diverse and changing world, the ability to adapt is critical for survival, and adaptability is tied to genetic variability to a large degree.

### 4.4.3 Motility

Plants don't need to move, because their source of energy is everywhere. Fungi, which devour mostly dead organisms, don't have to move because their source of food isn't going anywhere. Animals do need to move, because they have to find new producers when the local supply has been exhausted, or they need to seek out lower level consumers that serve as their prey. Whatever name our alien scientists would give to "plants," "fungi," and "animals," they would surely predict that some would need the ability to move about while others would not.

To the extent that motility is an advantage to any organism, natural selection would exert pressure for it to be increased in some of them. If our alien scientists were insightful enough to recognize that Earth's oxidizing atmosphere would probably make a high level of energy metabolism quite possible, they could make the logical deduction that some organisms might achieve a high degree of motility. They would likely conclude that, especially on land and in the air where the oxygen supply is ample, high levels of energy expenditure for movement could be attained.

Whether they would anticipate *all* the mechanisms for motility that have actually evolved – jet propulsion by cephalopods (squid and octopi), lateral body oscillation by fish, flapping wings by birds and butterflies, legs for walking, running, and crawling by animals, arms with legs for climbing by some animals, swim bladders for buoyancy in fish, wiggling by worms, and so forth – they would surely anticipate that a variety of mechanisms should have evolved over the history of life. Quite possibly they would be surprised to learn that among the animals, flight is the form of locomotion employed by the greatest number of species on Earth.

### 4.4.4 Sensory systems

Being able to sense and respond to the environment, both living and non-living, is a vital aspect of being alive. Even passive organisms like plants rotate their leaves toward the light and close their pores to hold in water when the air dries out. This implies the ability to detect light and moisture. Sensing the presence of a beneficial chemical substance, and avoiding one that's potentially dangerous is essential for the simplest microbe. No wonder that the chemical senses are the oldest and most widespread of all sensory abilities.

Among animals, we refer to those vital chemical senses as taste and smell. They may be part of the makeup of every living thing in the universe. The basic physics of energy transmission would predict what a number of the other sensory capabilities ought to be: vibration in the air or water (evolving into sound for complex animals); tactile sensations of various sorts – for touch, heat, cold, pain, and pressure; electromagnetic radiation, which different organisms would perceive as vision or heat, depending on the wavelengths prevalent and most relevant in their particular environments; and orientation in space, so that organisms could keep their balance in the gravitational field, tell the difference between up and down, and sense in which direction and how fast their bodies are accelerating.

Because of the diversity of habitats that our alien observers would know to exist on Earth, they would probably predict that all the physical features of the environment would be candidates for detection. Depending on the nature of life on their planet, they might be able to guess what cellular specializations for light, heat, and pressure detection would be like. In view of the considerable motility that many organisms on Earth would be presumed to have, they would probably assume that a complex system for taking in parallel streams of information, integrating and analyzing it in the light of experience (implying the capacity for storing information), and coordinating a response must exist. Whatever they would call it, we would recognize it as a nervous system.

Sedentary organisms can get by with fairly simple sensory systems, because their actions are limited to passive, defensive reactions. However, organisms that move about are encountering an ever changing array of incoming information, and adjusting their behavior accordingly. As the aspects of their environment become more numerous, or require higher resolution, their information processing systems have to become more sophisticated. And, of course, the larger the organisms are, the more body mass there is to be coordinated. So nervous systems or their equivalent are an inevitable necessity for active organisms. Once any form of life begins to move about, a nervous system is bound to emerge. When the amount of information that needs to be detected, integrated, and coordinated becomes great enough to require hierarchical processing and centralization, a brain will be the evolutionary result.

### 4.4.5 Cognition

It takes but a cursory review of life on Earth to reveal that a high level of sensory processing is associated with higher cognitive abilities. All the animals that we recognize as more intelligent tend to have extensive sensory abilities, with specialization in usually more than one of the senses. Perhaps not surprisingly, they have larger brains. However, the opposite is not necessarily true – not all organisms with remarkable sensory abilities are necessarily cognitively advanced. Moths have exquisite sensitivity for pheromones (air borne chemicals for reproductive attraction) but have no evident cognitive superiority over other insects. Some fish have elaborately developed brains specialized for electroreception, but they are not conspicuously intelligent. The relationship between sensory sophistication and cognition, therefore, is tenuous. All we can say with assurance is that cognition appears to be most highly developed in those organisms that move about in their environment, that confront changes in their environment, and whose behavior is dependent on learning and the processing of a lot of new, incoming information in real time. Large, active, organisms occupying biological niches in heterogeneous habitats like those on Earth would be expected to produce at least some forms with considerable cognitive abilities.

The value of cognition should not be overrated, however. Microbes and plants dominate the biomass of the Earth and have no cognition in any meaningful sense of the term. Sponges, clams, and mushrooms likewise have little in the way of personality, yet they've been fixtures of the biosphere on Earth from ancient times and will likely outlast all the more intelligent forms of life that share the planet with them. Our point here is that cognition is hardly the path to evolutionary success for the vast majority of life on Earth (or probably anywhere in the universe). However, it has arisen in some forms of life, under the peculiar conditions that characterize our planet. To the extent that those conditions exist elsewhere, cognition should be equally predictable. And cognition has consequences.

### 4.4.6 Technology

Under the right combination of circumstances, cognition can lead to technology, which transforms the ability of an organism to act on and control its environment. We know this because we as a species have done it. If we knew for sure that we were being watched from afar, we would know that it's happened elsewhere as well. The generic conditions that enable the emergence of technology are explored in detail in chapter 12. As a preview, we can consider why it happened among our species on Earth.

The circumstances leading to human technology can be traced retroactively through the conditions described in previous sections of this chapter. They include (a) a long history of evolution with time for critical transitions like the development of multicellularity and high-energy oxidative metabolism, (b) heterogeneous environments with extensive habitat fragmentation leading to great biodiversity, (c) an inexhaustible and ubiquitous form of energy sufficient for supporting complex ecosystems with large, active organisms at the top of the food chain, (d) a variable and dangerous world that places a premium on cognitive processing and social behavior, (e) ancestry in an environment (presumably arboreal) that promoted the evolution of grasping hands and a high degree of eye-hand coordination, and (f) evolution on the land, where a diversity of raw materials was readily accessible and energy from fire could be harnessed.

If our alien observers were aware of all these circumstances, they would likely assume the possibility that intelligent forms of life giving rise to technology were capable of evolving on Earth. If they were able to observe Earth with sufficient detail to know all that, they would more than likely know whether it had in fact happened, because they would be able to detect the radio and other artificially generated emissions that we've been sending into space for a hundred years now. And if they had been observing Earth long enough to see the radical changes in our atmosphere – the accelerated accumulation of greenhouse gases and the sudden appearance of mysterious compounds like chlorofluorocarbons – they would be certain that a techno-capable form of life exists on Earth. *They* would know for sure that they are not the only intelligence in the universe.

## 4.5 What alien observers could get wrong about life on Earth

We have contrived in this chapter a scenario to be the mirror image of our own situation. By asking what a distant intelligence at a level of technology comparable to our own could tell about us, then comparing the answers with the reality of life on Earth, we can gauge how good our speculations about life on other worlds might be.

The overall message of this chapter is that a remarkable amount of information about life on Earth could reasonably be deduced from a knowledge of the Earth's age and the broad outlines of its geochemical and geophysical features. But assuming that our alien observers' technological capacity really isn't a lot greater than our own, from a distance of 10 light years, they would be unable to confirm any details of their prediction. In fact, they might well get some things wrong. By looking at their possible mistakes, we can get a better gauge of our own potential for accuracy (or lack thereof) when speculating from the vantage point of Earth, with our level of knowledge and technology, about life on other worlds.

A lot depends on the nature of our observers. If they are carbon-based forms of life operating with water-soluble biochemistry confined in the water-insoluble bodies of individuals, they should have no trouble imagining our basic biological properties. However, if they were totally aquatic and had no knowledge of life outside the water, they would probably underestimate our terrestrial biosphere. If their world is low on oxygen, they would probably discount its importance and see it only as a toxin or highly flammable hazard inconsistent with life on Earth. Even if they recognized the possibility of life on land, they would probably underestimate the prevalence of flight as a means of locomotion. Therefore, they would likely have no idea of the prevalence of animals like birds and insects.

If our alien observers had arisen from more exotic conditions in relation to those on Earth – for instance, from a silicon-based chemistry at a much colder temperature in an non-aqueous medium, they would probably have a correspondingly difficult time imagining just how water-soluble, carbon-based polymeric chemistry could give rise to living systems, particularly within such a narrow thermal range that their theory would tell them would have to be the case.

Knowing the dynamic nature of Earth's wind and ocean currents, they might assume the existence of kinetotrophs deriving their energy from the motion of the atmosphere, ocean currents, or tides (which would be inferred from the presence of a moon in orbit around Earth). The reality is that no such organisms have been found, probably because of the superior efficiency of phototrophy and chemoautotrophy. This miscalculation exemplifies the erroneous conclusions that theory in the absence of empirical evidence can lead to.

Their own size would probably bias their view of what size could be attained by organisms on Earth. Their sensory abilities would certainly bend their focus toward looking for the same capabilities among the organisms of Earth. Their own evolutionary history would likely inform their speculations about the extent of evolution on Earth. Had their evolution consisted of two billion years of existence as unicellular organisms only, for instance, they would find the notion of an equally long formative period for life on Earth as totally credible. But if their evolution had been much faster, they might be puzzled at the lack of a more technologically advanced civilization on a planet as old as 4.5 billion years.

The bottom line is that the more an alien planet from which we were observed would be like our own, and the greater the similarity in our evolutionary histories would have been, the more likely our observers would be able to guess right about our nature. Turning this around, if whatever life *we* might find on other worlds is similar to us, our success in finding it might simply be due to the fact that we know enough to recognize forms of life like ourselves. It *could* mean that life fundamentally like us (water-borne, carbon-based) is the only, or at least the highly preferred, form of life in the universe. If, on the other hand, there are forms of life out there very different from our own, we will have to be very smart and very open minded to come close to correctly imagining what it might be like. And our failure to find life in our own image will by no means be evidence of its absence. We will need to bear this in mind as we probe the conditions and possible planetary histories of the other worlds we will survey, in our quest for those on which life may have evolved.

## 4.6 Chapter summary

With carbon-based biochemistry, water widespread and in liquid form as a solvent, and abundant sunlight, the planet Earth has all the characteristics deemed most favorable on theoretical grounds for the existence of life. To those advantages are added the fact that it has an atmosphere, continental masses with variable topography interfaced with both fresh and salty water, active geological activity, and, at 4.5 billion years of age, has been around for a long time.

We have empirical knowledge that a rich panoply of life has arisen, proliferated, and evolved through time under these conditions. The abundance of water and sunlight, along with the opportunity for chemoautotrophy, enables the turnover of an extensive biomass at the producer level. The heterogeneity of climates and topography fragment habitats so extensively that a great amount of biodiversity has resulted. The cyclic biogenic consumption and production of carbon, hydrogen, and oxygen, including high energy-yielding oxidative metabolism, support an extensive network of food webs at a number of trophic levels involving consumers with a great variety of lifestyles, sizes, and activity levels.

Ever increasing complexity in some forms has been the net effect of evolution, but the transitions toward complexity have taken a long time, and have always involved a small select few. Life on Earth remained prokaryotic in form – the simplest of all cells – for the first billion and a half years of its history, then took another billion years to even start putting cells together into macro multicellular organisms. The one sample of life we have is not sufficient to tell us whether such a long time was an inherent necessity, or just the way that the history of life unfolded by chance on Earth. It remains one of the least constrained guesses we can make about life on another world.

Once life emerged on land, the pace of both diversity and innovation quickened, probably made possible by a critical rise in the level of oxygen in the atmosphere. Life has seen a number of upheavals and transitions, usually quite suddenly, and increasingly suspected of being associated with catastrophic events like global volcanic eruptions or massive meteorite impacts. Thus far, however, each catastrophe has served to generate new opportunities.

Several conditions presented by both nature and history have combined to give rise to intelligence in a relatively small number of forms. The superior cognitive ability of humans, in combination with their unique evolutionary history and their access to raw materials and energy for construction of tools and machines, has provided the foundation for a degree of technological prowess that is extending their impact far beyond their immediate biological capabilities. The technology of this species has enabled it to start exploring worlds beyond its own, and to start seeking forms of life both in its own image, and in forms as yet unimagined.

## 4.7 References and further reading

1   Schopf, J. W. 1999. *Cradle of Life: The Discovery of Earth's Earliest Fossils.* Princeton NJ: Princeton Univ. Press.
2   Lineweaver, C. H. and Davis, T. M. 2002. Does the rapid appearance of life on Earth suggest that life is common in the universe? *Astrobiology* **2**: 293–304.
3   Nisbet, E. G. and Sleep, N. H. 2001. The habitat and nature of early life. *Nature* **409**: 1083–1091.
4   Whitman, W. B., Coleman, D. C. and Wiebe, W. J. 1998. Prokaryotes: The unseen majority. *Proc. Natl. Acad. Sci. U.S.A.* **95**: 6578–6583.
5   Cowen, R. 1995. *History of Life.* Boston: Blackwell.
6   White, T. D., Asfaw, B., Beyene, Y., et al. 2009. *Ardipithecus ramidus* and the paleobiology of early hominids. *Science* **326**: 75–86.

http://www.windows.ucar.edu/tour/link=/earth/earth.html
Excellent, comprehensive resource covering all the Earth sciences, in both English and Spanish.

http://tolweb.org/tree/phylogeny.html
A comprehensive overview of life on Earth.

http://evolution.berkeley.edu/evolibrary/home.php
A good source for many web sites and articles related to the history of life on Earth.

# 5 Frozen Desert

## Life beneath the cold surface and thin atmosphere of a planet like Mars

Looks can be deceiving. Every close-up picture of the surface of Mars has revealed what appears to be a barren, stony, still, and totally lifeless landscape (Figure 5.1). We do know the atmosphere is dynamic, since from a distance we can see that winds on Mars can get high enough to kick up global dust storms that last for weeks. For all we can tell, though, the surface lies frozen in lifeless immobility – and *has* for a very long time. But we would be wrong.

Astronomically speaking, Mars is the little sibling of the larger twins, Earth and Venus. Like all the planets in our Solar System, they formed about the same time, from the same accretionary process that turned them and Mercury into the relatively smaller, rocky planets closer to the Sun than the distant gas giants. From every indication, Earth and Mars shared a similar history over their first few hundred million years – both differentiating into metallic cores and mineralized crusts, bombarded with meteors large and small that delivered kinetic energy, water, and probably organic compounds to a surface still heated by high levels of inner radioactive decay, both generating atmospheres that held the heat in, keeping them warm enough for liquid water to bathe their hardened solid surfaces.

In time, however, their different sizes and distances from the Sun would set them increasingly apart. Earth would settle into a pattern of more gradual changes which , despite oscillating periods of global warming and cooling, would keep liquid water abundant (albeit possibly frozen at its surface for periods of time) under a fairly dense, nitrogen-rich atmosphere. Mars, on the other hand, would quickly cool, lose its atmosphere, and see its surface water freeze, then disappear. Life would germinate or be emplaced on Earth, then blossom in profusion. Whatever life may have arisen or been emplaced on Mars either disappeared or became so untenable on the surface that, if it remains at all, is probably consigned to long if not permanent existence beneath the ground, or survives close to the surface only through specialized adaptive mechanisms.

## 5.1 Peeling through layers of Martian mystery

Every advance in technology directed toward Mars has increased our understanding but deepened the mysteries surrounding the planet known in Roman mythology as the god of war for its blood-red tint as first seen by the naked eye of

L.N. Irwin and D. Schulze-Makuch, *Cosmic Biology: How Life Could Evolve on Other Worlds*, Springer Praxis Books, DOI 10.1007/978-1-4419-1647-1_5,

© Springer Science+Business Media, LLC 2011

**Figure 5.1** Surface of Mars. This view of the rocky, barren ground was taken looking southeast at near noon on 21 July 1976 from Viking Lander 1 in Chryse Planitia. Compounds of iron predominate in the Martian soil, giving the planet its characteristic reddish-brown color. Even the sky is pink, due to airborne dust of the same composition (NASA/JPL).

the ancients. When Percival Lowell trained his telescope on Mars and thought he saw an infrastructure constructed by intelligent beings to redistribute water presumed to be at its poles, Mars was imagined to host a civilization comparable to, if not evolutionarily beyond, our own.

When the optics for observing the planet had improved enough to reveal no such structures, the dry and barren landscape that came into view offered no hope of a civilized race, and even cast the existence of life itself on Mars into doubt. By the time the images from the *Mariner 4* flyby in 1965 had shown the surface to be pot-marked with the craters of a thousand bombardments, the survival of life on a planet more reminiscent of the lifeless Moon became hard to imagine.

But Mars is not the Moon, as *Mariner 9* and the *Viking* missions dramatically revealed. *Mariner 9* was the first orbital insertion around another planet. The

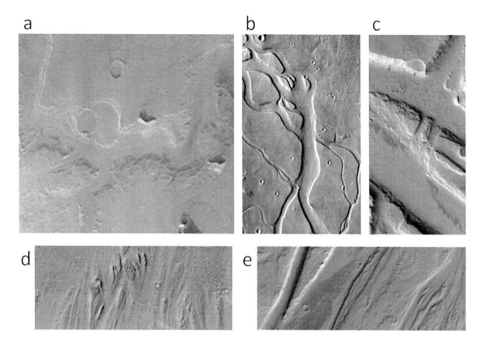

**Figure 5.2** Channels on the surface of Mars. (a) Early hints of strong erosional forces were first seen by the Viking orbiters, as in this 20 km wide channel near the edge of the Hellas Basin. (b) Fretted channels with flat floors and steep walls were revealed in higher definition by the THEMIS camera from the Mars Odyssey orbiter, as shown here by a channel about 3 km wide west of Elysium Mons. Note the teardrop-shaped outcrops indicating downstream flow of liquid. Other images captured by the Mars Orbiter Camera (MOC) aboard the Mars Global Surveyor revealed (c) intersecting channels 1–2 km wide a little east of Olympus Mons, (d) streamlined landforms indicating catastrophic flooding 15 degrees north of the equator near Elysium Mons, and (e) a channel several hundred meters wide east of Olympus Mons. (Images: NASA/JPL-Viking Orbiter (a), NASA/JPL/ASU-Mars Odyssey Orbiter (b), NASA/JPL-MOC, Mars Global Surveyor (c-e)).

remarkable feat was achieved on 13 November 1971, in the midst of a global dust storm that shrouded the planet in obscurity for weeks. When the dust began to settle, a point of land appeared where previously a giant crater had been suspected. Soon, four more points emerged from the cloud of dust, and gradually grew larger, suggesting not craters but mountains. These would turn out to be the giant shield volcanoes of the Tharsis Rise and Elysium Mons, soon to be recognized as the largest volcanoes in the Solar System. When the winds had totally abated, a giant canyon (Valles Marineris) two-thirds the breadth of North America was revealed. And perhaps most surprising of all, was the evidence that something like water or ice had scoured the surface, forming channels suggesting the movement of powerful forces across the land (Figure 5.2).

With the higher definition images of the surface provided by the Viking orbiters, another curiosity clearly became evident – the planet's northern and southern hemispheres vary greatly in their density of impact craters, with cratering much more prevalent and showing a greater range of ages in the south than in the north. A compelling explanation for this anomaly was suggested when the Mars Orbital Laser Altimeter aboard the Mars Global Surveyor orbiter confirmed that elevations in the northern hemisphere are lower on average than those in the southern hemisphere (Figure 5.3). A lower rate of cratering, with craters generally younger, and a lower elevation in the northern hemisphere, are exactly what would be predicted if the northern hemisphere had been covered with water. Barbel Lucchitta first suggested in print the presence of a northern ocean in 1985, and Timothy J. Parker proposed not one, but two ancient coastlines circling the globe soon thereafter. Later, evidence for several coastlines from a succession of oceans lasting into the Hesperian emerged from the work of Alberto Fairén [1]. Other explanations were possible, like recent and massive collections of sand at lower elevations in the north, or hemispheric submersion by extensive lava fields. But these were less satisfactory, especially after huge outflow channels tending northerly, and the pedestal platform of Olympus Mons suggestive of shield volcano construction surrounded by water or ice (Figure 5.6c), were revealed.

Discovery of the large outflow channels emptying into the northern lowlands, along with the clear evidence that the surface in the northern hemisphere is relatively young and devoid of the massive meteorite bombardment evident in the southern highlands, swung the pendulum of opinion gradually in the direction of accepting the existence of an Oceanus Borealis, or northern ocean [1]. But mysteries remain. If, at one time, there was enough water for a global ocean 400 meters deep, as calculated by Michael Carr [2], what happened to all that water? If the pedestal base of Olympus Mons was a sea-cliff over 5 kilometers high, as postulated by Henry Faul in a 1972 manuscript that he couldn't get published [3], what became of the sea? And furthermore, the large valley channels themselves look peculiar – more akin to collapse ("sapping") of underground reservoirs, or conduits for massive spring-fed or snow-melt discharges, than traditional rainwater runoff tributaries. And if there was a succession of oceans fed by large rivers with extensive dendritic tributaries, like those of the Amazon or Nile, the evidence of river systems that long and extensive has disappeared.

One part of the mystery was solved when massive amounts of water were discovered to be sequestered underground by the gamma-ray spectrometer on the Mars Odyssey orbiter [4]. The subsurface water-ice forms a substantial fraction – 70% and higher by weight – of the upper levels of the substrate as the poles are approached. This vast reservoir of ice is covered by a few cm of soil, giving the surface of Mars its bone-dry appearance. Models indicate that when Mars is tilted at a more extreme angle on its axis, the sheets of ice creep further away from the poles (Figure 5.7). Even then, the vast sheets of ice may remain obscured beneath the ever present layers of dust and soil that swirl about the planet. But Mars, we now know for sure, holds plenty of water still, out of sight

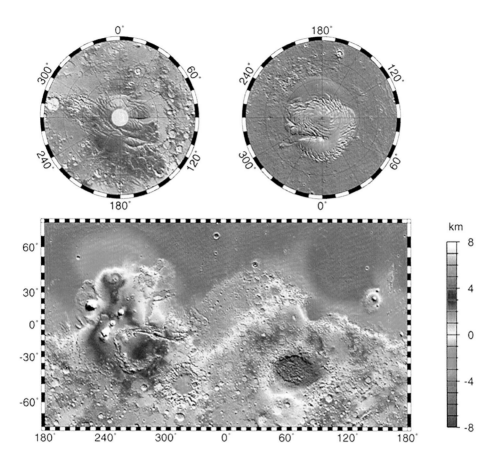

**Figure 5.3** Elevations on Mars. The altitude of the entire surface of Mars was mapped by the Mars Orbiter Laser Altimeter (MOLA) instrument aboard Mars Global Surveyor. Measurements are color-coded, with blue indicating the lowest and white indicating the highest elevations. Upper circles: Stereographic images of the southern (left) and northern (right) hemispheres. Lower rectangle: Mercator projection of surface from -70° to +70° latitude. Higher elevations are clearly concentrated in the equatorial Tharsis Rise and the southern hemisphere generally, with the exception of the large impact craters forming the Argyre and Hellas Basins. The horizontal gash at the eastern edge of the Tharsis Rise is Valles Marineris (NASA/JPL).

most of the time, but ready to erupt in large amounts onto the surface when provoked.

To account for the obvious episodic flow of massive amounts of water into what surely was one or more extensive bodies of water in the northern hemisphere, Victor Baker and his colleagues proposed that catastrophic outflows of ground water were periodically discharged into the northern basin, forming the massive outflow channels [5,6]. They gave their model the name, MEGAOUTFLO, an acronym for Mars Environmental Glacial Atmospheric

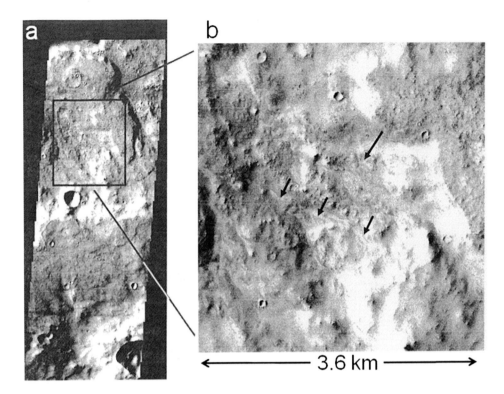

**Figure 5.4** Small-scale volcanism and erosion in the equatorial highlands. (a) This terraced strip of land about 4 km wide sloping northward in Terra Sabaea contains a number of deep and imperfectly rounded craters, the calderas of small volcanoes tens to hundreds of meters in diameter. The rough terrain is old, as shown by highly degraded and eroded features, but several of the small calderas appear to be relatively recent because their rims are high and still well formed. (b) At higher resolution old landforms scoured by drainage channels (arrows) and basins for collection of lava or water runoff can be seen. (Image taken by the THEMIS camera aboard the Mars Odyssey orbiter at 4.8°S and 46.6°E, courtesy NASA/JPL and the Mars Student Imaging Project, Arizona State University [7]).

OUTburst FLood Oscillations. They postulated that periodic episodes of internal geological activity, especially pulses of volcanic activity, were responsible for the occasional megaoutflow events.

The Tharsis and Elysium Provinces are spectacular examples of volcanic eruptions on Mars, but careful study of the surface at the higher resolutions now available betrays a long history of volcanism widely distributed across what is now referred to as the southern highlands. An unexceptional slice of land in Terra Sabaea (Figure 5.4). illustrates how widespread and possibly recent smaller scale volcanism on Mars has been.

The kinetic energy of bollide (meteorite) impacts is the other source of heat that could have triggered eruptions of ground water. Models have shown that the

a                             b

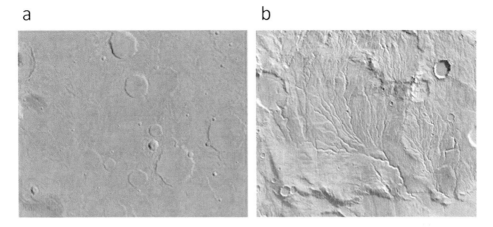

**Figure 5.5** Valley drainage networks in the southern highlands of Mars. (a) An ancient valley network of dendritic channels photographed from the Viking Orbiter (b) A similar network imaged at higher resolution from the Mars Odyssey orbiter. The drainage systems shown in both cases are ancient, as indicated by considerable degradation of some branches and the overlay of numerous impact craters (NASA/JPL).

very large impacts characteristic of the planet's early history could have delivered enough kinetic energy to produce scalding rains and liquid water runoff for decades to millennia, though such episodes were likely interspersed by millions of years of predominantly freezing temperatures [7]. Smaller impacts up to the present time would be capable of causing local melting. These perhaps account for the more Earth-like dendritic valley networks seen mainly in the southern highlands (Figure 5.5)

## 5.2 Overview of Martian planetary history

Each flyby, orbiter, and robotic mission to Mars has added critical information to our understanding of what has occurred to the Red Planet since its birth as the fourth rocky planet from the Sun in our Solar System. Interpretations of that information remain controversial, but in our opinion have achieved sufficient consistency for us to sketch out a general history of the geological evolution of the planet. Based on our own research and that of many other scientists, we offer the following outline of that history.

1. Mars was formed, like Earth, as a water-bearing rocky planet about 4.5 billion years ago. In the beginning, it was warmer and wetter than today, with bombardment from frequent meteorites delivering kinetic energy, and endogenous radiolytic decay providing internal heat sufficient to drive plate tectonics, keep water liquid, and generate a humid atmosphere. This lasted perhaps midway (500 million years) through the Noachian Eon.

2. Tectonic rearrangements led to partial subduction of the northern hemisphere under equatorial and southern regions, which were thereby elevated, with continuing heavy bombardment from meteorites, including massive impacts that created the basins of the Hellas and Argyre Planitias.

3. By the end of the Noachian (3.7 Gya) and perhaps a good bit earlier, the planet's core had cooled and solidified, reducing its heat output and eliminating its magnetic field. The mantle hardened and thickened, and the atmosphere was largely blown away by solar wind now bathing the surface unimpeded.

4. With its heat-retaining atmosphere gone and its generation of internal heat greatly diminished, Mars became cold. Water disappeared from the atmosphere due to the planet's low gravity and ongoing solar wind exposure, and sublimated away from the frozen surface. However, the porous basaltic crust created by extensive volcanism absorbed and retained much of the planet's water underground.

5. Since plate tectonics had stopped, great pressure from the planet's interior built up at focused hot spots. Eventually, magma erupted, creating the giant volcanoes of the Tharsis Rise and Elysium Mons – somewhat analogous to the build-up of the Hawaiian island chain (Figure 5.6a,b). However, the volcanoes on Mars are much larger because (a) the planet's lower gravity allows more massive buildups, and (b) the points of eruption stay stationary over hot spots since the plates are no longer moving.

6. These massive volcanic eruptions melted underground water reservoirs and caused catastrophic flooding sufficient to fill most of the northern hemisphere periodically. Olympus Mons was surrounded by water, or possibly ice covered water, long enough to assume the characteristic pedestal shape of shield volcanoes surrounded by an ocean (Figure 5.6).

7. As meteor bombardment subsided, occasional large impacts would briefly raise global temperatures, creating a transient atmosphere and brief periods of precipitation that could last between 1 and 1000 years. Runoff from this precipitation formed streambeds now visible in the highlands. Most of the time, however, Mars would be cold and dry between these brief "warmer, wetter" interludes.

8. Longer periods of warmth and humidity persist because of Mars' cycle of extreme obliquity in which the polar axis of the planet tilts to $45°$ or greater every 124,000 years or so, causing the frozen water and carbon dioxide at the poles to sublimate into a thicker, warming atmosphere before condensing as solids closer to the equator (Figure 5.7).

9. The obliquity cycles, occasional volcanic eruptions, and meteor impacts have caused infrequent, sparsely distributed eruption of ground water leading to ponding of water-ice in ephemeral lakes. Persistence of underground water continues to cause gully formation in craters and some dendritic or branching valley networks, associated with occasional sudden eruptions and flow of water onto the surface.

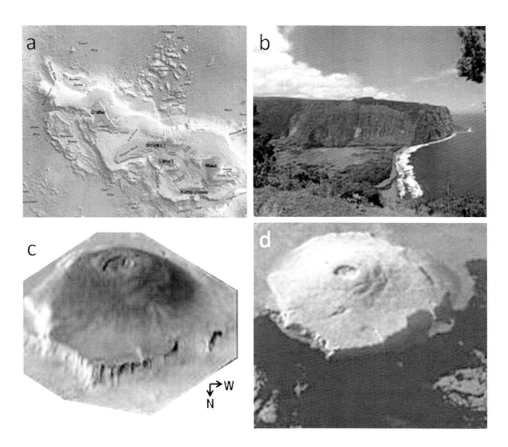

**Figure 5.6** Argument for a surface ocean northwest of Olympus Mons, by terrestrial analogy. (a) Aerial view of the Hawaiian island chain, showing pedestal structure of volcanic islands surrounded by ocean. (b) Waipaio Valley, Island of Hawaii, showing characteristic gentle slope from volcanic peak to ocean, where lack of firm support and water erosion result in steep sea cliffs. (c) Reconstructed view of Olympus Mons, emphasizing pedestal structure on the northern and western edges (NASA/JPL, Photojournal) (d) Conception of Olympus Mons, bordered by an ocean to the northwest at an earlier time (Art by Kees Veenenbos, with permission, based on data from the NASA/JPL/MOLA Science Team).

The picture that emerges, then, is of a water-rich planet early growing cold and freezing over, unable to sustain water above ground because of the low vapor pressures and freezing temperatures of the atmosphere, but able to soak the water into underground reservoirs in the porous basalt beneath the surface.

Ocean-scale inundations and widespread glaciations occurred periodically for the first billion years of Mars' planetary history. Then, with decreasing frequency, occasional outflows and brief periods of rain and snow carved drainage valley networks, recycling water to the surface, while long-term oscillations in the tilt of its axis and the eccentricity of its orbit caused Mars to

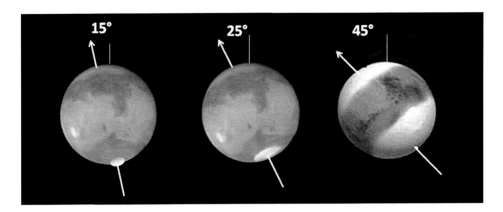

**Figure 5.7** Effect of obliquity on humidity and ice distribution. The polar axis of Mars oscillates between a tilt of about 15° to over 40° every 124,000 years. In a cycle of about 1 million years, the tilt becomes even more extreme, from about 25° to possibly 60°. During periods of low tilt, the middle of the planet is heated by sunlight most of the time, so frozen water and $CO_2$ accumulate at the poles. As the tilt increases, longer and more intense summer sunlight at the poles causes the frozen water and $CO_2$ to sublimate, making the atmosphere denser and spreading the moisture further from the polar regions into thinner layers that blanket the surface closer to the equator. The figures above represent Mars at the minimum (15°), current (25°) and maximum (45°) tilt of its present cycle (Adapted from images by NASA and Jim Head/Brown University).

redistribute water-ice from the poles toward more equatorial reservoirs at periodic intervals, recharging underground aquifers and sustaining an ample amount of water to the present day. In the less than half a century of time – a blink of the eye, geologically speaking – in which we have had realistic images of the actual climate and surface conditions on Mars, we have been witness to an inter-fluvial, inter-glacial period of quiescence, which over a longer span of time is deceiving.

Mars has been to Earth as a volatile child prone to periodic emotional outbursts is to a larger sibling who, despite occasional upheavals as well, maintains a more placid demeanor most of the time. The history of both planets was tumultuous in the beginning. Earth then settled in to a relatively more stable water-world enveloped in a protective atmosphere with sustained plate tectonic activity to recycle its minerals and recharge its air. But Mars grew cold, lost most of its atmosphere, and entered into a period of increasingly infrequent geological upheavals and occasional eruptions of water to the surface, while retaining its moisture mostly underground, most of the time. Earth has been relatively warm and wet for its entire existence. Mars started out that way, but probably turned into a frozen desert early on, with periodic outbursts that belie the cold barren view we have of it today.

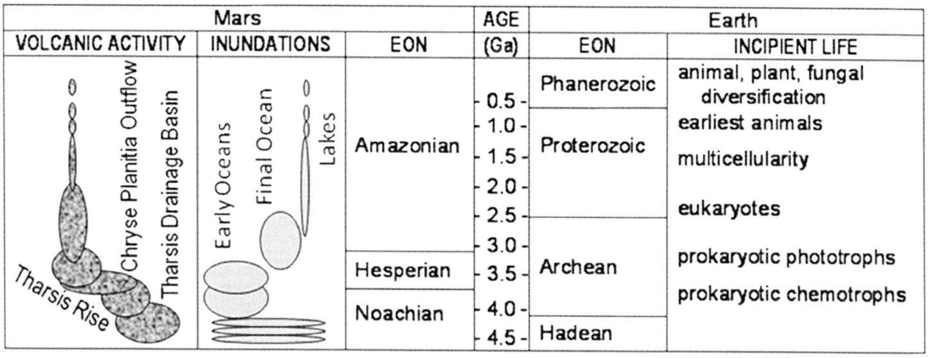

| Mars | | | AGE | Earth | |
|---|---|---|---|---|---|
| VOLCANIC ACTIVITY | INUNDATIONS | EON | (Ga) | EON | INCIPIENT LIFE |
| | | Amazonian | 0.5 | Phanerozoic | animal, plant, fungal diversification |
| | | | 1.0 | Proterozoic | earliest animals |
| | | | 1.5 | | multicellularity |
| | | | 2.0 | | |
| | | | 2.5 | | eukaryotes |
| | | | 3.0 | | |
| | | Hesperian | 3.5 | Archean | prokaryotic phototrophs |
| | | Noachian | 4.0 | | prokaryotic chemotrophs |
| | | | 4.5 | Hadean | |

**Figure 5.8** Schematic timeline of geological history on Mars, with comparison to Earth. Column 1 shows major volcanic events, with width representing magnitude and height the temporal duration. Column 2 shows timeline and estimated magnitude of water after probable recurrent inundations and vaporization events associated with ongoing bombardment during early "warmer, wetter" periods. Column 4 shows time in billions of years prior to the present. Column 6 indicates the time at which key biological innovations appeared on Earth. (Adapted from Fairén et al [1] and Schulze-Makuch et al [9]).

## 5.3 Reconstructing a plausible evolutionary history for putative life on Mars

A couple of hundred million years – the most conservative estimate of how long water stood on the surface of Mars – is long enough for life to get going, if Earth is any guide. And a billion years, which some of us feel is more likely for the existence of surface water-ice on Mars, could well be sufficient for a considerable diversity of biological forms to arise.

Whatever evolution of life occurred on Mars was front-loaded in time, to be sure. The histories of life on Mars and Earth, assuming similar beginnings, would have to be mirror-images of one another. Earth, settling into a stable and relatively unchanging water world early on, provided a constant environment with little pressure to change, once the revolutionary mechanism of photo-synthesis evolved. It would take 2.5 billion years for the oxygen byproduct of photosynthesis to accumulate to the point of enabling a radical diversification of macrobiotic forms. Mars, by contrast, experienced dramatic upheavals that recurrently altered its climate, surface, and atmosphere in its early history, challenging whatever life first emerged with strong pressure to adapt to dramatic, often catastrophic, changes. Then, like the calm after the storm, the planet settled into a much quieter, relatively stable stretch of 2.5 billion years that has seen much less geological action, and probably little if any biological innovation.

What would life have looked like on a youthful Mars, and what would be left of it today? We have summarized in Figure 5.9 a schematic time-line for our vision of what the major evolutionary transitions could have been. In the discussion that follows, the numbered steps refer to the transitions shown in Figure 5.9.

### 5.3.1 Rise of the autotrophs

When life arises *de novo* anywhere in the universe, we assume is does so as protocellular aggregates, gradually enclosing increasingly complex metabolic interactions that feed on the chemical nutrients in their environment, progressively acquiring the ability to replicate themselves with reliability, in accordance with the principles described in chapter 2. While pangenesis could have seeded Mars with pre-existing forms of life from elsewhere, including Earth, we will here assume that *de novo* protocellular aggregates were the founding form of life on Mars.

For life to be self-sustaining in the long run, it has to be able to assemble its own larger, energy-rich molecules for energy storage pending utilization. It has to manufacture its own food, in other words, which is the meaning of autotrophy. We assume that the first autotrophs on Mars, as almost surely on Earth, were chemoautotrophs, using chemical exchange reactions that sucked small bits of energy from the inorganic compounds available for use in assembling larger, energy-storing molecules. Given the abundance of water on Mars, and the great advantage of carbon-based macromolecular constructions in water, we are assuming that life on Mars would be carbon-based and therefore dependent on principles of organic biochemistry similar to those used by life on Earth.

The transition from protocellular aggregates to simple chemoautotrophs *(step 1)* would have signaled the onset of fully living cells, most probably in the early oceans of Mars. Even if the water were largely frozen, the concentrating power of liquid inclusions in ice could have provided a formative environment for the earliest chemoautotrophs. Photoautotrophy, or the ability to use the energy of sunlight to assemble food molecules, could also have arisen early in the history of life on Mars *(step 2)*.

We think it more likely that this came later than the appearance of the first chemoautotrophs, simply because photosynthesis as we know it on Earth requires more subcellular structural specializations; but we can't say for sure. In any event, 20 to 100 million years after its formation, after the planet had cooled enough for the oceans to persist, autotrophy of one if not both forms could have been well established.

### 5.3.2 Phototrophic diversification

With the Sun growing slowly brighter as the conversion of hydrogen into helium made it denser at the core, life driven by sunlight should have prospered on Mars as long as it could stay near the water's surface. Even in an ocean frozen over, microbial phototrophs would be able to grow and reproduce while embedded in ice inclusions. Then, when global warming or a large meteorite impact melted the ice layer, the cells would have been released for dispersal into the transiently liquid oceans or lakes.

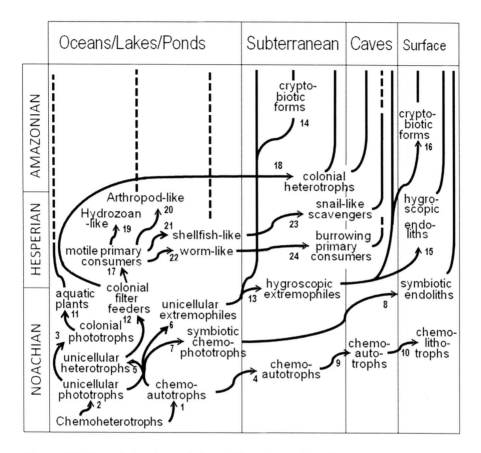

**Figure 5.9** Scenario for the evolution of life on Mars through time in different environments. Plausible steps in the major transitions that life could have undergone are shown schematically at the approximate time in planetary history (vertical axis) in the indicated principle environments (horizontal axis). Solid lines represent evolution or persistence of living forms. Dashed lines represent persisting fossil remnants of forms that have become extinct.

Where bodies of water persisted for long enough periods, the aggregation of phototrophic cells into colonial aggregates – what we would call primitive plants, like green algae, perhaps – would be expected *(step 3)*. Invasion of the land on a young Mars would even have been possible, had the atmosphere been substantial enough to moderate the influx of solar radiation and keep global temperatures above freezing part of the time. Ground cover akin to primitive mosses on Earth would thus have been conceivable.

If the early colonial phototrophs were able to evolve into more complex forms *(step 11)* – something like early plants to us – it most likely would have occurred in water, where temperature fluctuations would have been buffered and kept warm enough for metabolism to continue. In time, however, as the surface of the planet dried out and life was forced underground in order to survive, photoautotrophy would have become untenable. Fossilized plant-like organisms would be the only remnants of this episode in the life history of Mars that would be found on the planet today.

### 5.3.3 The heterotrophic succession

With autotrophs established, the stage would be set for the evolution of heterotrophs able to feed on the autotrophs. Hence, the first microbial consumers would have been able to evolve *(step 5)*. Meanwhile, the earliest chemotrophs may have been evolving more extreme adaptations to what was probably growing acidity and salinity in the oceans, and dropping temperatures globally *(step 6)*. These microbes would have been analogous to the acidophilic ("acid-loving"), halophilic ("salt-loving"), and psychrophilic ("cold-loving") microbes that inhabit their respective extreme environments on Earth (bearing in mind that the word "extreme" is used here in comparison to what is normal on Earth, not Mars).

Being the inventive process that it is, evolution on Mars may well have given rise to a symbiotic marriage of convenience between an early phototroph and an early chemotroph, resulting in a hybrid form of life able to use both sunlight and chemical energy from its substrate *(step 7)*. Such an organism would have been well pre-adapted for taking up residence inside rocks, persisting in that form in or near the surface today *(step 8)*.

### 5.3.4 Colonial heterotrophs ensue

As population densities of autotrophs and primordial heterotrophs rose in the ancient oceans and lakes of Mars, selective pressure for a more efficient feeding mechanism would have been strong. This is the circumstance that would have favored aggregation of microbial cells into colonial forms acting in concert to haul in more food, more effectively *(step 12)*. Sponges on Earth are little more than a confederation of cells specialized for filter feeding, by passing food-bearing water in and out of their loosely organized interiors. Such an organism may have been the first form of life that could be regarded as multicellular on Mars. Motility is an ancient property of life on Earth. Some bacteria and many unicellular protists have the ability to propel themselves through the medium they inhabit. Being stationary and filtering out the food that comes your way is an effective life style, as persistence of sponges to this day demonstrates. But being able to move about, either avoiding the filter feeder or seeking a meal that isn't coming your way, provides a further advantage in the struggle for existence.

There is thus every reason to assume that motility would have evolved early among the living organisms on Mars *(step 17)*. And if the organism were already prone toward multicellularity, the evolution of mobile, multicellular organisms would seem to have been inevitable. They may have started out just as a small motile stage in the life cycle of an otherwise sessile form, like the pleural stage of certain hydroids on Earth. However they arose (if they in fact did), once swimming about the waters of Mars, grazing over the underside of her ice-covered oceans, or plodding along the beds of her lakes and seas, these motile ancestors could have been the forerunners of the Martian biodiversity to come.

### 5.3.5 Offshoots of colonial life diversify

Once the motile stage of an otherwise stationary organism had evolved, selective pressure would have favored its remaining in the mobile phase of its life cycle, living as a primary consumer on the plant-like organisms and microbes that by then were populating the aquatic environments of Mars. Mobility would then have been a portal into a variety of life styles. Figure 5.9 pictures an evolutionary radiation into organisms something like soft-bodied jellyfish *(step 19)*, a Trilobite-like crustacean *(step 20)*, a molluscan-type animal with a calcified shell *(step 21)*, and a representative of one of the most successful body forms for life on Earth, an elongated worm-like structure *(step 22)*. These are purely speculative, of course, and are meant to illustrate only that some form of diversification would have been natural for this stage of the evolutionary process – not that these were in fact what appeared in the late Noachian or Hesperian seas of Mars.

Could this much evolutionary change have occurred within a billion years on Mars? After all, by that same length of time, eukaryotes had not yet evolved on Earth, much less anything multicellular (Figure 5.8). The answer is, we don't know. But we do need to avoid thinking that the pace of evolution would have been no faster necessarily on Mars than on Earth. Mars was undergoing convulsive vacillations in its geophysical makeup, hence probably in its climate and atmosphere, over the first billion years of its existence, a period of time in which Earth was settling in to a steady, prolonged world of global oceans and low-lying land masses. Pressures for directional selection would have predominated on Mars, while stabilizing selection would have been the stronger tendency for life on Earth, once the late heavy meteorite bombardment tapered off. So Martian seas swarming with a variety of macrobiotic forms of life as the end of the Hesperian and the crises of evaporating seas approached, are thus well within the realm of plausibility.

### 5.3.6 A succession of subterranean retreats

The crisis, however, did come. The sequence of events that would lead to desiccation of the surface of Mars, save for periodic outbursts of surface flooding, were by the early Amazonian (or even Hesperian) leaving the planet increasingly devoid of the aquatic habitats that had presumably spawned the early proliferation of biodiversity on Mars. Water would persist on the planet, but underground only. For most forms of life on Mars, this meant adapt to a subterranean life style or perish. And many would have perished.

Soft-bodied consumers were probably among the first to go, as the buoyancy of open water was essential for maintaining their morphology. In fact, animal-like organisms and plant-like forms of any sizable dimensions would have been unlikely to adapt to the confining spaces of the subsurface soil.

For microbial forms, the question of space was not critical, but adaptation to microenvironments that were cold, dry, and probably quite salty would have been the challenge. Descendants of the earliest chemoautotrophs *(step 4)* could have survived through appropriate extremophilic adaptations. One of the most important was probably the ability to absorb the increasingly declining levels of moisture in the atmosphere *(step 13)*, so hygroscopic extremophiles (organisms able to absorb water from the air) may have become prominent members of the subsurface biota. One of us (Schulze-Makuch), with our colleague, Joop Houtkooper, has suggested [10] that the hydrogen peroxide found in greater abundance on Mars than on Earth could be a key adaptation to the extreme cold and dry air of Mars, for two reasons: (1) It's very hygroscopic, drawing in the very limited amount of water vapour available in the atmosphere; and (2) the freezing point of hydrogen peroxide solutions can reach lower than -56°C, or 70°F below zero, and even then the solutions just get firm instead of forming ice crystals that disrupt cellular structures. Thus, hydrogen peroxide is both hygroscopic and cryoprotective – two key adaptations for surviving on Mars today.

Microbes with adaptations such as these may, in turn, have given rise to cryptic forms that persisted on or near the surface in a state of suspended animation, capable of being revitalized only by rehydration upon the occasional reappearance of water on the surface *(step 23)*. Rotifers provide an example of such organisms on Earth. Of the larger forms, those with worm-like body shapes may have been able to make the transition *(step 24)*. Worms of various sorts constitute by far the bulk of the subterranean animal biomass on Earth.

One form of quasi-unitary organism that may have survived underground could have been something like the slime molds and fungi that permeate certain soil habitats on Earth. They could have descended from aquatic colonial heterotrophs *(step 18)*, retaining their colonial organization and simple morphology well suited for permeating the confining spaces between soil particles underground. A member of the fungal family, *Armillaria*, has been found to extend over an area of nearly 10 square kilometers beneath a forest floor in Oregon, absorbing nutrients from the roots of the plants it enshrouds [11].

### 5.3.7 The sanctuary of caves

Mars has lots of caves. We know this because it has lots of volcanoes, and volcanoes create lava tubes that lead to caves. As lava flows across a landscape, the outer layers of lava cool first, forming a solid shell, while liquid lava continues to flow and occasionally drain completely out of the surrounding shell of solidified rock, leaving a hollow tube. Because of the lower gravity on Mars, its lava tubes are much bigger than on Earth. Indeed, we can see a number of what appear to be collapsed lava tubes in the high-definition images now available from the surface of Mars. Given the certainty of underground water and the growing evidence for marine stratigraphy on Mars, caves formed by mineral dissolution like our limestone caves on Earth may also be present well beneath the surface.

Seaside or lakeside caves on Mars would have taken in directly whatever aquatic life was present. As the seas and lakes dried up, water may have lasted longer in the caves, providing a respite from the extinction likely occurring on the outside. How much water remains in the caves of Mars cannot yet be known, but the presence of groundwater suggests that some humidity may yet prevail in the cave habitat. It is thus possible that microbial mats consisting of chemotrophs *(step 9)*, and consumers of a fungal-like nature *(step 18)* could cover the floors of caves on Mars today, conceivably even supporting higher-order consumers like snail-like scavengers *(step 23)* as well as a host of detrivores.

Caves on Earth provide a privileged sanctuary, where forms of life uniquely adapted for the perpetual darkness, constant temperature, and altered atmosphere of caves evolve in sometimes bizarre directions, away from the ancestors that first invaded the cave habitat. By the same token, caves on Mars may be host to a variety of living organisms no longer found in the much harsher environment outside. Here could still live mats of microbes not found on the outside, and snail-like scavengers that might graze upon them. The food web would have to be based on chemotrophic producers [12], and the greatest potential for survival of this type of sequestered community would be in caves deep underground or well sealed off from the atmosphere, where temperatures would not be frigid and some humidity would be preserved. The caves of Mars could, in fact, be the most favored habitats for the persistence of life on the planet. We just won't know till we get there.

### 5.3.8 Living stones
If anything survives in a perpetually active form on the surface of Mars today, it's likely to be encased in stone. Endolithic ("stone dwelling") organisms are well adapted to an environment scoured by desiccating winds and radiation. Simple chemoautotrophs living off the inorganic chemical energy in their substrates would be natural inhabitants *(step 10)*, particularly those with well-developed hygroscopic abilities *(step 15)*. Phototrophs or symbiotic photochemoautotrophs close enough to the surface to harvest sunlight through their mineral coverings *(step 8)* would likewise find the endolithic life style a strategy for survival.

### 5.3.9 Cryptobionts
Under the conditions prevailing on Mars today, it is difficult to imagine any active forms of life on the surface other than endolithic organisms, and the "activity" of even those would be limited to the very slow rates of growth and metabolism that life inside rocks would permit. There could, however, be cryptobiotic microbes descended from hygroscopic extremophiles that lie near the surface, ready to be revitalized by brief episodes of rehydration *(step 16)*. Even macroscopic organisms the size of spores could lie in wait for those infrequent respites from the unrelenting cold and dust-dry conditions that now prevail the vast majority of the time.

### 5.3.10 Fossil remnants and life unseen
The fact that no camera landed on Mars has yet detected the slightest hint of anything alive is the consequence of the likelihood that (1) there really isn't anything living on the surface of Mars, other than endolithic forms that would

be very hard to detect without microscopic examination, and (2) life on Mars has retreated totally to subterranean habitats. Assuming that Mars was well populated in its youth with microbes of many sorts, their descendants could still be living in profusion in the underground aquifers that hold most of the planet's water. Likewise, colonial heterotrophs like underground fungi and slime molds could still be found wherever moisture is retained.

If our assumptions about planetary history are correct, however, there was a time when life could have flourished with considerable diversity in and beneath the surface waters of Mars. Among those organisms could have been forms encased in calcified shells or other types of exoskeletons that maintained their shape after death. Others may by fortuitous circumstance have been preserved by a mineralization process. Therefore, there could be fossils to be found wherever water stood for a considerable length of time. Most of the robotic exploration of Mars has been conducted by landers (Viking 1 and 2, and Phoenix) and rovers (Sojourner, Spirit, and Opportunity) that have been placed where water at one time was abundant. They haven't found any identifiable fossils yet, but an alien rover placed randomly in a dry lake bed on Earth probably wouldn't either. Until humans walk the surface of Mars, and probably not until they dig beneath its surface, will we likely find fossil evidence of the life that we think might have been there.

## 5.4 A putative Martian biosphere

In Table 5.1 we summarize what a list of generic organisms that have ever lived on Mars would look like. Note that it's a composite of life that presumably existed on Mars in its early history, with that which may survive today. The two biospheres differ considerably.

When Mars was young and water was pooled on the surface probably for millions of years, life could have become fairly diverse, with well-developed aquatic and subterranean food webs and levels of complexity approaching what the oceans of Earth may have looked like near the dawn of the Cambrian explosion. No form of life on Mars was likely very large, as life on Earth remained microscopic until the accumulation of an oxygen-rich atmosphere made much more efficient metabolism possible; and Mars has probably never had an oxygen-rich atmosphere. But colonial and multicellular organisms were well within the realm of possibility. The terms, "arthropod-like," "mollusk-like," and so forth as used in Figure 5.9 and Table 5.1 are intended to suggest only that organisms something like jointed animals with hard exoskeletons, or sessile filter-feeders with mineralized shells would have been plausible inhabitants of the habitats presumed to have been available. Until a robot or human unearths a fossilized form of macro-organism on Mars, any thoughts of what that organism would be like in detail, of course, are purely speculative.

Whatever the macrobiota of early Mars may have consisted of, it by now most likely is only a weak echo of what once existed. As water disappeared from the

**Table 5.1 Summary of putative organisms on Mars.**

This lists the minimal set of organisms that may have lived or may still be found in living or fossil form on Mars.

| TROPHIC LEVEL | ENERGY SOURCE | GENERIC ORGANISM | EARTH ANALOG |
|---|---|---|---|
| Producers | reduction of iron | iron aquatic chemoautotrophs | *Deferribacter* |
| | reduction of iron | iron chemolithotrophs | *Shewanella* |
| | reduction of $CO_2$ | methanogenic chemoautotrophs | *Methanobacterium* |
| | light | fresh water phototrophic bacteria | cyanobacteria |
| | light | marine phototrophic bacteria | *Roseobacter* |
| | light | fresh water plant[1] | water lily |
| | light | marine plant | kelp |
| | light and inorganic chemistry | symbiotic photochemotrophs | *Symbiodinium* |
| Primary Consumers | chemo- and photoautotrophs | microbial heterotrophs | protozoa (*Amoeba*) |
| | chemo- and photoautotrophs | acido-halophilic heterotrophs | brine shrimp |
| | lithoautotrophs | endolithic heterotrophs | boring thallophytes |
| | fresh water plant | fresh water herbivore | tadpole |
| | marine plant | marine herbivore | sea snail (*Aplysia*) |
| Secondary Consumers | microbial heterotrophs | hygroscopic heterotrophs | endolithic fungi |
| | acido-halophilic heterotrophs | hygroscopic heterotrophs | endolithic fungi |
| | various heterotrophs | colonial fungus-like heterotrophs | slime mold |
| | various heterotrophs | colonial filter-feeders | sponge |
| | various heterotrophs | cnidarian-like consumer | sea anemone |
| | various heterotrophs | arthropod-like consumer | shrimp |
| | various heterotrophs | mollusk-like consumer | clam |
| | various heterotrophs | worm-like consumer | roundworm |
| Tertiary Consumers | colonialfungus-like heterotrophs | snail-like scavenger | snail |
| | marine filter feeders | marine grazers | starfish |
| | worm-like consumer | carnivorous fungi | nematophagus fungi |
| Decomposers | fresh water detritus | microbial heterotrophs | *Paramecium* |
| | marine detritus | benthic megascavengers | lobster |
| | subsurface detritus | burrowing decomposers | earthworm |
| | subsurface detritus | subsurface fungi[2] | mold |
| | surface detritus | surface microbes | mushroom |

[1] any multicellular phototroph
[2] any multicellular organism that ingests externally digested food by absorption

surface, the subterranean habitats where water could persist would have been unable to sustain anything more than microorganisms and possibly a few highly specialized larger forms. What remains today most likely consists of microbes surviving in underground water reservoirs, some cryptobiotic forms capable of rejuvenation when the soil becomes transiently rehydrated, and possibly some endolithic microbes on or near the surface. As noted in the previous section, some life could still exist in caves on Mars as well – the nature of which, however, is hard to guess.

## 5.5 Ecosystems on Mars

Ecosystems on Mars today probably bear only a faint resemblance to what they once were. We assume that all surviving life on Mars, except for a few endolithic forms, is sequestered underground. There is a good chance then that the biosphere beneath the ground consists exclusively of simple, microbial forms. If so, producers would presumably be restricted to chemoautotrophs, and consumers would be simple heterotrophs comparable to bacteria, protists, or fungi occupying similar habitats on Earth.

Since we cannot exclude the possibility that underground lakes or pools of liquid water deep inside caverns could still exist, we can't rule out the possibility that larger, more complex organisms could still survive. Assuming the low productivity available from chemoautotrophs at cold temperatures, the biomass of the producers must be low, and therefore capable of supporting very minimal amounts of biomass at the consumer level.

Earlier in the history of Mars, the picture may have been quite different. With millions of years to evolve in liquid oceans under a denser, more humid atmosphere, fully developed ecosystems might well have appeared. Figure 5.10

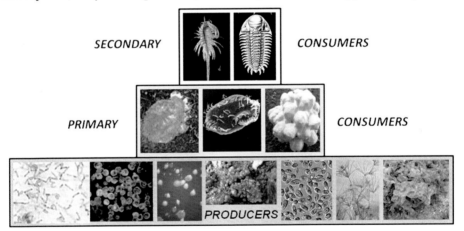

**Figure 5.10** Trophic structure of Martian ecosystem. Producers would consist of chemotrophs and phototrophs, analogous (from left to right) to various archaebacteria, symbiotic bacteria, extremophiles, brown algae, green algae, fresh water plants, and seaweed. Primary Consumers would include microbial heterotrophs analogous to slime mold, protists, and colonial filter-feeders. Secondary Consumers would include small animal-like predators and scavengers analogous to the brine shrimp and trilobite pictured here. In the absence of further information about more complex life at the top of ancient aquatic food chains on Mars, speculation about tertiary consumers would be unfounded. All organisms shown are examples of life on Earth for generic reference, and should not be taken to represent the actual appearance of present or past life on Mars. The horizontal breadth of each level indicates that global biomass decreases with each successively higher trophic level (Photographs by J. McKenna (brown algae), Kristian Peeter (seaweed), djpmapleferryman (brine shrimp)).

illustrates what the trophic structure of such an ecosystem could have looked like in generic terms, using forms of life on Earth as analogs for the actual organisms populating Mars. Evolution should have been able, in theory, to proceed to the point where at least a secondary consumer trophic level was supported. Tertiary consumers can certainly not be ruled out, but organismic form and function become so speculative at higher levels of complexity, that restricting our model of a Martian ecosystem to a three-tiered trophic structure is prudent until some form of life on Mars is definitively confirmed.

## 5.6 Biotic communities on Mars

Early in the planet's history, the marine biotic community was most likely the dominant if not only life-supporting habitat on Mars. Life would most likely have arisen in the early oceans of Mars, and would have been able to diversify and move about most freely in that environment. A fresh-water origin for life on Mars cannot be ruled out, but the shorter duration and greater susceptibility to freezing of the smaller pools of fresh water would probably have limited the ability for life to evolve to a significant degree of diversity there. Most fresh water forms would probably have been transitioned fairly quickly to the marine habitat, where biodiversification was easier and more likely, in any event.

Invasion of the land at an early stage by microbes or simple plant-like organisms cannot be excluded, particularly since glaciers and isolated pockets of frozen water may have provided spaces in liquid inclusions or beneath an ice cover for cold-adapted forms to slowly grow. Their diversification would have been even more restricted than in fresh water, however.

From the beginning, subterranean moisture would have provided a stable, protected environment for the proliferation of at least microbial forms of life. This is a biotic community that conceivably survives unbroken from its origin over three billion years ago. Assuming this habitat has been stable and unchanging (except for local periodic disruption by volcanic activity or bolide impacts), stabilizing selection would be expected to have kept the surviving forms in a relatively ancestral state. Much of the subsurface biosphere of Mars may not look much different today from what it was like three billion years ago.

Cave life represents the most imponderable aspect of Martian biology. There is no doubt that caves abound on Mars, from the very fact that lava flow has been extensive and that lava tubes are an inevitable result of those flows. Indeed, surface images of Mars provide many instances of what appear to be collapsed lava tubes. And given the lower gravity on Mars, they must be much larger than on Earth. One can imagine the Tharsis region being riddled with lava-generated caves as it encroached into the northern sea. Then, those caves on the northwestern perimeter of the rise must have been flooded and occupied by whatever life was then existing in the ocean. When the ocean receded, was the water in the caverns retained? If so, was life in the water self-sustaining, and might it still be there today? Might there be other caverns, carved deep underground by erosional forces like those that formed the great caverns in sedimentary deposits on Earth, or by processes unfamiliar to us but natural on Mars? These are fascinating questions that most likely will not be answered until humans have the opportunity to go cave exploring for the first time on our neighboring world. Whatever life there is or ever has been in the caverns of Mars most likely represents the most unique and specialized biotic community on the planet.

## 5.7 Earth analogues of Martian habitats

A variety of terrestrial analogs mimic the environmental near-surface conditions of Mars. The most relevant surface analog environments are the Atacama Desert in Chile and the Dry Valleys of Antarctica. The Atacama is the driest desert on Earth. The Dry Valleys of Antarctica are not as dry, but simulate better the very cold temperatures on Mars. Both analog environments are used intensively for studies supported by NASA and the European Space Agency (ESA). For example, the Viking life detection tests were duplicated with soil from both deserts and did not reveal any signs of life. Yet, closer inspection revealed that neither the Atacama Desert nor the Dry Valleys of Antarctica is sterile.

The Atacama Desert receives only a few millimeters of precipitation per year on average, and sometimes no precipitation whatsoever. The average rainfall in the core area of the Atacama Desert is less than 1 millimeter per year. The desert is 50 times drier than California's Death Valley and extends for over 100,000 square kilometers. Not only is the Atacama Desert extremely dry and mostly cold, but it also contains perchlorates, chemical compounds that were recently discovered on Mars by the Phoenix mission and which are consistent with the presence of life. Some locations in the Atacama Desert receive marine fog that provides sufficient moisture for algae and lichens, which live underneath rocks (a hypolithic life-style). In other, even dryer areas of the Atacama Desert, microbes make a living by attracting life-sustaining water directly from the atmosphere via hygroscopic compounds such as halite and other salts.

With regard to temperature, the Antarctic desert's high altitude dry valleys may be the best terrestrial analog for the surface of Mars. Air temperature remains below -20°C from early April to August, and even during the peak of the austral summer, air temperature does not rise above 0°C. The climate is very dry, permanently lacking surface liquid water. Most of the snow, which is the only form of precipitation, is either blown into the valleys or sublimates without melting.

The year-round subzero temperature, combined with low water availability, results in a true desert. Plant life is nonexistent. Even rock lichens, which are common in the maritime Antarctic, are rare and occur only on exceptionally well protected rock surfaces. Soil microbial counts are exceedingly low and most of the life present appears to live within rocks, mostly beneath sandstone surfaces. This type of life style is termed cryptoendolithic and may be a model for life on Mars. On Earth, this habitat supports a variety of eukaryotic algae, fungi, and prokaryotic cyanobacteria. Survival is possible in the sandstone because the pore spaces provide the physical space, sunlight can penetrate the outer layer of the translucent sandstone, and rock surfaces, especially those that are sloped and north-facing, are warmed by insulation. In summer, rock temperature can even reach up to +15°C.

Recently, hypolithic cyanobacteria have been discovered that colonize the lower surface of dolerite clasts, a magnesium and iron-rich igneous rock similar to basalt. The main source of water to the community of cyanobacteria is large snow packs. During the ten days in late January of 2005 when the discovery was made by Henry Sun of the Desert Research Institute, snow melting lasted only 2–3 hours daily, around noon. During the rest of the day, the ground was frozen solid. Microscopic observations revealed that the dominant cyanobacterium, *Oscillatoria* sp. has gliding motility. Apparently, the hypolithic cyanobacteria are able to actively migrate under the clasts during the 2–3 hour window during which metabolism is possible. The motility also explains why the organisms can use opaque basalt for protection – they can come out to catch the sunlight and then retreat to a damper environment.

In deserts, cyanobacteria grow under half-buried stones principally for moisture. Soil covered by a stone remains moist much longer than does bare, unprotected soil. The colonization target are mostly translucent stones because they also transmit sunlight, in addition to the moisture benefit, but this example shows that basalt-like rocks can also be a habitat for life. And indeed, most of the Martian surface is made up of basalt rocks, though sedimentary rocks such as sandstones are also present on the Martian surface in some areas.

Another analog environment for possible life on Mars is the lava tube cave. Lava tubes provide the most plausible current cave environment for Mars, as they are clearly visible in many places. We can see them, map them, and ultimately send missions to them. Other types of caves may well be present but not yet detected. On Earth, many caves provide a radically different environment for life than does the overlying surface. Caves offer protection from desiccation, ultraviolet radiation, and weather. The interior of caves can provide energy sources distinct from the biomass above ground, serving as a more suitable model for a potential Martian subsurface ecosystem. Caves also provide access to bedrock compounds that can serve as chemolithotrophic energy sources, and/or act as collecting chambers for the buildup of reduced gases coming from below. Prior work by Penny Boston from New Mexico Tech and colleagues has shown the wide variety of materials metabolizable by cave organisms in oligotrophic circumstances [12]. Common in many basaltic lava flows, lava tubes on Earth contain extensive visible evidence of microbial growth and associated mineral precipitation. These lava tubes experience a wide variety of climatic influences from ice-volcano interactions in Iceland to hot sand-floored lava tubes in Saudi Arabia. Some lava tube caves on Earth, like those in the Mojave Desert with very low precipitation, may mimic conditions on Mars particularly well.

## 5.8 Characteristics of life on Mars

Our speculations about what life may be like on Mars today are bound to be more reliable than our notions of what life may have been like on a younger planet harboring a level of biodiversity long since lost. That is because we can be reasonably confident that the only life currently surviving on Mars, other than endolithic microbes and cryptobiotic beings in deep suspended animation, must be residing in the subsurface of the planet, where life is slower, simpler, and highly confined. If by chance some survivors still lurk near the surface, they would have to have developed specialized adaptations as suggested in section 5.9. This doesn't spare us the need, however, to offer thoughts on what life may have been like at those earlier, more robust, times. So again, we will be distinguishing between Martian life now, as opposed to the past.

### 5.8.1 Metabolism

In the beginning, metabolism would have been a hit-or-miss, trial-and-error testing of various biochemical pathways, as it must have been on Earth when conditions on the two planets were similar. But the metabolic machinery of life on Mars would have been heavily influenced early on in a way that metabolism on Earth was not, by the onset of perpetual winters that have most likely kept the planet very cold from early in its history. Metabolism on Mars has always probably been a case of chemistry in slow motion.

Carbon dioxide and compounds of iron, sulfur, and perhaps other metals have been readily available for energy-yielding reactions from the beginning. Likewise, sunlight has been an ever present alternative form of energy for autotrophy. Other compounds, like hydrogen peroxide, may have played special roles on Mars, and forms of biochemistry unknown to us cannot be ruled out. What we can rule out, it appears, is the evolution of highly-efficient organic oxidations made possible by an abundance of oxygen. If ever there was very much oxygen on Mars, it apparently became sequestered in metallic oxides long ago. Once the disappearance of surface water marked the end of photosynthesis (except maybe to a minor degree for endolithic phototrophs), the ability to generate oxygen by splitting water – if that was the nature of photosynthesis on early Mars – would have vanished. There appears to be an oxidant on the surface of the planet still, perhaps hydrogen peroxide, but empirical observation shows that it hasn't been sufficient to raise the level of atmospheric oxygen to a significant degree.

In summary, the combination of pervasive low temperatures and the lack of a concentrated oxidizing agent have probably meant low metabolic rates for any organisms that have ever lived on Mars.

### 5.8.2 Reproduction

In the large proportion of Martian life that has always been microbial, the same rules of simple reproduction that apply to microbes on Earth have probably applied on Mars. That means simple fusion, with the possibility of some conjugation (fusion for the exchange of genetic material) on occasion. The dividing times would probably be longer than in microbes on Earth because of the consistently lower temperatures.

On early Mars when life may have evolved to colonial and multicellular levels of complexity, budding would probably have been a common innovation for the larger forms. Since biologists still aren't sure what all the factors are that tip the balance for or against sexual reproduction (and it goes either way in different circumstances), it's hard to predict what role sexual reproduction may have played. We do know that sex is an ancient adaptation for reproductive success on Earth. If used, it would probably have started out like on Earth, as life cycles consisting of alternations between generations consisting of one or multiples of genetic material – assuming that genetic information was carried in discreet molecular entities analogous to chromosomes. If and when mechanisms evolved for syngamy, or the fusing of gametes (cells bearing copies of genetic material) from more than one parent, passive dispersal through the water would have enabled even sessile organisms to exchange genetic material.

Except for a few large animals, aquatic organisms use external fertilization, since water protects against dehydration of the gametes and provides a convenient dispersal mechanism. There is little reason to assume, then, that internal fertilization and the more elaborate morphological and behavioral adaptations that are required for it would have been a characteristic of Martian organisms.

As water became increasingly rare, and mostly colder and drier climates started oscillating with transient warmer, wetter periods, a premium may have been placed on rapid reproduction and maturation of those forms subjected to desiccation and the need to go into some form of hibernation. An alternation of generations, featuring the formation of spores that preserved the capacity for regeneration after increasingly long periods of cryptobiotic quiescence as the periods of dry climatic conditions grew lengthier, seems likely. Whether evolution could have produced spores able to survive for millions of years between the episodic return of water is hard to tell. All we know for sure from looking at life on Earth is that evolution can lead to some pretty hardy innovations.

### 5.8.3 Motility

Amoeboid movements or flagellar-like propulsion of microbes through moist channels underground may represent the limit of motility on Mars today. Conceivably, some small worm-like forms of multicellular life deep beneath the surface could wiggle slowly through the substratum; or if liquid water is found deep inside caves, some swimming organisms might still survive in such a highly sequestered habitat. For life restricted to subsurface soil, however, mobility would clearly be limited.

Earlier, however, when water habitats abounded and biodiversity may have flowered, motility would have been an advantageous strategy, especially for consumers. Indeed, we propose that radiation of complex forms was probably spurred by the advent of motile stages of the life cycle of a simpler colonial ancestor (Figure 5.9, *step 17*). Some of those descendants may have been very motile, like crustaceans, or may have remained sessile, like shellfish. Slower moving forms, like echinoderms, are equally plausible.

A difference between the motility of Martian organisms and those on Earth would probably have been a much slower pace for the former. This would have been the consequence of a lower rate of metabolism, as discussed above.

### 5.8.4 Sensory Systems

Sensory systems of sessile organisms are limited to simple trophic (approach or avoidance) responses to chemistry, touch, and, where available, light. Trophic responses to chemical signals appear to be universal for living cells on Earth, for the simple reason that chemistry is a matter of life and death. Contact can play an important role as well, where social interactions among cells are important, as they often are. The earliest Martian organisms can therefore be assumed to have had the capacity to detect "odors" and touch. Phototrophs can also be assumed to have been sensitive to light.

Dedicated sensory systems did not appear until more complex organisms evolved, with a need to integrate incoming information. The more mobile the organisms were and the more coordination their motor activities required, the more complex the sensory systems would have grown. If we assume, as Figure 5.9 does, that organisms as complex and active as crustaceans could have evolved on Mars, then sensory systems of comparable complexity to that of crustaceans are likely to have evolved as well. Given the slower pace of life that was likely the case on Mars, the sophistication of sensory systems may have been correspondingly less than that of analogous organisms on Earth, but only in a matter of degree.

### 5.8.5 Cognition

Cognition bears a direct relationship to sensory capability and complexity of motor control. To the extent that both have most likely been limited in the course of evolution on Mars, cognitive capability – hence intelligence – is not likely to have arisen to a very measurable degree. A conservative but warranted conclusion is that intelligence has never been greater in any organism on Mars than it has been in a marine arthropod on Earth. However, lacking certain knowledge of the degree of biodiversity and complexity that unfolded in the oceans of Mars within its first billion years, we resist the temptation to make absolute pronouncements about the level of intelligence attained by any form of life on Mars.

### 5.9 What could be wrong with this picture?

As with all our speculations based on limited information, our vision of how life could have evolved on Mars could be off in either direction.

On the one hand, we may have underestimated the scope and diversity of life on Mars. This would most likely be the case if Mars has been wetter and warmer for a greater span of its planetary history than most experts think. If a northern ocean existed continuously as late as the early Amazonian, that would mean that a marine biosphere could have had well over a billion years in which to develop and mature. While the pace of evolution on Earth during its first billion years can be described as relatively slow, the stronger selective pressures generated by more drastic and volatile changes over a comparable period on Mars could have spurred rapid innovation and led to the rise of macrobiotic complexity much faster there than here. The apparent lack of highly-efficient oxidative metabolism may have limited the size and diversity of organisms, but they could well have exceeded what we have envisioned in our scenario.

On the other hand we could also be overestimating the ease with which life could have arisen on Mars, or – even if having begun – the durability with which it could have survived. If Mars has in fact been mostly cold and dry almost from the beginning, with its protective atmosphere blown away very early on, life may have had a much tougher time getting established there than on Earth. As previously noted, life on Earth took a long time to evolve to a substantial degree of diversity and complexity – at least a billion years to reach the complexity of the eukaryotic cell. If that is an intrinsically limited pace at which evolution can occur, there simply wasn't enough time for life of any form to get very far. Our scenario for the evolution of life on Mars could thus be an overestimate that, taken to the extreme, simply never happened at all.

Against this maximally negative assessment stands the radically different view that life on Mars has in fact already been found [15].

## 5.10 Life may have been discovered on Mars already

"Today, rock 84001 speaks to us across all those billions of years and millions of miles. It speaks of the possibility of life." That possibility – that life had been discovered to have existed on Mars – seemed near enough to reality for President Bill Clinton to announce it in the words above in 1996.

"Rock 84001," or ALH84001 in NASA terminology, was the first meteorite found in the Allen Hills of Antarctica in 1984. Originally formed in the crust of Mars, cracked enough by an unknown force 3.6 billion years ago to let some water seep in, then blasted off the surface of Mars by another powerful impact, it drifted through space for nearly 16 million years before crashing onto the Earth's surface about 40,000 years ago. It was delivered to NASA's Johnson Space Center in Houston, where David McKay and his colleagues put it through a rigorous set of chemical and optical observations. The results of their tests led them to the conclusion that, in the aggregate, the evidence pointed to discovery of the remnants of microbial life on Mars [13]. They based their argument on a collection of circumstantial evidence, including the detection of organic compounds (polycyclic aromatic hydrocarbons, or PAHs) known to be breakdown

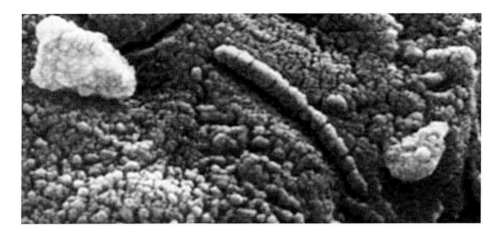

**Figure 5.11** Putative fossil microbe from Mars. The segmented, rod-shaped structure is no larger than the smallest complete organisms (nanobacteria) known to exist on Earth. It was suggested by David McKay and his colleagues [13] to be the fossil of a microbe that lived 3.6 billion years ago on Mars. Others have claimed that it could simply be an inorganic mineral precipitate (NASA).

products from dead organisms, and other biochemical properties linked to microbial metabolism, along with the stunning visual imagery of what appeared to be a segmented, rod-like fossilized microbe (Figure 5.11)

Another, earlier, advocate for evidence already in hand that life has been detected on Mars was Gilbert Levin, lead investigator in the Labeled Release Experiment conducted by the two Viking landers in 1976. This experiment consisted of inoculating a sample of the Martian soil with water and a nutrient solution containing glucose with $^{14}C$, a radioactive isotope of carbon. Had any organisms been present to feed on the added nutrients, they presumably would have broken down the organic compounds, releasing $^{14}C$-containing carbon dioxide (assuming energy metabolism as known on Earth). The radioactivity counters in the instrument panels of both Viking landers did in fact detect a time-dependent release of $^{14}C$-labeled gas that was not seen from soil samples pre-heated to sterilizing temperatures as controls [14]. This could be interpreted as a metabolic signature of a living organism [16]. And another procedure, the Pyrolytic Release Experiment, did show a significant amount of organic synthesis – an observation difficult to reconcile with an inorganic process.

The majority opinion, however, among scientists who have examined the evidence closely has favored non-biological explanations for the observations from both ALH84001 and the Viking landers. Every piece of evidence offered by McKay in favor of the residue of a living organism in the meteorite from Mars has been countered by an argument for a non-biotic origin for that evidence, although McKay's claim is fairly compelling that the most straightforward explanation for all the arguments together is past microbial life. In addition, the water-hydrogen peroxide adaptation hypothesized for Martian organisms [10] does provide a biological explanation for the Viking observations. Nevertheless the majority opinion in the scientific community remains that the argument for life on Mars is, at best, not yet proven. The reality is that the argument cannot, and will not, be settled until more evidence becomes available.

## 5.11 Chapter summary

Mars was formed, like the other rocky planets of the inner Solar System, with sufficient energy to cause differentiation into a core, mantle, and crust, with heat left over to drive plate tectonics and feed a frenzy of volcanic activity that covered the planet with porous basalt. Among the geological agitation early in its history was some process that created a great northern basin that held an ocean of water, periodically retreating and expanding, for somewhere between 200 million and a billion years.

But the small size of Mars brought radiogenic heating to an early end, dispelling its protective magnetic field, and exposing its atmosphere to the blistering solar wind. Its heat- and humidity-retaining atmospheric envelope thus blown away, the planet grew cold, and the water either sublimated into the thin air or retreated into the porous recesses of the substrate.

Much of our speculation about the evolution of life on Mars depends on how warm and how wet it was, and for how long. In this chapter we have adopted a model midway between the assumption of a planet growing cold and dry shortly after its formation, and one which has seen warmer and wetter conditions for a considerable portion of its history. We assume that Mars did grow cold fairly early, but that water remained on the surface, albeit possibly frozen over, for perhaps a half a billion years or so, periodically reappearing in lesser quantities thereafter [17].

A half a billion years of liquid water, with mineral substrates and sunlight available for driving the energy reactions that living cells require, would have provided sufficient time and resources to generate a biosphere based on a host of producers and at least two trophic levels of consumers. Most of the biodiversity would have occurred in those early oceans. Some may have continued locally as long as water of either a fresh or salty composition remained on the surface. And descendant forms of life could have been sequestered in caverns where they still could exist.

The flowering of biodiversity on Mars, however, almost surely came to an end once retention of water on the surface became untenable. At that point, the major biosphere on Mars would have retrenched to underground habitats, including the floor of caves, and possibly some endolithic and hygroscopic forms at or near the surface. A million years on average is a long time to wait for surface moisture to rehydrate cryptobiotic forms of life, but evolution is extremely inventive, so such organisms might still lurk near or beneath the surface as well.

Whatever life evolved on Mars, were we able to watch it in action, would appear to move in slow motion by our standards. Everything from growth rates to size to motility would be restrained by the relatively low metabolic rates imposed by the cold and lack of strongly oxidizing conditions. In the absence of high degrees of mobility, sensory systems would probably never have achieved a great deal of sophistication. Cognitive abilities would have been correspondingly limited as well, so the prospect of intelligence ever having evolved on Mars beyond that of ancestral invertebrates on Earth is small.

Our speculations about the diversity of life and the trajectory it might have taken on Mars could be in error in either direction. Perhaps life was more diverse and robust than we have imagined. Or maybe it never existed on Mars at all. We think on balance that the truth most likely lies somewhere in the middle, aware that some would argue that life on Mars has already been demonstrated by a preponderance of the evidence overall.

## 5.12 References and further reading

1   Fairén, A. G., Dohm, J. M., Baker, V. R., et al. 2003. Episodic flood inundations of the northern plains of Mars. *Icarus* **165**: 53–67.

2   Carr, M. H. 1996. *Water on Mars.* Oxford: Oxford Univ. Press.

3   Kargel, J. 2004. *Mars – A Warmer, Wetter Planet.* Chichester, UK: Praxis Publishing.

4   Levrard, B., Forget, F., Montmessin, F., et al. 2004. Recent ice-rich deposits formed at high latitudes on Mars by sublimation of unstable equatorial ice during low obliquity. *Nature* **431**: 1072–1075.

5   Baker, V. R., Strom, R. G., Gulick, V. C., et al. 1991. Ancient oceans, ice sheets and the hydrologic cycle on Mars. *Nature* **352**: 589–594.

6   Dohm, J. M., Ferris, J. C., Baker, V. R., et al. 2001. Ancient drainage basin of the Tharsis region, Mars: Potential source for outflow channel systems and putative oceans or paleolakes. *J. Geophys. Res.-Planets* **106**: 32943–32958.

7   Lozano, G., Carballo, G., Adcox, A., et al. Volcanism and erosion in Sabaea Terra: Implications for planetary history of a new image from the Mars Odyssey orbiter. *European Workshop on Exo-Astrobiology. Mars: The Search for Life.* Madrid, Spain: European Space Agency; 2003. pp. 233–234.

8   Segura, T. L., Toon, O. B., Colaprete, A., et al. 2002. Environmental effects of large impacts on Mars. *Science* **298**: 1977–1980.

9   Schulze-Makuch, D., Irwin, L. N., Lipps, J. H., et al. 2005. Scenarios for the evolution of life on Mars. *J Geophys Res – Planets* **110**: E12S23.

10  Houtkooper, J. M. and Schulze-Makuch, D. 2007. A possible biogenic origin for hydrogen peroxide on Mars: the Viking results reinterpreted. *Int. J. Astrobiol.* **6**: 147–152.

11  Ferguson, B. A., Dreisbach, T. A., Parks, C. G., et al. 2003. Coarse-scale population structure of pathogenic Armillaria species in a mixed-conifer forest in the Blue Mountains of northeast Oregon. *Canadian J. Forest Res.* **33**: 612–623.

12  Boston, P. J., Ivanov, M. V. and McKay, C. P. 1992. On the possibility of

chemosynthetic ecosystems in subsurface habitats on Mars. *Icarus* **95**: 300–308.

13  McKay, D. S., Gibson, E. K., Jr., Thomas-Keprta, K. L., et al. 1996. Search for past life on Mars: possible relic biogenic activity in martian meteorite ALH84001. *Science* **273**: 924–930.

14  Levin, G. V. and Straat, P. A. 1977. Recent results from the Viking labeled release experiment on Mars. *J. Geophys. Res.-Planets* **82**: 4663–4667.

15  Schulze-Makuch, D. and Darling, D. 2010. *We Are Not Alone: Why We Have Already Found Extraterrestrial Life.* Oxford: OneWorld Publications.

16  DiGregorio, B. 1997. *Mars: The Living Planet.* Mumbai: Frog Books.

17  Morton, O. 2002. *Mapping Mars: Science, Imagination, and the Birth of a World.* New York: Picador.

http://photojournal.jpl.nasa.gov/targetFamily/Mars
Repository for thousands of images from the myriad of robotic missions to Mars.

http://marsprogram.jpl.nasa.gov
The Jet Propulsion Lab's website devoted to Mars exploration.

http://www.google.com/mars
An enlargeable topographic image of Mars, color-coded for altitude.

http://mars.astrobio.net/pressrelease/56/water-on-mars-not-so-ancient-after-all
A short informative article on the discovery of extensive evidence for surface water on Mars in the past and subterranean water reservoirs on Mars today.

http://www.space4case.com
Home page for visionary digital art of ancient and present-day Mars and other worlds by Kees Veenenbos.

# 6
# Hell Fire and Brimstone

*The possibility of life on a runaway greenhouse world like Venus*

On 15 December 1970, the Soviet lander, Venera 7, radioed good news and bad news back to Earth. The good news was that it had landed softly enough on the surface of Venus to still be intact and talking, the first man-made object to do so from the surface of another planet. The bad news was that, at a temperature of 470°C and an atmospheric pressure 90 times greater than on Earth, it wasn't going to last long. Indeed, it died 23 minutes after touching down.

The fate of Venera 7 was not totally unexpected. Eight years earlier, Mariner 2 had become the first spacecraft from Earth to encounter another planet, in a flyby that took it within 35,000 kilometers of Venus. Mariner 2 had sensed that the surface was hot, and detected a thick, continuous cloud layer reaching 73 kilometers into the sky, suggesting a dense and heavy atmosphere. The foreboding nature of Venus was confirmed in 1967 when Venera 4 parachuted successfully into the atmosphere, but stopped operating at an altitude still 25 kilometers above its landing site. Belying its beauty in our nighttime sky, it has become clear that Venus is the closest thing to hell there is in our Solar System [1].

## 6.1 Nature of Venus

Venus is nearly the same size as Earth, with nearly the same elemental makeup. Otherwise, Earth and its twin are different worlds entirely.

The second of the four rocky planets out from the Sun, Venus has almost, but not quite, become locked into gravitational synchrony with the Sun; its day (one rotation on its axis) is longer than its year (one orbit around the Sun). Despite facing sunlight for 117 Earth-days at a time, the daytime and nighttime surface temperatures on Venus are nearly identical, at a little over 460°C (860°F). This is due to the high heat conductance of an atmosphere many times denser than Earth's, and high winds toward the poles that quickly spread the heat evenly around the planet (Figure 6.1). Despite the strong winds near the top of the atmosphere, however, the uniform temperature below means no wind at the surface; and the absence of water boiled away long ago means no rain, so there is not much weather to speak of – just a suffocating, oppressive hothouse at an unrelenting, crushing pressure.

L.N. Irwin and D. Schulze-Makuch, *Cosmic Biology: How Life Could Evolve on Other Worlds*, Springer Praxis Books, DOI 10.1007/978-1-4419-1647-1_6, © Springer Science+Business Media, LLC 2011

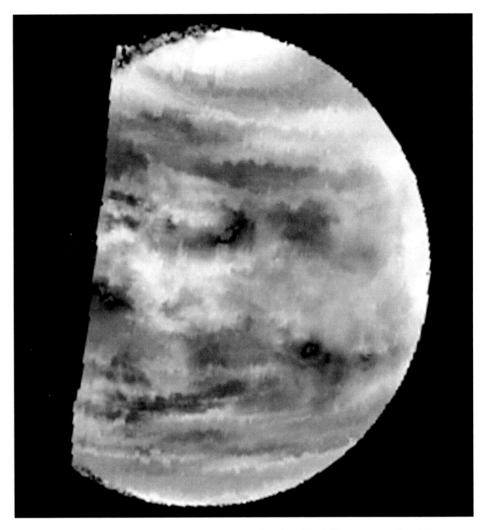

**Figure 6.1** Venus. This infrared image was taken by the Galileo spacecraft as it approached Venus on 10 Feb 1990, modified to indicate what the cloud layers would look like from reflected sunlight. Towering puffs are seen at the equator, while winds as high as 240 km/h (150 mph) string the clouds out into linear streaks near the poles (NASA/JPL).

### 6.1.1 Atmosphere

The hellish heat trapped beneath the clouds of the jewel-like image we see from Earth as our morning or evening star is the result of a runaway greenhouse effect [2] in which the dense atmosphere of carbon dioxide, with thick clouds of sulfuric acid at its ceiling, absorbs the infrared radiation from the Sun that penetrates the clouds but can't get out, as well as the heat emitted from the planet's interior (Figure 6.2).

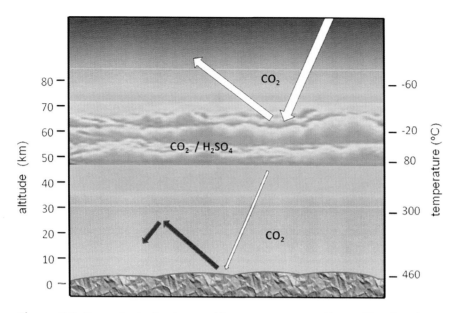

**Figure 6.2** Atmospheric structure and its consequences on Venus. The planet's massive inventory of carbon dioxide ($CO_2$) rises to about 70 km (43 miles) above the surface. Clouds of sulfuric acid ($H_2SO_4$) nearly 30 km (18 miles) thick form the atmosphere's ceiling, and reflect about 80% of the Sun's radiation. The 20% that gets through the clouds, along with infrared energy emitted from the surface, is absorbed by the dense $CO_2$ at lower levels. Thus, near the surface the temperature is hot enough to melt lead, but above about 50 km, the temperature is cool enough for the stability of organic compounds (Art by Louis Irwin.)

## 6.1.2 Topography

Thanks to the armada of spacecraft sent to Venus by both the United States and former Soviet Union, and especially to the detailed radar mapping of the surface by the Magellan Mission in the early 1990s, we have a detailed view of what the surface of Venus would look like if we could peer beneath the clouds. Using color coding for elevation, with cool colors for low and warm colors for high altitudes, the topography of the surface of Venus centered at $0°$ longitude, is shown in Figure 6.3a.

As in the case of Mars and the Earth, broad basins capable of holding water at one time are evident. Unlike Mars and Earth, however, Venus is characterized by a larger number of smaller basins, scattered rather randomly around the planet, including the hemisphere on the opposite side from the one pictured here.

There are only three large upland areas. Ishtar Terra (Figure 6.3b) in the far north comes closest to being continental in size. At its western border it rises 3.3 kilometers above the plateau at its base, with fairly steep cliffs reminiscent of the pedestal formation around Olympus Mons on Mars. This elevated plain supports mountains as high as the Himalayas, though fewer of them. About 20 smaller upland regions are scattered among broad plains across the planet.

a
b

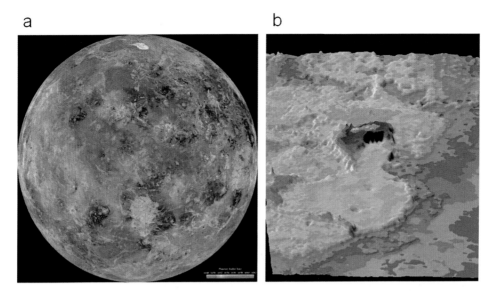

**Figure 6.3** Topography of Venus. Elevations based on radar data from the Magellan orbiter and supplemented by altimetry data collected by the Soviet Venera and US Pioneer spacecraft, color coded to represent low elevations in blue and high elevations in red and white. (a) Global view of the eastern hemisphere. (b) Enlarged view of Ishtar Terra, a plateau the size of Australia that rises 3.3 km above the surrounding lowlands. Note the cliff-like appearance of the western (near) edge. The tallest mountain in the center of the image is Maxwell Mons, the highest point on the planet at an elevation of greater than 10 km (32,000 feet) above the lowlands (NASA/JPL/USGS).

Channels indicative of fluid flow are found on Venus. Many of these channels exhibit the meandering features and channel topography typical for water-carved rivers on Earth (Figure 6.4). But the Venusian channels generally have gentler, broader curves than the drainage channels seen on Earth and Mars, and are often obliterated over parts of their length by infill from younger geological processes

### 6.1.3 Volcanism
Evidence of volcanic activity is pervasive on Venus [3]. Eighty-five percent of the surface consists of lava-covered plains and the volcanoes that paved them. The tallest volcanoes, like Maat and Sapas Mons, are typical shield volcanoes (Figure 6.5a). Others, like Gula Mons, appear to have formed from thicker lava, because of their steeper slopes (Figure 6.5b). Then there are the pancake hills – smaller, flattened domes seen on no other planet (Figure 6.5c).

Volcanic features on Venus are clearly recognizable as such, but they do have properties rarely seen elsewhere. Often the volcanoes lack calderas; the lava fields extend over vast areas, often for hundreds of kilometers across the surrounding plains; and the structures, like Gula Mons and the pancake hills, are odd in shape by comparison with those on Earth and Mars. Different types of lava, indicative

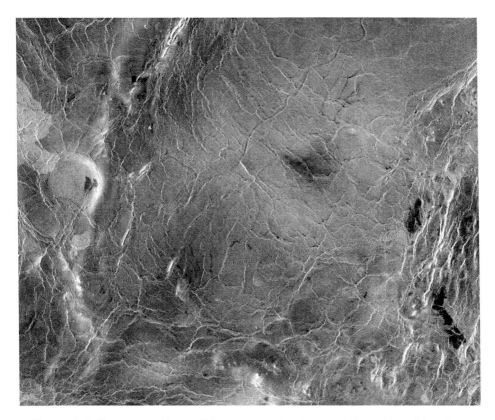

**Figure 6.4** Channels on Venus. This compressed radar mosaic image from Magellan covers an area 460 km (285 miles) wide. Coursing from lower left to upper right is a 600 km long segment of a gently meandering channel about 1.8 km wide. Originally discovered by the Soviet Venera 15 and 16 orbiters to be at least a thousand km in length, this is the longest channel known on Venus, making it lengthier than the Nile. Gently sinuous channels like this are frequently seen on the lava plains of Venus, often cross-cut by other channels, fractures, or ridges at nearly perpendicular angles, suggesting an older origin than the more wrinkled features that intersect them (NASA/ JPL).

of different compositions and liquid content at the time of their emission, are evident. Their vast extension suggests greater spreading than on other planets, perhaps because of the higher temperatures and greater pressures prevailing at the time of their eruption.

Impact craters on Venus are far fewer than on Mars, despite planetary exposure for just as long. The absence of small craters reflects the filtering effect of the dense atmosphere on smaller meteorites, but the relative lack of even large craters must reflect extensive resurfacing in the geologically recent past. Many planetary scientists think that an extensive series of eruptions resurfaced much of the planet about 500 -700 million years ago.

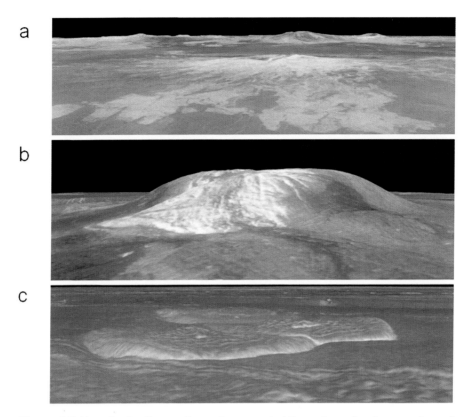

**Figure 6.5** Venusian landforms. Computer-generated three-dimensional perspectives of notable geological features on the surface of Venus. The vertical scale is exaggerated 5 ×. (a) Sapas Mons, in the center of the image, as seen from a distance of 527 km. This volcano is 400 km (250 miles) across and rises 1.5 km above the surrounding plain. Its lava spreads over a wide area for hundreds of km. An even taller volcano, Maat Mons, rises 5 km above the plain on the horizon in the background. (b) Gula Mons, a bell-shaped volcano rising 3 km from its base, spreads lava fields over a hundred km distant from the point of eruption. Note the steeper sides, unlike those characteristic of many shield volcanoes, presumably due to a more viscous form of lava. Dark terrain indicates a rougher surface. (c) Pancake lava domes, averaging 25 km (15 miles) in diameter, a little less than 1 km in height. These structures, unlike any other in the Solar System, may represent a slow outpouring of viscous lava that solidified first on the surface, then was stretched by continuing intrusion from below, forming fracture lines across the domes (NASA/JPL).

### 6.1.4 Tectonic features

Features resembling plates, though much smaller than the tectonic plates on Earth, as well a crumpled terrain, rift valleys, faults, and other evidence of tectonic activity indicate a planet that at one time had a mobile, fractured crust and plate tectonics. Where tectonic features exist, though, they appear to be much older than more recent and extensive volcanic activity. The general consensus is that tectonic activity on Venus stopped some time ago.

## 6.2 Planetary history of Venus

Location, location, location. While not quite everything, as in real estate, a planet's position in its solar system determines to a large extent its fate. Venus is perhaps the poster planet for this principle. Otherwise, there would be no way to explain how Earth and Venus, so close in age, size, and composition, could end up so differently.

On theoretical and empirical grounds, we believe that Venus was, like Earth, a cooler, wetter world early in its history. Deuterium to hydrogen ratios in its atmosphere indicate that tremendous amounts of water have been lost. The morphology of some of the channels suggest that water flowed once across its surface. The spread of lava is so extensive that it must have been facilitated by a higher water content. And the amount of carbon dioxide present now, in the opinion of most planetary scientists, could not have generated the extent of the runaway greenhouse effect we see on Venus without a major contribution from water.

Reconstructing a planetary history for Venus is a challenge, because Venus, more than any other planet, has destroyed its own past. As recently as several hundred million years ago, volcanic activity of such catastrophic proportions occurred that most of the surface was restructured, burying a past that itself may have buried most of the evidence of what the planet was like in earlier times. But planetary scientists are a clever lot. By piecing together bits of information from here and there, making reasonable assumptions, applying sound theory, and using logical analogies, they have offered explanations for quite a few pieces of the historical puzzle. What follows is our summary of what seems like the most reasonable sequence of events for Venus, from its origin to the present day.

1. Venus was formed by accretion, like all the planets of the Solar System, about 4.5 billion years ago.
2. It formed in orbit around the Sun at a distance, and with a mass, that differentiated into a metallic core, solid silica-rich mantle, and rocky crust, like Mercury, Mars, and especially Earth, which it closely resembles in size. The conditions of its formation, plus the bombardment from water-bearing comets and meteorites that was intense at the young age of the Solar System, produced a rocky planet with a substantial amount of water.
3. Though one-third closer to the Sun than Earth, the Sun was about a third less intense when the planets first formed, so temperatures on the surface of Venus were low enough to condense water in the liquid state. The formation of clouds would have reflected sunlight, helping to maintain a lower surface temperature, so that standing bodies of water, possibly oceanic in size, most likely existed on the surface, for several hundred million years, or perhaps even longer.
4. The outflow of internal heat derived from radiogenic decay and absorbed from the kinetic energy of bolide impacts, in combination with lubrication from the available water, led to plate tectonics in the crust. However, the

evidence suggests that such activity may have occurred on a smaller scale, or with smaller (sub-continental) plates, than on Earth.

5. As the Sun gradually grew more intense, water began to evaporate from its surface liquid stores at a rate faster than those stores could be replenished by rain. Water vapor acted as a strong greenhouse gas, trapping radiant heat from the surface and raising global temperatures further, accelerating the evaporation of water, while preventing the condensation that would lead to rain.

6. Under such runaway greenhouse conditions, whatever oceans existed, along with all standing bodies of water, evaporated. Once the water was gone, carbon dioxide present in the atmosphere and outgassed from the interior had no way of being dissolved. Also, with water no longer available for lubrication, plate tectonic activity ceased. The $CO_2$-carbonate-silicate cycle that has kept the $CO_2$ content at a low level in the Earth's atmosphere was disrupted on Venus. Lacking a way for $CO_2$ to be removed from the atmosphere, it collected there, acting as a further greenhouse gas to raise the surface temperature even higher.

7. Volcanic activity continued to emit sulfur dioxide into the atmosphere, which added to greenhouse warming and reacted with the water to form sulfuric acid. This hygroscopic compound proceeded to remove more water from the atmosphere. In addition, the slightly lower gravity of Venus compared to that of Earth, and the higher elevation of the cold trap above the hot surface allowed water to rise higher in the atmosphere, where ultraviolet radiation caused photodissociation of the hydrogen and oxygen. The lighter hydrogen then escaped to space and the oxygen became immobilized in oxidized minerals, leaving the atmosphere consisting mainly of $CO_2$ and sulfuric acid clouds.

8. The cessation of plate tectonics eliminated the horizontal convection of heat beneath the crust. With heat outflow now restricted to vertical convection unable to escape for long periods of time, intense pressure would build up at stationary spots, eventually erupting catastrophically into a dense hot atmosphere that kept the lava in liquid form as it spread out over great distances. The last, globally massive outburst probably occurred 500–700 Mya, leading to extensive resurfacing of the face of Venus.

9. Each eruption disgorged more sulfur dioxide into the atmosphere. The resulting sulfuric acid clouds grew so dense, in fact, that they reached the point where today they reflect 80% of the Sun's thermal energy. But this has been too little too late. The 20% that does get through, with the continuing radiation of infrared heat from the interior, keeps the surface at about 460°C, warm enough to give molten lava unusual flow characteristics and mold the surface into one of the Solar System's more unusual landscapes.

## 6.3 A possible evolutionary history for putative life on Venus

There is good reason to believe that conditions were present for the origin or introduction of water-borne, organic (carbon-based) life early in the history of Venus, Earth, and Mars, all three [4]. To the extent that those conditions were similar, the life spawned within them could have been similar, and possibly even related. But like the geophysical histories of the planets themselves, the histories of their biospheres must have taken radically different turns. On Earth, temperatures stabilized enough to keep water liquid; and plate tectonics in concert with the hydrological cycle stabilized the atmosphere long enough for complex forms of life to evolve and eventually colonize nearly every conceivable habitat. On Mars, any life had to adapt to increasing cold and diminishing bodies of water, leading to underground retreat where any remaining life remains sequestered. On Venus, the challenge was increasing warmth rather than cold, as the atmosphere thickened, trapped heat from a Sun growing increasingly bright, causing the seas to disappear, and baking whatever life remained.

For a time, though, life may well have flourished on Venus. In considering the trajectory it may have taken (Figure 6.6), we need to distinguish between the biospheres possible when water was present, the changes in life that rising temperatures in the water and on land would have required, and finally, the fate of the Venusian biosphere as water disappeared and the surface temperature rose to the highest, globally-distributed level known in the Solar System.

### 6.3.1 Life in the age of water on Venus
Some time soon after the formation of Venus, when its upper crust had cooled sufficiently to allow liquid water to remain, conditions should have been excellent for the origin of life at the confluence of water and land, or in the seas that formed in the planet's low-lying basins. Organic molecules delivered from space or synthesized from simple precursors exposed to lightning or other geophysical processes could have provided the energy for the earliest protocells. In time, the boundary structures and metabolic complexities of persistent, living cells would have evolved into self-perpetuating lineages of microbial life. Or possibly forms of life already underway from an origin on Earth or Mars could have been transported to Venus in those early eons when interplanetary transfer of materials was more frequent.

Either way, the earliest forms of true life on Venus were most likely chemoautotrophic. Given the abundance of sunlight bathing Venus, the possibility that light could have been harvested for energy from the start cannot be discounted. Photoautotrophic life as we know it on Earth, however, is far too complicated to be a candidate for a truly ancestral bioenergetic process. That something less elaborate but sufficient to drive the synthesis of high-energy compounds – a primitive, crude, and inefficient form of photosynthesis – could have been the source of energy for the earliest cellular life on Venus is a possibility. Or, it could have been thermosynthesis, the utilization of heat energy, suggested as a precursor of photosynthesis by Anthonie Muller [5].

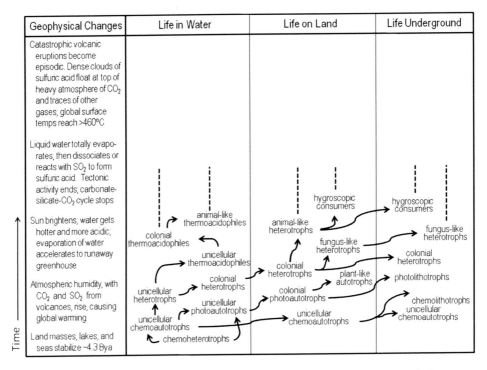

**Figure 6.6** Scenario for the evolution of life through stages of environmental change on Venus. Plausible steps in the major transitions that the different generic forms of life could have undergone are indicated for the geophysical conditions present. Time, without units since the actual timing of the transitions is not known, is represented on the vertical axis, with three principle habitats displayed horizontally. The possibility of life in the fourth principal habitat – the atmosphere – will be considered in Chapter 7. Solid lines represent evolution or persistence of living forms. Dashed lines represent fossil remnants, presumed to disappear when the lines end.

Whichever was first to provide the energy for living systems on Venus, light and chemistry would both likely have emerged as the major bioenergetic sources early on.

The early seas of Venus should have been swarming with chemotrophs and phototrophs alike, while water was still abundant, temperatures were still mild, and tectonic activity was dissipating internal energy and keeping an interchange of volatiles like carbon dioxide, sulfur dioxide, and nitrogen moving between the atmosphere and substrate. Planetary differentiation and the volcanic activity that has presumably been a hallmark of Venus from the beginning would have produced a humid atmosphere that moderated the harshness of conditions on land.

Life may have moved fairly soon out of the water, onto and underneath the dry surface. Photoautotrophs would have been particularly prone to grow in complexity outside the water, where they could soak up sunlight and absorb nutrients from both the soil and the air. Organisms that we would call plants may have spread across the land fairly early on.

Once autotrophs of whatever stripe became available for food, the evolution of consumers to feed on them would have followed. Heterotrophs would have evolved both in the sea and on the land, initially as unicellular consumers but progressing to colonial forms, then possibly to well-integrated macrobiotic organisms.

It would be reasonable to assume that some of the consumers would have been stationary and parasitic or saprophytic (living on other organisms without harming them), akin to fungi, while others would have been more motile and animal-like. Of course, all generic forms, whether microbial, fungus-like, or animal-like, would have included detrivores to decompose dead organisms, as well as predators of living prey. At this early stage, life on Venus may well have resembled in complexity, if not the specific character, of the biospheres on Earth at the beginning of the Archean, and on Mars at the mid- to late-Noachian (see Figures 5.8 and 5.9 in chapter 5).

### 6.3.2 Life as Venus became hotter and the water evaporated

Given the warming temperatures (sustaining higher metabolic rates and faster life cycles), along with the changing environments, directional selection may have driven evolution at a faster pace than on either Earth or Mars, as the rocky planets were beginning to diverge. Three billion years ago, Venus may have been home to the most complex and diverse biosphere in our Solar System.

But as temperatures on Venus continued to rise, water kept evaporating, and the seas became acidic as more $CO_2$ and $SO_2$ dissolved in them. Under those conditions, directional selection for greater tolerance to heat and acidity would have become the dominant evolutionary theme.

Whether unicellular, colonial, or macrobiotic, organisms would have survived in direct proportion to their ability to stand the heat and, in the water, deal with the growing acidity. Since we know that extremophiles can live on Earth in highly acidic water up to its boiling point, we can assume that such adaptations would have been able to evolve on Venus. Furthermore, we have no reason to doubt that life would have persisted as long as water stood anywhere on the surface and still permeated the soil. Even after the water was gone from everywhere except the atmosphere, some very hardy hygroscopic consumers may have been able to hang on for a while.

In the end, heat and desiccation would have combined to kill off all the surface and subsurface life on Venus. By the time the end came, all forms of life would have been very different from their ancestral origins. Unlike on Earth and Mars at a comparable point in time, where ancestral remnants of early, simple forms of life would still have been extant, directional selection would have driven life on Venus to its extremophilic, and ultimately terminal, variations.

### 6.3.3 The possibility of life below the surface of Venus

*6.3.3.1 Carbon-based life*

The life that would have originated or been brought to Venus would almost surely have been carbon-based, given the presence of liquid water and the great advantages of carbon-based biochemistry in water as a solvent. How long carbon-based life could have existed beneath the surface of Venus would have depended largely on how long any moisture would have been available. Hygroscopic organisms could have survived the increasing desiccation, but once all water was gone, their fate would have been sealed fairly rapidly. This is because the loss of water would likely have accelerated quickly under runaway greenhouse conditions, rather than occurring over a sufficiently long period of time for directional selection to promote the evolution of long-living cryptobiotic forms.

As on Mars, an unknown step in the retreat of life from increasingly harsh conditions on the outside, would have been the state of life in caves on Venus. Caves derived from lava tubes, certainly, and possibly from other geological modes of formation must have been a planetary feature. Numerous images of the Venusian surface today are best interpreted as collapsed lava tubes. So where there were caves, they would surely have been invaded, and would have supported at least chemoautotrophs and whatever consumers could feed on them. Even in these sequestered habitats, though, the temperature would eventually have reached levels that would have driven water away. Once the water was gone and the temperature was hot enough to destroy carbon-based biomolecules, caves would have offered no protection.

One other possibility deserves consideration. Far beneath the surface, water could persist in local reservoirs in a supercritical state, well above its boiling point but under such pressure that it can't boil away. The persistence of any form of life under such conditions is hard to imagine for two reasons. First, water in the supercritical state has properties very foreign to the characteristics that make it a good solvent for carbon-based biochemistry. Secondly and more critically, molecular conformations and even chemical bonds are so unstable at such extreme temperatures that complex macromolecular structure – a hallmark of the chemistry of life – is simply unsustainable.

*6.3.3.2 Silicon-based life*

Some polymeric forms of silicon are stable at very high temperatures. Silicones (organosilicon polymers with a silicon-oxygen backbone) can exist in stable form at temperatures well above the point of destruction for purely organic polymers. Silicates, which make up much of the rocky portion of a planet's crust, are stable at even higher temperatures. Therefore, it's fair to ask if life could have survived on or beneath the surface of Venus by changing into a silicon-based form. This possibility seems to us highly unlikely. The reason is that life would have come from a different direction – from a carbon-based, water-borne origin. To evolve into a form of life based on a fundamentally different biochemistry would represent a transition far greater than any known to have taken place in the history of life on Earth.

Yet another possibility is the prospect of a later, independent origin of silicon-based life on Venus as the thermal and barometric conditions on the planet approached their current state. While a theoretical possibility, if the requirement for liquid is truly essential for life, as we suspect, a suitable solvent would have to be found that would be compatible with silicone and could persist in stable pools, for this possibility to be a reality. Of the candidate liquids that might meet these criteria on present-day Venus, only silicates in the form of liquid magma seem like realistic possibilities, and there is a real question of whether even polymeric silicones could remain stable at the temperature of magma. If some form of silicon-based biochemistry had evolved on Venus, its features would likely be so exotic that they would be hard to recognize. Nothing resembling the form of a living organism has ever been found among the extensive deposits of lava on Earth. We therefore regard the prospect of silicon-based life on Venus so speculative that informed consideration of its possible features are not possible at this time.

## 6.4 The prospect of finding fossil evidence of life on Venus

No fossils are found in basaltic formations on Earth. This confirms the theoretical expectation that the tremendous temperatures reached in magma destroy all remnants of carbon-based organisms. Thus, no fossils could reasonably be expected to be found in any basaltic formations on Venus. Since most of its surface is basaltic, paleontology on Venus would likely be a frustrating pursuit.

If local spots were to be found where sedimentary remnants of ancient seas lie exposed among the lava-paved surfaces surrounding them, conceivably miner-alized fossils could still exist. Wide basins at low elevations, well away from any evident volcanic emmisions, would be the place to look. Given the difficulty of even robotic missions to the surface of Venus, though, the search for fossils there is not likely to be anyone's astrobiological priority for a long time.

## 6.5 Ecosystem possibilities for life on Venus

Spurred by warm temperatures and heterogeneous habitats, evolution concei-vably could have spawned a biosphere on Venus at least as diverse and complex as that envisioned for a youthful Mars, if water persisted on Venus for several hundred million years. In that case, the biotic inventory should have been no simpler nor less diverse than that on Mars (Table 5.1 in chapter 5).

As conditions progressively worsened for organic life, diversity would have declined. Whatever consumers had evolved to feed on phototrophs in the seas and on the land would have disappeared as soon as their source of food died out. Even chemoautotrophs and their consumers would have succumbed eventually to the heat and desiccation. As in all mass extinctions in the history of life on Earth, when a part of the ecosystem falters, the collapse is widespread and quite rapid. We are tempted to suspect that, with conditions worsening exponentially, the end to life on the surface of Venus came quickly.

## 6.6 Characteristics of life on Venus

To the extent that life on Venus may have been comparable in complexity and diversity to that on early Mars, so too would the characteristics of that life have been analogous. This generalization needs to be qualified in two ways, however.

First, while Mars grew colder, Venus got warmer. For living processes, which are fundamentally based on chemical interactions, this would have meant that everything worked faster. So if life on Mars proceeded in slow motion, on Venus it would have been moving at a fast-forward pace. In cases where the motion was literal, as in organismal motility, energy consumption would have become a rate-limiting factor. At some point, acceleration of metabolic rates may have become detrimental, as energy stores could not keep up with the reaction rates that the high temperatures made mandatory.

Secondly, and perhaps as a partial counter to the above, there appears to have been a greater prospect for oxidative metabolism, for a time, on Venus. This is because the atmospheric inventory of oxygen would have risen as water photodissociated in the upper atmosphere. However, unless there had been a biogenic way to generate oxygen, such as a photosynthetic mechanism like the one used by plants on Earth, oxygen consumption by living organisms would have been unidirectional, and therefore short lived, since the source of oxygen (water) was a non-renewable resource. This is where knowledge of photosynthetic processes on other worlds would be invaluable, in giving us clues about how common the evolution of oxygen-generating forms of photosynthesis are.

## 6.7 Possibilities for life on exoplanets like Venus

The lessons of Venus, whatever they are, may have broader applicability than we would have imagined just a few years ago. That is because hot planets, though probably in the minority, are clearly not uncommon in the universe.

In the first place, planetary histories like that of Venus should not be unusual, given the widespread availability of water, and the formation of rocky planets at a distance from their central stars where atmospheres and liquid water can lead to greenhouse conditions. Indeed, the fate of life on Venus is likely the fate of life on Earth, later if not sooner. So in looking at Venus in the past, we are seeing Earth in the future.

In the second place, our previous simple view of how solar systems are generally structured, with small rocky planets in the interior and gas giants remotely, has come unraveled with the discovery of so many very large exoplanets orbiting so close to their central suns. Without knowing the atmospheric and planetary compositions of those planets, the thermal conditions on them cannot be known; but it's safe to say that many of them must be warm in the extreme. A history of life like we have envisioned for Venus may be a history that has been recurrent for life on many other worlds.

Given our appreciation for a greater variety of planetary positions, conditions, and histories than we once thought to be the case, we need to consider the possibility that some worlds have retained high temperatures from the start. On worlds such as those, under conditions we have a hard time imagining but cannot refuse to consider, silicon-based life, or biochemical systems incorporating organosilicon components and more heat-tolerant solvents, may be a possibility.

## 6.8 Chapter summary

Venus began as Earth's near-twin and Mars' sibling, with liquid water washing over rocky shores. Blessed with warmth, abundant sunlight, and presumably the same inventory of organic nutrients available to its siblings, the prospects for life were bright in the early days of the planet's existence.

Too bright, as it turned out. With the Sun growing more intense over time, Venus was a little too close to handle the increasing flux of energy. Water evaporated from her seas, turning the atmosphere into a greenhouse trap for heat that boiled water away even faster, till none was left. Lacking water to dissolve the carbon dioxide, that greenhouse gas accumulated in the atmosphere as well, piling to a height of about 70 kilometers with a ceiling of sulfuric acid, and pressing down on the surface with 90 times the weight of the atmosphere on Earth. Life on Venus became, almost certainly, intolerable.

Until its demise, however, life on Venus may have flowered to a level of complexity and diversity that exceeded that on either Earth or Mars. If anything, life may have been quicker as well as more widespread, on Venus than on her sibling planets, especially if oxidative metabolism, which was a theoretical possibility, could have taken hold.

When the good times ended, they probably ended quickly. Runaway cycles are so named for a reason; the pace of change accelerates toward the end. With heat rising from below as well as accumulating above the surface, retreat to caves and below the ground would have offered no reprieve. The only habitat free from the suffocating heat of the surface and lower atmosphere is the layer of sulfuric acid clouds about 55 kilometers in the air. There the temperature of a mild summer afternoon on Earth can be found. The chance that life could have retreated to that niche, and may still exist there, will be considered in the next chapter.

### 6.8 References and further reading

1   Grinspoon, D. H. 1997. *Venus revealed: a new look below the clouds of our mysterious twin planet.* Cambridge, MA: Perseus Publishing.
2   Kasting, J. F. 1988. Runaway and moist greenhouse atmospheres and the evolution of Earth and Venus. *Icarus* **74**: 472–94.
3   Head III, J. W. and Basilevsky, A. T. 1999. Venus: surface and interior. In: P. R. Weissman, M. L.-A. and T. V. Johnson (eds). *Encyclopedia of the Solar System.* New York: Academic Press; pp. 161–189.
4   Owen, T. C. 2000. The prevalence of earth-like planets. *Acta Astronautica* **46**: 617–620.
5   Muller, A. W. J. 1995. Were the first organisms heat engines? – a new model for biogenesis and the early evolution of biological energy- conversion. *Progr. Biophys. Molec. Biol.* **63**: 193–231.

http://photojournal.jpl.nasa.gov/targetFamily/Venus
Repository for thousands of images of Venus.

http://www.nasa.gov/worldbook/venus_worldbook.html
Compact fact sheet on Venus, with brief historical review of robotic missions to the planet.

http://www2.jpl.nasa.gov/magellan
Home page for the Magellan mission that mapped Venus.

# 7 Suspended Animation

## Life in the clouds of a dense atmosphere on planets like Venus or the gas giants

The skies are cloudy on seven of the eight planets in our Solar System. They are thick and global on Venus, Jupiter, and Saturn. Uranus has hazy clouds, while those on Mars and Neptune are thin and wispy. Partly cloudy would be a more accurate description for the clouds on Earth, which are never global but can be locally thick and extensive. Because we know that the Earth's atmosphere is filled with airborne life, including microbes that circulate at high altitudes in the clouds above the Earth, we need to consider the possibility that life in the clouds is a reality on other worlds as well.

Atmospheres come in two varieties. Those of the outer gas giants in our Solar System are known as "primary" atmospheres because they hold, presumably, a mixture of gases that made up the primary nebulae from which the planets were formed – mainly hydrogen, helium, nitrogen, and small molecules like ammonia and methane. The inner rocky planets (except for Mercury) are surrounded by "secondary" atmospheres, generated by ongoing geophysical processes like outgassing that have occurred since the planets were first formed. As we have already seen, the history of a planet has a lot to do with the probability that it now or ever has harbored life. Given the very different planetary histories that distinguish primary and secondary atmospheres, we will deal separately with the possibility that some forms of life could be harbored in them.

Life in Earth's secondary atmosphere forms a continuum with life on its surface, as dealt with already in Chapter 4. Since life on Mars has very likely been subterranean for quite some time, and the clouds on Mars are so high and thin, we think it highly unlikely that any forms of life are found up there. That leaves Venus, which is another story in many ways.

## 7.1 Prospects for Life in the Clouds of Venus

In the previous chapter, we discussed how the surface and atmosphere of Venus made it the greenhouse heat trap it is today, with temperatures and pressures that render life on or beneath its surface highly unlikely. A closer look at its thick atmosphere and complicated cloud structure, however, reveals a highly

L.N. Irwin and D. Schulze-Makuch, *Cosmic Biology: How Life Could Evolve on Other Worlds*, Springer Praxis Books, DOI 10.1007/978-1-4419-1647-1_7,
© Springer Science+Business Media, LLC 2011

differentiated environment that just might include a habitat where life could thrive.

### 7.1.1 Composition and characteristics of Venusian clouds

The atmosphere of Venus can be divided into four general zones (Figure 7.1). From the surface up to an altitude of about 38 kilometers (24 miles), the air is clear and composed overwhelmingly of carbon dioxide ($CO_2$). The next 10 kilometers consist of a thickening haze of sulfuric acid vapor with a diminishing density of $CO_2$. The third zone is composed of thick clouds of sulfuric acid in a

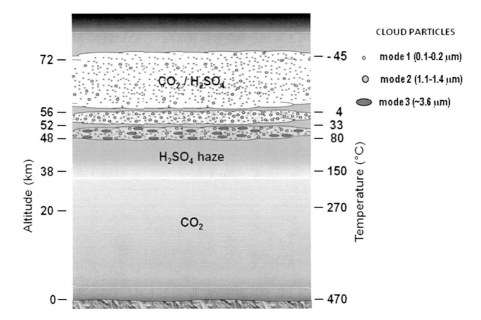

**Figure 7.1** Schematic structure of the Venusian atmosphere. Carbon dioxide ($CO_2$) and sulfuric acid ($H_2SO_4$), with trace amounts of water, nitrogen, carbon monoxide, sulfur compounds, and inert gases comprise the atmosphere of Venus. Four zones can be distinguished. The lowest consists mainly of $CO_2$ and is clear. With increased elevation above 38 km, the air becomes hazy with $H_2SO_4$. The third zone consists of thick clouds of $H_2SO_4$ about 20–25 km in height. The fourth zone is a layer of diminishing $H_2SO_4$ haze and $CO_2$ extending to an altitude of over 80 km. The cloud layers consist of three sublayers, distinguished by the size and composition of the $H_2SO_4$ aerosol droplets that compose them [1]. Mode 1 and 2 droplets are spherical, while mode 3 droplets are non-spherical. The lowest of the three cloud layers is about 4 km thick, and consists of all three droplet sizes. The middle cloud layer is also about 4 km thick, but made up of only mode 1 and 2 droplets. The upper cloud layer is the thickest at about 16 km, and consists of a lighter mix of mode 1 and 2 droplets. Droplets are shown disproportionately large for clarity. Temperature data were collected by the Magellan orbiter on 5 October 1991 from the atmosphere and extrapolated to the surface, as reported by Jenkins et al. [2] (Art by Louis Irwin).

$CO_2$ matrix, made up of three distinguishable layers, extending to an altitude of a little over 70 kilometers (about 45 miles). The fourth and highest zone consists of a diminishing haze of sulfuric acid ice crystals and $CO_2$.

### 7.1.2 Properties conducive to life in the clouds of Venus

*7.1.2.1 Temperature and pressure*
At the bottom of the cloud deck on Venus, the air temperature is about 80°C (Figure 7.1). It continues to decline to 4°C at about the top of the middle cloud layer. Over this same range, atmospheric pressure varies from 1.1 to 0.3 times the pressure at sea level on Earth. Thus, the two lower layers of the cloud deck on Venus harbor temperatures within the liquid range for water, at or below pressures similar to those in Earth's atmosphere.

The lowest cloud layer varies in temperature from 80° to 33°C – a range that encompasses normal and thermophilic microbial life on Earth. Normophilic to psychrophilic microbes could comfortably inhabit the middle cloud layer, at about 30° to 4°C. Thus, organic macromolecules could be stable at the temperatures and pressures found in the lower cloud layers of Venus.

*7.1.2.2 Water, acidity, and organic chemistry*
Water is sparse in the atmosphere of Venus, but is found at a few hundred parts per million in the lowest cloud layer. Below the clouds, the temperature quickly becomes too high for water to stay liquid, and above the middle cloud layer, from about 58 kilometers upward, water is frozen into ice crystals. Eventually, water molecules at an altitude that high are dissociated by ultraviolet light or cosmic radiation.

Sulfuric acid is liquid over a very broad range, from 10° to 337°C at 1 bar of pressure. Thus it floats as liquid droplets through the atmosphere, from a few kilometers above the surface to about the middle layer of clouds, at an altitude around 55 kilometers [2]. From about 30 to 48 kilometers above ground, the droplets start forming a haze. At about 48 kilometers in altitude, the sulfuric acid starts condensing into droplets large enough to constitute opaque clouds. The lowest cloud layer holds the largest droplets, about 3.6 μm in length, as they appear to be non-spherical. This is about the size of the water droplets that make up the clouds on Earth.

Sulfuric acid is very hygroscopic, so it absorbs any water available. Consequently, the droplets are best described as solutions of concentrated sulfuric acid, with a pH of 0 or lower. This equates to an acidity equivalent to battery acid. While chemically destructive to most organic macromolecules, extremely acidophilic microbes are found to exist on Earth at levels of acidity this high.

None of the probes equipped to detect organic chemicals has done so, to a significant degree. If organic compounds exist in the clouds of Venus, they must be very sparse. However, even though Earth's atmosphere and especially surface is rich in life, particularly microbial life, Earth's atmosphere is poor in organic compounds when measured from space.

*7.1.2.3 Sources of Energy*
There is no lack of energy in the clouds of Venus. Sunlight is strong, of course. Though much is reflected from the upper reaches of the cloud deck, enough light gets through to illuminate the surface, so even the lower cloud layers receive a fair amount of light. Winds are strong too, ranging from 50 meters per second (about 112 mph) at the floor of the lowest cloud layer to 200 meters per second (about 450 mph) at the top of the middle layer – fast enough to circle the planet in 3–4 days [3]. If by chance any forms of life on Venus had evolved the ability to harness kinetic energy, there would be plenty of it. Lightning has also been detected in the atmosphere of Venus. Though too powerful and erratic for use by living organisms, lightning could contribute to the production of organic compounds that could end up in the metabolic pathways that any organisms would have to have.

Besides the traces of ethane detected by Pioneer Venus 2, and whatever other small amounts of organic compounds might be generated by photolytic or electrical activity in the upper atmosphere, smaller molecules theoretically capable of engaging in chemical reactions that could yield energy are present. Sulfur dioxide, for instance, which has been outgassed from the interior for billions of years, could react with hydrogen and carbon monoxide to yield hydrogen sulfide and a good bit of energy:

$$SO_2 + H_2 + 2CO \rightarrow 2\ CO_2 + H_2S + energy \tag{1}$$

Sunlight could be used to power a reverse version of that reaction:

$$CO_2 + 2\ H_2S + sunlight \rightarrow CHOH + H_2O + S_2 \tag{2}$$

Something in the clouds of Venus absorbs ultraviolet light. Sulfur compounds and photopigments, among other molecules, have this ability. One such compound is a ring-like structure of 8 sulfur atoms. This molecule could act as an ultraviolet sunscreen, as a pigment that couples absorbtion of UV light to high-energy chemical bonds in a unique photosynthetic-like reaction, or as a converter of UV light to lower frequencies more useful for photosynthesis by another molecular mechanism [4]. An intriguing observation is the fact that hydrogen sulfide, a product of the energy-yielding reduction of sulfur dioxide in the first equation above, is more concentrated in the atmosphere just below the lowest cloud layer, consistent with the possibility of its biogenic production through a chemoautotrophic process in the lower clouds. Whether any of these reactions actually occurs as part of the metabolism of living organisms in the clouds of Venus is unknown. The key point is simply that the presence of sulfur compounds in both an oxidized ($SO_2$) and reduced ($H_2S$) form suggests a chemical disequilibrium that could be harvested for energy, if the appropriate molecular machinery has evolved to do so – making them biosignatures for life itself.

*7.1.2.4 Habitat stability*
Unlike the ephemeral nature of clouds on Earth, the sulfuric acid clouds of Venus are much larger and longer lasting. Models suggest that the sulfuric acid droplets

that make up the clouds last for months, compared to just days in clouds on Earth. Thus, a mode 3 droplet in the lower cloud layer of Venus could remain intact and suspended in the air – a floating drop of aquatic microhabitat, home to an aggregate of microbes over many generations of a typical microbial life cycle. Mode 3 particles, a few microns across and within the typical size range of microbes on Earth, could also represent individual Venusian microbes encoated in sulfur compounds [4].

### 7.1.3 Challenges for life in the clouds of Venus
Appealing as the notion of airborne islands of sulfuric acid housing a family of microbes on a three- or four-day circumnavigation of Venus may be, reality demands that we admit to some challenges for a life style like this.

For starters, the acidity would be a formidable obstacle. As previously noted, there are some forms of microbial life on Earth that thrive at near-boiling temperatures in the acidic water of hot springs. They surely require special adaptations, and even at that, not many are known to live at a pH of 0.

The relative lack of water is another problem. Assuming that any life living in the clouds actually does inhabit liquid droplets that consist of a mixture of water and acid, this is certainly not insurmountable. They are probably very hygroscopic, though, soaking up every stray molecule of water they come across.

Organic reactants are rare as well. While organic compounds can be cobbled together from various oxidized forms of carbon (CO, $CO_2$, CHOH), nitrogen, and ammonia, with the help of ultraviolet light or perhaps lightning, the tell-tale signature of organic molecules beyond trace amounts has consistently failed to be detected by the many probes sent to Venus. If the 2 ppm of ethane detected by Pioneer is the high water mark for organics in the clouds of Venus, the prospect of a carbon-based biochemistry there may be dim.

Buoyancy in air is ultimately a problem for any organism because its density is greater than the air it inhabits. The density of droplets of sulfuric acid, however, is sufficient to keep organisms easily afloat. That means, of course, that organisms are restricted in size to the space available in their airborne aquatic housing.

### 7.1.4 Possible trajectories for the evolution of life in the clouds of Venus
Notwithstanding the challenges described above, the possibility that even a sparse biomass could subsist on the chemical resources available in the clouds of Venus opens our minds to an exciting alternative form of alien life, and raises some intriguing question. Is it really possible that some forms of life could permanently reside in the clouds? How could that have come to pass? How could life born and raised on the surface of a planet take up permanent residence in its clouds?

In the previous chapter, we made the argument that Venus may well have harbored a richer and more diverse biosphere early in its history than either Earth or Mars. The oceans of Venus at the outset may have swarmed with life (Figure 6.6 in chapter 6). As the planet warmed, the atmosphere would have grown progressively humid, with fog and low-lying clouds a common occurrence. The mist, like the water from which it arose, would have been acidic. The relentless pressure of directional selection, as more and more carbon dioxide and sulfur dioxide dissolved to turn the water into carbonic acid and sulfuric acid, would already have given rise to acidophilic microbes. As the mist whipped up from the waters, and rose into clouds of longer duration, the acidophiles would have been just as comfortable in the airborne droplets of acid as they were in the water where they originated. The earliest persistent life in the clouds of Venus would likely have been acidophilic chemoautotrophs and photoautotrophs. As heterotrophs evolved to feed on the autotrophs in the waters below, they too gradually would have joined the microbial populations in the skies. The transition of life from water to clouds is shown schematically in Figure 7.2.

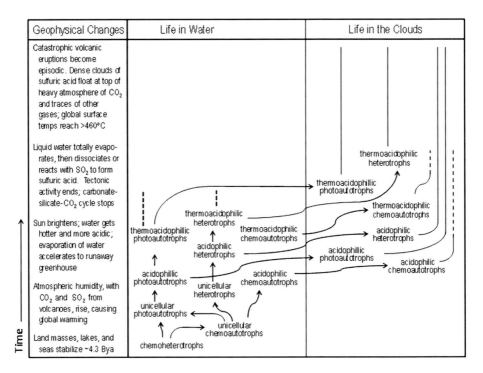

**Figure 7.2** Scenario for the evolution of life in the clouds through stages of environmental change on Venus. Plausible steps in the major transitions that the different generic forms of life could have undergone are indicated for the geophysical conditions present. Time, without units since the actual timing of the transitions is not known, is represented on the vertical axis, with two principle habitats displayed horizontally. Only forms of life directly relevant to ultimate habitation in the clouds are shown. Solid lines represent evolution or persistence of living forms. Dashed lines represent evolution toward extinction.

At first, evaporation would have buffered the rise in atmospheric temperature, keeping the water temperatures just moderately warm. But in time, as the greenhouse effect gathered steam, so to speak, the waters would have become progressively warmer. Again, directional selection would have favored thermophilic adaptations, so a new generation of organisms more tolerant of the heat than their ancestors would have come to dominate the dwindling lakes and seas on the surface. They, in turn, would have risen into the clouds like the normophilic ancestors before them, once the clouds themselves became warm enough for the comfort of microbes grown used to a hot habitat. Meanwhile, the earlier generation of normophilic microbes would have risen to higher levels of the atmosphere, where temperatures were cooler than the warming air below. A simpler scenario would be for thermoacidophiles to evolve from simple acidophilic ancestors already established in the clouds.

By the time water was gone entirely from the surface, the clouds would have harbored two distinct microbial populations. The older, normophilic populations would be residing higher up, at cooler temperatures. The more recent, thermophilic arrivals or descendants, would occupy the lower clouds, closer to the greenhouse heat of the ever thickening atmosphere.

We don't know when the present structure of the cloud layers on Venus became fixed, or why, exactly, there are two distinct layers, each about 4 kilometers thick, forming the floor of the cloud stack. For reasons that are unclear, but probably have to do with the particular size and mass of the sulfuric acid droplets composing them, in combination with the interaction between pressure, temperature, and altitude, the two layers happen to vary in temperature over precisely the range that normophilic and thermophilic organisms on Earth find comfortable. The lowest cloud layer ranges from about 33° to 80°C. The higher one varies from about 30° down to the approximate freezing point of sulfuric acid solutions.

Because the two cloud layers are separated by a space of clear air, each represents a distinct thermal zone. Thermophiles would own the lower level, while normophiles would live in the one above. Like populations on different islands, one in the South Pacific and the other in the Arctic Ocean, they may have been separated long enough to be quite different, above and beyond their different thermal adaptations.

This scenario assumes that thermophiles and normophiles would have risen to the clouds from founding populations already divided by thermal adaptation before they left the surface waters of their origin. This is the trajectory shown in Figure 7.2. An alternative scenario would have the thermophilic adaptation arising later, from normophiles only after they had achieved a lasting airborne presence. Which of these scenarios is historically more accurate would depend on the path the atmosphere took in arriving at its two thermally distinct lower cloud layers. If the temperature of the atmosphere remained mild until well after the seas were all gone, a normophilic origin for all the life in the clouds is more likely.

### 7.1.5 Ecosystem possibilities in the clouds of Venus

The simplest trophic structures that could be assumed from the evolutionary transitions shown in Figure 7.2, would be a foundation of producers and a smaller biomass of consumers feeding on them. Life in the clouds would probably impose a severe limitation on the evolution of the size of the consumers, however, since weight would be a distinctive disadvantage in an airborne habitat. Given the restricted size of the droplets that each micro-ecosystem would inhabit, there wouldn't be much room for larger organisms of the type that usually make up secondary and higher levels of consumers.

We can't really say how diverse the producers and consumers would be. Diversity on Earth is minimized in harsh habitats, and since the habitats in the clouds of Venus are harsh by Earth's standards (which is fair enough, since life presumably arose on Venus under Earth-like conditions), biodiversity in the clouds of Venus is probably low. The survival of only a very few forms of life, exquisitely adapted as either producers or consumers to the lowest (warm) or next-to-lowest (cooler) layer of clouds may be the only life to be found on the planet. Hot springs on Earth typically consist almost exclusively of autotrophs. The same may well be true in the clouds of Venus, where an assemblage of chemotrophs and phototrophs could simply recycle the abiotic nutrients that are available. This would be consistent with indications of an atmosphere somewhat in chemical disequilibrium.

## 7.2 Prospects for life in the atmospheres of gas giant planets

There is no reason to believe that the four giant planets encased in deep layers of hydrogen and other gases in our Solar System are atypical of gas giant planets throughout the universe. Thus, by looking at each of them in a little more detail, we can probably make the best estimate possible at the current state of our knowledge of the plausibility of finding life among them.

Life in the atmosphere of Jupiter or other gas giants is not a new idea. Carl Sagan and Edwin Salpeter suggested in a provocative abstract in 1976 [5] that photosynthetic organisms using methane instead of $CO_2$ as a carbon source could populate the clouds of Jupiter. Organisms in the form of "thin gas-filled balloons" could float at a stable altitude. They went on to imagine small powered "hunters" that could seek one another out and coalesce into "sinkers" and "floaters." The floaters could even achieve large sizes, conceivably from meters to kilometers in diameter.

The image of flocks of organisms floating through the skies of the gas giant planets, darting and dodging to avoid predatory hunters, themselves the targets of prowling megafloaters of Deathstar dimensions, is certainly evocative. Now more the 30 years after such notions were "floated" into the scientific community, are we in a position to evaluate their plausibility? Let's begin by reviewing what we know about the gas giant planets in general.

Conventional theory predicts that the outer gas giants will be fundamentally different from the inner rocky planets of our Solar System. Accretion of rocky material distant from the sun, the argument goes, allows the lighter, more abundant elements to accumulate around the denser core without being blown away by the solar wind. As the mass feeds on itself, drawing more gasses like hydrogen, helium, ammonia, and methane, toward the gravity well of the growing planet, the pressures become strong enough in the interior to press the gas into a liquid, and ultimately into the metallic form where electrons swarm free of their neutrons. Above the core, the transitions from metal, to liquid, to gas are indistinct.

If we were to approach the planet from the outside with pressure gauge and thermometer in hand, we would first encounter clouds at various heights, depending on their chemical composition, and bone-chilling temperatures. Plunging further into the gaseous atmosphere, we would find it growing denser, with pressures rising and temperatures warming. Eventually we would be swimming instead of flying, as gas turns to liquid under crushing gravitational forces, but we wouldn't know it because the temperatures by this time would have become hot enough to incinerate us. If our chemical remains kept plunging toward the planet's center, they would eventually disintegrate to individual atoms, then finally have their electrons stripped away.

Each of the transitions would be gradual, with fuzzy boundaries at best. Until the rocky core of a gas giant is reached, there are no distinct surfaces – liquid, solid, or metallic sea. And the temperatures and pressures at that rocky interface with liquid metal are so incredibly high that life as we have defined it is inconceivable. What remains for speculation is whether some form of life could exist somewhere among those higher layers, including the clouds at the top of the planet.

### 7.2.1 Composition of the gas giants
All the gas giants are composed (beyond their solid cores) mostly of hydrogen and helium, with variable amounts of organic molecules and biochemical precursors like methane, ammonia, ammonium hydrosulfide, nitrogen and water. By knowing the relative amounts of these chemical constituents, and the sizes and densities of the planets, good models of the interior structures of each of the gas giants can be formulated. Models such as these for each of the four gas giants in our Solar System are compared in Figure 7.3. Not everyone's models are the same, but there is broad general agreement that the two smallest – Uranus and Neptune – differ significantly from the two largest – Jupiter and Saturn – in having a much higher proportion of methane, ammonia, and water. The difference is pronounced enough to give rise to a fairly thick layer of these compounds beneath the mantle of Uranus and Neptune that hardly exists on Jupiter and Saturn (darker blue layer, Figure 7.3). All four differ in the nature of their clouds. Thus, we'll briefly summarize the atmospheric properties of each in preparation for evaluating the possibility that life could be found among any of them.

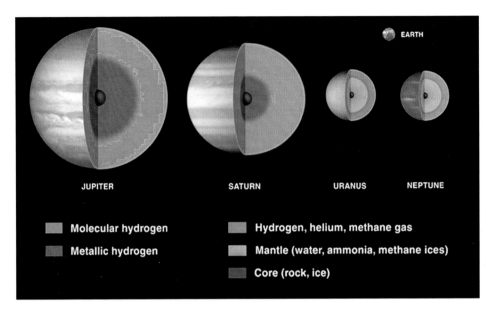

**Figure 7.3** Internal structure of gas giant planets. The four gas giants of our Solar System are thought to consist of a rocky core encased in a thick layer of hydrogen mixed with variable amounts of other gases. Planets are shown in their approximate relative sizes (NASA, courtesy of nasaimages.or).

### 7.2.1.1 Jupiter

Jupiter's atmosphere is 90% hydrogen, with helium and traces of nitrogen, ammonia, methane, sulfur gases, and a few other organic molecules making up the rest. Once thought to have water in its lower atmosphere, the Galileo probes and orbiter found almost none. The composition of Jupiter is closer to that of the Sun than any other planet.

Beneath the highest, wispy clouds of ammonia race thick billowy clouds of various hydrocarbons and their derivatives, each difference in colored zone or belt the result of different chemistries and condensation temperatures, which measure about -145°C (-230°F) at the top of the clouds. Deeper into the cloud layer, room temperature (21°C, or about 70°F) is reached at a depth where the pressure is about 10 times Earth's air pressure at sea level. Water would be stable as a liquid under these conditions, but no water clouds have been detected.

Winds on Jupiter are strong, especially at the equator where the faster clouds move at around 110 meters per second (250 mph). Solar heating at the equator, in combination with heat sinks at the poles, thermal energy emitted from the interior, and the fastest rotation rate of any planet in our Solar System (just under 10 hours per day) make for a very dynamic and turbulent atmosphere (Figure 7.4 left).

### 7.2.1.2 Saturn

Saturn's atmosphere is over 96% hydrogen, with helium and trace gases making up the small remainder. Its upper atmosphere is thought to consist of three layers: ammonia at the top, ammonium hydrosulfide below that, and water-ice clouds lower still. The thin clouds of ammonia on top absorb ultraviolet light. The resulting heat causes the ammonia to form a smoggy haze, which obscures the cloud layers below and gives the planet an indistinct laminated appearance (Figure 7.4 right). It also heats the upper layer of ammonia more than the cloud layers immediately below, but with increasing depth the temperatures start rising again.

Saturn's tilt of almost 27 degrees means that each of its poles faces the Sun more directly than its equator for close to seven years at a time. This results in warmer temperatures at the pole than at the equator during the summer when that pole is tilted toward the Sun. During the southern summer in 2004, for example, temperatures near the top of the cloud layer ranged from -140°C at the equator to -123°C at the south pole. A little lower in the atmosphere, temperatures were -188°C at the equator and -182°C at the south pole [6]

This complicated thermal profile, coupled with the rapid rotation of the planet and the different temperatures at which the chemical constituents of the clouds condense, gives rise to the horizontal bands of visible clouds that encircle the planet. At the equator, they are stronger than on Jupiter, with speeds of 500

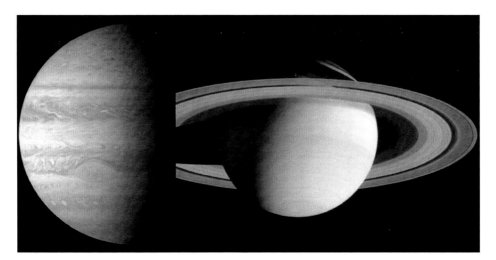

**Figure 7.4** Cloud patterns on Jupiter and Saturn. Different chemical compositions absorb different wavelengths of light, resulting in visible bands that travel around the planet at different velocities, altitudes, and in some cases, direction. (a) Jupiter's lighter "zones" are higher and cooler, while its darker "belts" are warmer and lower. A great deal of turbulence is evident, especially at the boundaries of zones and belts, which tend to move in opposite directions. The prominent Great Red Spot is a hurricane several Earth diameters across that has been raging for centuries. (b) Saturn's bands appear more numerous and smoother, though this is largely an artifact of the hazy upper atmosphere that blurs visibility of the clouds below. Closer inspection reveals turbulence and ephemeral storm cells, just like on Jupiter (NASA/JPL/Space Science Institute).

meters per second (1,200 mph). Zonal and altitudinal differences in wind velocity create tremendous sheer forces, generating turbulence and huge storms at certain latitudes

### 7.2.1.3 Uranus

The atmosphere of Uranus consists of roughly 83% hydrogen, 15% helium, 2% percent methane, and trace amounts of ethane and other gases. The much higher content of methane on Uranus absorbs red wavelengths, and gives rise to a blue haze that obscures the lower cloud levels (Figure 7.5a). Using other wavelengths, clouds of different composition come into view (Figure 7.5b). Recent observations suggest a much more dynamic atmosphere than the placid picture first observed close up by Voyager 2. Winds have been measured up to 218 meters per second (over 490 mph), and blotches indicative of storm systems have been observed [7]. Voyager 2 measured a temperature of –214°C near the top of the cloud layers.

### 7.2.1.4 Neptune

Neptune's atmospheric composition is similar to that of Uranus, with about 3% methane and trace amounts of other gases, but with even more helium (18%) and less hydrogen (79%). Likewise appearing blue to the human eye because of the large methane content, Uranus and Neptune look the same at a distance, but closer images reveal more visible white clouds (Figure 7.5c,d). Temperatures in the upper cloud layers vary in the vicinity of -200°C.

The winds on Neptune are the highest known in the Solar System, reaching 600 meters per second (1,340 mph) at the equator, faster than the speed of sound on Earth. Voyager 2 and subsequent observations have revealed a very dynamic atmosphere on Neptune, with large scale atmospheric phenomena appearing and disappearing over a matter of days.

Neptune's inner atmosphere may hold an ocean of liquid water over the rocky nucleus, almost the size of the Earth, perhaps an immense liquid-water ocean mixed with silicates, methane, and ammonia, which extends up to the atmosphere of hydrogen, helium, and methane. The ocean must be extremely hot (and in a supercritical state), but likely remains liquid as a consequence of the high pressure [8].

## 7.2.2 Conceivable habitats for life on the gas giants

A frustrating lack of solid information about the atmospheres of the gas giant planets makes speculation about life in their midsts difficult. We do know enough to appreciate that all of them have complicated structures, with chemistries and temperatures that vary with latitude, depth, and seasons, making all of them highly dynamic – even violent by Earth-based standards.

Solar radiation is much less intense at the distance of the gas giants, especially as far away as Uranus and Neptune, but it remains strong enough to affect weather patterns. Heat at lower altitudes and higher pressures is an abundant source of energy. Above all, the kinetic energy of winds way beyond that of the

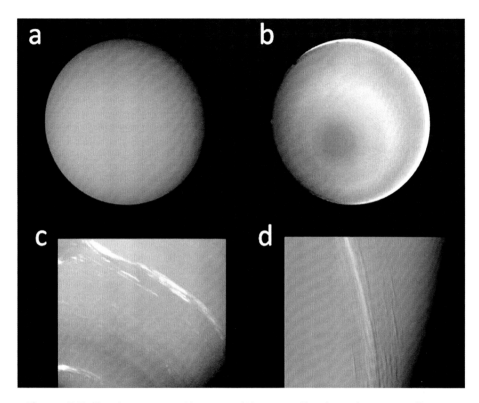

**Figure 7.5** Cloud patterns on Uranus and Neptune. Clouds on the two smaller gas giants are much thinner and less distinct than on Jupiter and Saturn. (a) This true color image was taken as Voyager 2 approached Uranus in 1986. The blue color results from absorption of red wavelengths by the planet's relatively high concentration of methane (b) The same image of Uranus, false colored to enhance subtle differences that reveal a banding pattern parallel to the plane of rotation. The circles appear concentric because Uranus is tilted at an angle of greater than 98 degrees; so it rotates like a spinning top lying on its side. The south pole was pointing nearly straight at the Sun when Voyager 2 took this picture (c) True color image of Neptune from Voyager 2, showing long, wispy clouds parallel to the direction of rotation from the polar region to high latitudes. (d) Two hours from closest approach to Neptune by Voyager 2, this image shows clouds 50–200 km wide and 30–50 km high (NASA/JPL).

strongest cyclones on Earth is ever-present in the upper atmospheres of all the gas giant planets.

We can say with some confidence that there is a region somewhere between 0.1 to 10 bars where the temperature falls within the range for liquid water. On Neptune, liquid water may exist under much higher pressures and temperatures at a deeper level. We can also say that precursors for biomolecules are found within, above, and below this region. Their concentration overall is low, though the possibility that organic or even biochemicals could be locally aggregated into microorganisms is a theoretical possibility.

Clouds are dense on Jupiter and Saturn, and locally dense on Neptune. Beneath the methane haze near the top of the atmosphere of Uranus, clouds probably float as well. Temperatures within them appear to be quite cold, though, as required for the condensation that makes up the clouds. Lower, where atmospheric densities and temperatures become high enough for liquid water, little water is found. No region comparable to the bottom layer of clouds with mode 3 particles on Venus is known for any of the gas giants.

### 7.2.3 Assessing the plausibility of life in the atmosphere of the gas giants

Purely from an environmental point of view, the existence of life in the atmosphere of gas giant planets is theoretically plausible. There is opportunity, in the form of abundant energy and the presence of chemical precursors for biomolecules. And there are habitats, albeit poorly defined and unbounded, in the form of regions where temperatures and pressures are not incompatible with macromolecular stability. What is lacking is historical circumstance.

By historical circumstance, we mean a clear vision of where whatever life is there would have come from. The argument for life in the clouds of Venus is predicated on the firm possibility that the planet hosted a substantial biosphere on its surface that served as the ancestral origin for forms that became progressively adapted to the severe constraints of airborne life. No such primordial biosphere can be envisioned for planets that aggregated under conditions that generated a continuum between extreme cold in a vacuum to intense heat under great pressure. Even if a transient period existed during which life could have formed, it seems highly questionable that any form of life could have survived the growth in pressure and temperatures that are characteristic of a fully formed gas giant but are so incompatible with macromolecular stability.

Jupiter and Saturn in particular seem unlikely launching pads for life, on the basis of their paucity of organic precursors alone. They are not totally absent, of course. Trace hydrocarbons, nitrogen compounds, and sulfur complexes are found in the atmospheres of both planets. That these molecules could ever attain a critical concentration sufficient to engage in complex chemistry is unclear. The case could be made somewhat easier for Uranus and Neptune, because there, at least, methane and ammonia are major constituents. Throw just a little oxygen into the mix and amino acids under the right conditions could emerge. But oxygen is in short supply on the gas giants, and even if it weren't, amino acids themselves are far from being alive.

The gas giants today contain no boundaries or interfaces where self-organizational complexity could take hold. And the regions where temperatures and pressures could in theory maintain macromolecular stability are torn by winds so violent, radiation so strong, and upheavals so frequent that the establishment and maintenance of boundary conditions, as required by our definition of living systems, is hard to imagine. These severe constraints, when combined with the absence of any recognizable biomarkers that can't be explained equally well by abiotic processes (like the presence of methane), lead us to the conclusion that life anywhere in the atmosphere of our gas giant planets is unlikely.

That said, we will qualify our lack of enthusiasm for life on gas giants in three ways:

### 7.2.3.1 Gas giants elsewhere may surprise us

Already we know enough about exoplanets to realize that some of them are gas giants that differ significantly from our own. Even if ours turn out to be typical for the majority of gas giants in the universe, they will by no means be limiting cases. Planets very different from the gas giants in our Solar System are known to circle other stars. Some of them more massive than Jupiter have been found in orbits less than an astronomical unit (the Earth's distance from the Sun) from their central star. Double and triple gas giants circling around one another at close range are probably out there as well, not to mention all the planetary systems that surely have formed around double and triple stars.

What has been the history of their formation? If, as we argue extensively in this book, history is destiny for life in the cosmos, we can hardly rule out the possibility of life on planets whose history we have not yet deciphered. Could a Jupiter-sized planet arise close to a central star in a way that would have provided a window for the origin of life? Unless we can say no with conviction (and it's hard to see when that would be), we can't close the door on the possibility of life in the atmosphere of gas giant planets somewhere.

### 7.2.3.2 What about "life" outside conventional definitions?

One cannot fail to be impressed with the energy and chemistry that swirl about the upper atmosphere of a gas giant like Jupiter. Indeed, a time lapse movie of the stratosphere in motion on Jupiter suggests the entire planet itself is a brooding, roiling crucible of something that, for want of a better term, looks to be "alive." It's not that hard to get carried away by inspirational imagery of a Great Red Spot churning about the planet for at least three centuries, maybe even spawning small offshoots of itself in something akin to fission, if not true reproduction. In the end, though, these are nothing but deceptive imitations of certain characteristics of life, surely generated by highly dynamic but totally abiotic processes.

Nonetheless, is it possible that in all that vast, swirling vortex of activity, there might be some complicated organic chemical network – some extensive interaction of precursors and products that formulate, then break down, molecules in a more complicated way than conventional atmospheric science can explain? Even if not constrained by the boundaries of the thin-walled balloon-like "floaters" imagined by Sagan and Salpeter – organisms a kilometer across that could hold their contents through hurricane force winds lasting for years – maybe something like floating clouds of complex chemistry, churned into life by the tremendous forces of wind, temperature, and pressure unimaginable on Earth, could exist.

Even if they do, if we are true to our definition of life as set forth in Chapter 2, we would have to exclude such phenomena from our pantheon of life in the universe. To be without a hard-and-fast physical boundary automatically disqualifies even the most lavishly dynamic pool of activity from being considered "alive." Add to that the absence of any precise mechanism for replication, and the test for being a living entity has clearly been failed. But that doesn't mean that interesting entities don't roam the upper reaches of the atmosphere on gas giant planets, where energy abounds and the chemistry is far from simple – phenomena beyond easy explanation as purely physical processes, if not quite fully alive in a formal sense.

### 7.2.3.3 What about the immigration of life from another world?
We have concluded that on Venus, life in the clouds is distinctly possible because it probably existed below the clouds to start with. We have come to the opposite conclusion about the likelihood of life in the atmospheres of the gas giants, because we have seen no clear way it could have gotten there. What if this assumption is wrong?

The possibility that the inner rocky planets could have shared life from a common origin early in their history, spread among them by interplanetary transport (panspermia), has been discussed previously. Panspermia would solve the problem of life's non-origin on the gas giants, were it to have happened. If some form of life had managed to hitch-hike its way on a wandering asteroid from a distant point of origin to the neighborhood of a gas giant planet, thereupon falling into its fertile atmosphere and seeding it with life ever after, the problem of origins would be solved.

Generally, this has been regarded as an unlikely scenario for two reasons. First, the outer planets are so remote from the one planet we know that harbors life, and even the closest one (Mars) that might have, that the survival of a hitch-hiking progenitor seems very dubious. Secondly, even a surviving form of life would most likely be ill pre-adapted for life under conditions at its destination so different from its origin.

There is, however, one possible loophole is this line of logic. That is the possibility that panspermia occurred from a source not that distant, on a world where conditions may have been not that removed from the habitat in the clouds of the gas giant to which it is transported. If life exists on Io, a mere 0.003 of an astronomical unit from Jupiter, it would exist in a radiation environment at temperature extremes with a recognizable chemistry not that unlike some combination of conditions in the atmosphere of Jupiter. If life existed on Titan, a measly 0.008 of an astronomical unit from Saturn, it would come from a satellite well endowed with a dense, cold, and organic atmosphere hardly foreign to pockets of comparable conditions in the atmosphere of Saturn. A fairly localized version of panspermia conceivably could do the trick.

But what is the likelihood that life really has arisen on these smaller, distant satellites at the outpost of our Solar System? Before turning to that subject, we will briefly consider the possibility of life in the clouds of exoplanets.

### 7.3 Prospects for life in the atmospheres of exoplanets

Between the dimensions of relatively small, rocky planets like Venus with secondary atmospheres, and the gas giants with their primary atmospheres, lie exoplanets orbiting too close to their central star to hold a primary atmosphere of hydrogen and helium, but large enough and warm enough to generate a dense, secondary atmosphere. On exoplanets such as these, conditions might differ from the extremes we find in the clouds of either Venus or our own gas giants.

As a case in point, let's look at the planetary system known as Gliese 581. At least four planets are currently known to orbit a small M star, or red dwarf, at a distance of 20.7 light years from Earth (Figure 7.6). The innermost planet (581e) is the smallest, with 1.9 Earth masses, and appears to be far too close to the red dwarf to maintain an atmosphere, much less liquid water. The second innermost planet (581b) is the largest of the four, at over 15 Earth masses. It might be large enough to hold an atmosphere, but at a distance of only 0.04 AU (much closer

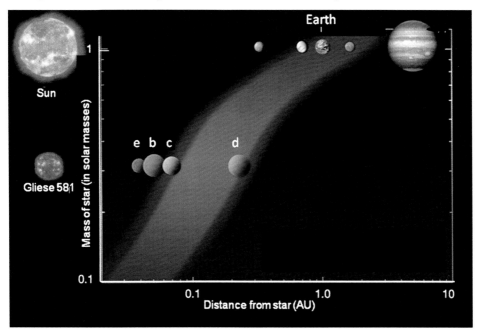

**Figure 7.6** Comparison of size and position of planets in the Gliese 581 system with those in our Solar System. Because of its much lower mass and energy output, Gliese 581 (a red dwarf star) could allow rocky planets to orbit much closer without loss of water or atmospheres. The two outermost planets (581c and d) may be large enough and sufficiently distant to hold an atmosphere and possibly even liquid water, the meaning of the "habitable zone," indicated by the light blue band. All the planets are shown in their approximately correct relative sizes, but are disproportionately large compared to the Sun (Image courtesy of the European Southern Observatory).

than Mercury is to our Sun), would very likely be a runaway greenhouse planet like Venus, if it did.

Gliese 581c is further out and smaller – between 5 and 10 Earth masses. At 0.07 AU from Gliese 581, it might be barely far enough away and large enough to retain not only an atmosphere but liquid water, if the atmospheric pressure and cloud albedo (brightness) are great enough. Finally, the fourth planet, Gliese 581d, is between 7 and 14 Earth masses at a distance of 0.22 AU from its central star. It might be heavy enough and barely warm enough to retain a dense, moist atmosphere, and possibly some liquid water, on the surface of rocky continental masses, or beneath the frozen surfaces of extensive oceans.

Were life to be present in the waters or the rocky surfaces of these two outer planets in the Gliese 581 system, the chances that some organisms might dwell for extended, or even permanent, periods of time in the airborne habitat of their clouds is a distinct possibility. As usual, the requisites for survival would be an energy source (presumably satisfied by light or radiant heat from the red dwarf not that far away), conditions allowing for the stability of physical boundaries, and a mechanism for reproduction. Our assumption would be that life would have originated on the surface, and migrated to the atmosphere, as envisioned on Earth and Venus, making life in the clouds of rocky exoplanets fundamentally analogous to life in the clouds on Earth. If the atmospheres are much denser but not much more dynamic, and if they are much more richly endowed with chemical nutrients as well as electromagnetic energy, the organisms in them might be larger, and the biotic communities richer and more complex, than in the atmospheres on Earth (and possibly Venus). Under those conditions, airborne communities on those distant worlds could approach the complexity that we see in arid land or benthic ocean habitats on Earth.

## 7.4 Chapter summary

Every planet except Mercury, and several of the larger moons, in our Solar System have atmospheres ranging from thin and tenuous to very dense. Given the availability of energy and the appropriate chemical necessities, there is no reason in principle why those habitats couldn't accommodate life. This chapter first considers the possibility that the densest atmospheres in our Solar System – those of Venus and the Gas Giants – could harbor life that, in all likelihood, cannot exist beneath their clouds. Then we briefly explore the possibility that exoplanets intermediate in character between cloud-enshrouded rocky planets and gas giant planets could also host forms of airborne life.

Venus is encased in an atmosphere at least 80 kilometers (50 miles) thick. At its lower levels, temperatures and pressures are too extreme for the stability of macromolecules. But a blanket of clouds over 20 kilometers thick, composed primarily of sulfuric acid droplets, rings the planet in perpetual motion. Two distinct layers, each about four kilometers tall, float at the floor of this global cloud cover. The bottom-most layer consists of droplets in three different sizes, the largest of which is comparable to the aerosol drops that make up the clouds on Earth. Temperatures range from about 80° to 33°C (176° to 91°F), and the atmospheric pressure is close to that of sea level on Earth. The next layer up is cooler, at about 30° to 4°C (86° to 37°F), with pressures comparable to that of tall mountain tops on Earth. These are conditions under which microbial life comfortably exists on our own planet.

The challenge for life in the clouds of Venus is formidable. Obstacles include a very dry habitat, except for the water absorbed within the droplets of sulfuric acid, the high acidity of the droplets, their lack of organic molecules in significant concentrations, and the relentless violence of hurricane-force winds. However, assuming that life flourished in the surface waters of Venus before they boiled away, directional selection could have pre-adapted microbes to stand the heat, acidity, and aridity they would eventually have to survive in the sky. A reasonable speculation is that normophilic microbes gradually came to reside in the cooler clouds that make up the higher of the two bottom cloud layers, while thermophiles came to occupy the lowest, warmer layer. A simple ecosystem consisting (most likely) of phototrophic and chemotrophic producers, subsisting on chemical cycles maintained partially by each, co-habiting within the larger droplets in the two lower cloud layers, is by no means implausible.

The chance for life to originate and take hold in the atmospheres of the gas giants is much lower, in our view. While the upper atmospheres of Jupiter, Saturn, Uranus, and Neptune lack the high acidity of clouds on Venus, and may have a slightly richer mix of organic compounds, the radiation environments are strong, the water content is low to non-existent, and the winds are even more powerful. There are narrow bands in the atmosphere of each planet where temperatures and pressures are compatible with macromolecular stability; so in theory a biosphere could be contained there. The principle problem with life on the gas giants is envisioning how it could arise in the first place, absent the sharp boundaries and tolerable combinations of temperature and pressure that we think are essential for the origin of life.

At our current limited state of knowledge about gas giants generally, especially of those we now know to exist in other solar systems that differ radically from our own, it would be premature to declare life an impossibility in their atmospheres. It could be that chaotic pools of chemical interactions swirl about somewhere in the heights or depths of their vast atmospheres, beguilingly complex but short of "being alive." However, unless a plausible scenario for seeding by panspermia – perhaps from life arising on one of their nearby satellites – can be invoked for the four gas giants in our own Solar System, we rate the chance that life could exist anywhere within them as very low.

Exoplanets with conditions intermediate between those of cloud-covered rocky planets like Venus and the vast primary atmospheres of gas giants are now known to exist. They present another possibility for the existence of airborne life. If living organisms are found beneath their clouds as well, they could simply present a variation on the theme of life as it occurs on Earth – within its waters, on and beneath its surface, and in the air as well.

## 7.5 References and further reading

1   Grinspoon, D. H. 1992. Venus revealed: a new look below the clouds of our mysterious twin planet, Perseus Publishing, Cambridge, MA.
2   Jenkins, J. M., Steffes, P. G., Hinson, D. P., et al. 1994. Radio occultation studies of the Venus atmosphere with the Magellan spacecraft. *Icarus* **110**: 79–94.
3   Counselman, C. C., 3rd, Gourevitch, S. A., King, R. W., et al. 1979. Venus winds are zonal and retrograde below the clouds. *Science* **205**: 85–87.
4   Schulze-Makuch, D., Grinspoon, D. H., Abbas, O., et al. 2004. A sulfur-based survival strategy for putative phototrophic life in the Venusian atmosphere. *Astrobiology* **4**: 11–18.
5   Sagan, C. and Salpeter, E. E. 1976. Particles, environments, and possible ecologies in the jovian atmosphere. *Astrophys. J. Suppl. Ser.* **32**: 624.
6   Orton, G. S. and Yanamandra-Fisher, P. A. 2005. Saturn's temperature field from high-resolution middle-infrared imaging. *Science* **307**: 696–8.
7   Hammel, H. B., Pater, I. D., Gibbard, S., et al. 2005. Uranus in 2003: Zonal winds, banded structure, and discrete features *Icarus* **275**: 534–535.
8   Baker, V.R., Dohm, J.M., Fairén, A.G., Ferre, P.A., Ferris, J.C., Miyamoto, H., and Schulze-Makuch, D. 2005. Extraterrestrial hydrogeology. *Hydrogeol. J.* **13**: 51–68.

http://www.solstation.com/stars/gl581.htm
An informative site about the Gliese 581 exoplanetary system.

http://nineplanets.org
An excellent compilation of data on all the planets and their satellites.

http://www.nasa.gov/worldbook/venus_worldbook.html
A thumbnail sketch of Venus and the robotic missions sent there.

http://www.nasa.gov/worldbook/jupiter_worldbook.html
A thumbnail sketch of Jupiter and the robotic missions sent there. Links to the other gas giants can be found here as well.

http://photojournal.jpl.nasa.gov/targetFamily/Venus
Repository for scores of images and topographical representations of the surface of Venus. Links to images of the gas giants can be found here as well.

# 8 Deep and Dark

*Aquatic life in perpetual darkness on an ice-covered water world like Europa*

Europa is on everyone's short list of destinations for finding life in the Solar System beyond Earth. This is because water is found there in abundance, almost surely in liquid form, protected from the harshness of space by a thick layer of ice. But the ice is restless, cracking continually and moving about in rafts and slabs that reflect the push and pull of gravity from Europa's huge parent, Jupiter, and its larger nearby siblings, the moons of Io, Ganymede, and Callisto. And gravitational flexing is only the start of the story. There's a tenuous atmosphere of oxygen, fed by the sputtering impact of Jupiter's powerful stream of radiation that splits the water molecules at the surface into their constituent hydrogen and oxygen atoms. A slight distortion of Jupiter's electromagnetic field at Europa suggests the presence there of either a metallic core, or a salty ocean beneath the ice, or both, that rotate at different rates (asynchronously) from the planetoid's surface. This could mean either a source of internal heating from radioactive decay in the core, or strong global currents of water, or both. Combine that with the possibility that thermal vents at the bottom of Europa's deep ocean and emissions of compounds like hydrogen sulfide and organic molecules like methane from the mantle could provide more energy options as well as the building blocks for polymeric chemistry, and all the requisites for life are at hand.

## 8.1 Nature of Europa

When Galileo first sighted the moon of Jupiter that Simon Marius would name Europa, it was barely more than a second dot in the sky beside its parent planet. Not until close-up images of Europa were transmitted back to Earth from the Voyager 2 flyby in July 1979, was the utterly unique nature of Europa revealed. Here was a satellite slightly smaller than Earth's moon, but so very different: the smoothest and brightest surface in the Solar System, almost devoid of craters, but crisscrossed with long lines and arcs like a cracked egg shell (Figure 8.1a).

We think we know what the inside of this globe must be like. Since we know its size (volume) precisely, and have been able to calculate its mass from the pull

L.N. Irwin and D. Schulze-Makuch, *Cosmic Biology: How Life Could Evolve on Other Worlds*, Springer Praxis Books, DOI 10.1007/978-1-4419-1647-1_8, © Springer Science+Business Media, LLC 2011

a                                    b

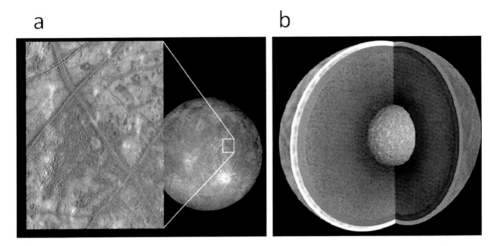

**Figure 8.1** Europa, surface and interior. The prototypical icy satellite, Europa is the ultimate water world, albeit completely frozen over. (a) Under constant gravitational flexing, Europa's exterior is continually being ground into the smoothest surface in the Solar System. The close-up view reveals the consequences of this dynamic activity, in the form of criss-crossing lines and fractures. A single rare crater can be seen in the lower right of the global view. The false-color image highlights an equatorial concentration of minerals on the surface. (b) Cutaway model of Europa's interior, revealing a metallic core surrounded by a rocky mantle, overlain by a global ocean frozen at its surface (NASA/JPL/ University of Arizona).

of gravity that bends the path of spacecraft flying by it, we know its global density. Since we also know the density of water, we can calculate that water-ice is insufficient to account for its entire bulk, so it must have a solid mineral interior. Knowing the density of silicates and metals like iron and nickel, we can approximate the size of the interior, assuming a rocky, silicate mantle surrounding an iron-nickel core, as shown in Figure 8.1b.

From its spectrographic signature, its relatively smooth surface, and its very high albedo (brightness), we can tell that the surface is solid ice. Solid, but not static, as the ever-shifting plates of ice slide past one another, collide, and reform in a chaotic jumble. Long lines reflecting surface fractures, and quite possibly eruptions from below, dissect the globe, but seldom in perfectly straight lines (more often as gentle arcs), and frequently at acute angles to one another (Figure 8.2a). Faulting is extensive, mainly of the slip variety, as frozen plates on the surface slide past one another, causing complex geometric formations (Figure 8.2b) and sometimes filling with upwelling liquid from below that freezes into widening trenches over time (Figure 8.2c). In the so-called "chaos" regions, plates of ice raft about one another like the scrambled pieces of a jigsaw puzzle (Figure 8.2d). Finally, the very low number of craters, compared to the number on nearby Ganymede and Callisto, are signs of a surface that has recurrently, or at least recently, been repaved.

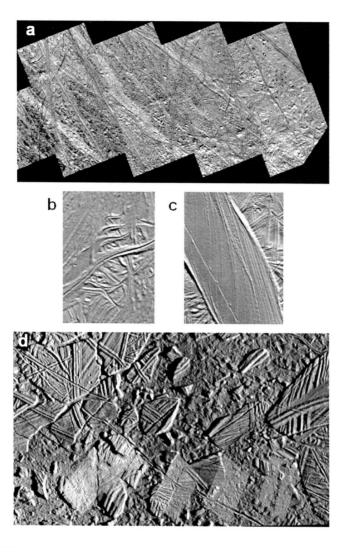

**Figure 8.2** Surface features of Europa. (a) An extensive network of brown double ridges extends across this 800 by 350 km mosaic in the northern hemisphere. Younger ridges are superimposed on older ones, both probably created by outflow of mineral-containing water from beneath the smooth blue layer of ice that forms a shell encasing a global ocean. Colors are enhanced to bring out detail. (b) A slip fracture, in which plates of ice slide along a fault line in opposite directions has created a series of sigmoidal ridges, generated at recurrent intervals. (c) Another slip fault has resulted in a widening gap, filled in with water erupted from below which rapidly freezes on exposure at the surface, forming a lane over 12 km wide. (d) This feature, known as "chaos," reflects the turbulent thrust and pull of gravitational forces that shuffle frozen rafts of ice around one another. The relative lack of craters in all images, and the obvious and frequent surface distortions, point to a very dynamic and geologically young surface. All pictures were taken from the Galileo orbiter (NASA/JPL/University of Arizona).

All of the above properties are exactly what would be predicted for a frozen water shell stretched and pulled first in one direction, then another by strong gravitational forces, exerted by bodies both inside (Jupiter and Io) and outside (Ganymede and Callisto) the orbit of Europa. Add to this the possibility that the continual squeezing and stretching generates heat, and the probability increases that the ice is melted below the surface. One of the strongest indicators of a liquid ocean beneath the surface is the induction of anomalies around Europa in the strong electromagnetic field generated by Jupiter. The easiest explanation for this effect is a mobile, electroconductive medium inside Europa, such as a salty, liquid ocean in motion [1]. Indirect evidence for interior rotation faster than the rotation of the surface adds strength to this argument. Finally, a number of surface features, like the long fracture lines bounded by minerals (falsely colored orange in Figure 8.1), heaved up perhaps from below, and the broken shards of ice that raft about chaotically, are all consistent with the presence of a dynamic underlayer of liquid, salty water serving as a floatation platform for the tectonic movement of ice plates at the surface.

Another feature known only for a few other bodies in the Solar System is the presence of an oxygen-containing atmosphere. To be sure, Europa's atmosphere is wispy thin – only hundredths of a percentage point of Earth's air, and much lighter than the atmosphere on Mars. With little atmosphere to filter it out, radiation splits water at the ice surface into hydrogen (which readily escapes) and oxygen. By "gardening" the newly released oxygen back into the ice (and perhaps dissolving it into occasional liquid eruptions) at the surface, oxygen could slowly work its way into subsurface contact with other chemicals, susceptible to oxidation. The air on Europa is so thin, though, that it provides virtually no protection against either radiation or meteorites. Thus, radiation continues to generate oxygen (with most of the released hydrogen escaping quickly into space), and the lack of craters despite the absence of protection from meteorites reinforces the view that the surface must be dynamic and under constant renewal.

## 8.2. Planetary history of the Jovian satellites

The four major moons of Jupiter share a common beginning, but have diverged in dramatic ways dependent on their size and relative positions. Before discussing Europa (and Io in the next chapter), we will first describe how they all arrived at their current condition.

1. When the Solar System formed from a massive cloud of gas and dust about 4.5 billion years ago, 99.8% of the mass gathered into the Sun at the center of the spinning, protostellar disk. The remaining 0.2% made up all the planets and their satellites in orbit around the Sun.
2. At least seven-tenths of the residual material was incorporated into the formation of Jupiter and its satellites, making it the most massive planetary system orbiting the Sun.

3. Of that residual material, 99.98% went into Jupiter itself; the remainder being divided mostly among four major moons: Io, Europa, Ganymede, and Callisto.
4. The moons formed from metallic and silicate compounds, in the presence of water and volatile gases. The metals and silicates formed solid cores. While not massive enough to hold onto the lighter gases, each captured an increasing volume of water, thereby decreasing their densities, with increasing distance from Jupiter.
5. Io, the moon nearest to Jupiter, was too close to retain an atmosphere in the intense radiation field to which it was subjected and the massive bombardment it probably incurred so close to Jupiter's gravity well during its early formation. These same factors probably drove away whatever water may have condensed there initially.
6. Europa, the second moon out from Jupiter, was far enough away to retain enough water to cover the globe entirely, but at a distance from the Sun that caused it to freeze over completely at the surface. Tidal flexing and possibly radioactive decay in the core generates enough heat to keep the subsurface liquid, however, and to cause ongoing resurfacing that obscures the degree of meteorite bombardment to which it has been subjected.
7. Ganymede, the third moon out from Jupiter, is the largest satellite in the Solar System. Its proportional water composition is greater even than Europa's, but internally generated heat appears to be much lower, and consequent resurfacing is not nearly as extensive as Europa.
8. Callisto, the furthest major moon out from Jupiter, is the third largest satellite in the Solar System. Its internally generated heat is even lower than that of Ganymede, resulting in a geologically inactive and ancient surface.

**Figure 8.3** The moons of Jupiter. (Left to right) Io, the most volcanically active planetary body in the Solar System, is continually resurfaced by lava eruptions. Europa is covered with a global shell of water, frozen but highly fractured at its surface. Ganymede, the largest satellite of any planet in the Solar System, experienced geological activity in the past but appears to be quiet, with a relatively old surface, at present. Callisto has the lowest density, hence the highest water content, of all four satellites. It appears to be geologically inactive, with a highly cratered, ancient surface. Evidence for liquid oceans beneath their frozen surfaces is strong for Europa, and suggestive for Ganymede and Callisto. Io appears to have very little water at all (NASA/JPL).

## 8.3 Conditions for life on Europa

Europa's resonant orbital period of one circuit around Jupiter for every two by Io, combined with a resonance with Ganymede of two Europan circuits for each one by Ganymede, meant that Io and Ganymede would alternately come close to, then move away from Europa. This guaranteed that Europa would be squeezed and pulled by enough tidal flexing to keep heat pumping out of its interior. This, along with some radiogenic decay, at least early in its history, appears to have resulted in a liquid ocean beneath its icy covering.

A quick review of the conditions necessary for life reveals why Europa would be a prime destination for the Solar System tourist searching for life on other worlds. For starters, there is more water on Europa than on Earth, and given all the favorable advantages of water as a solvent for living systems, the need for a solvent is satisfied at the outset. Since evidence tilts increasingly in favor of water in liquid form beneath the icy crust, a number of habitats for life become readily apparent, including the vast ocean itself, and the interface of the water with the mantle at the bottom and the ice ceiling at the top – boundary habitats teeming with life on Earth.

And if water is the solvent, chances are that carbon is the most likely backbone for the polymeric chemistry that life requires. Direct evidence for organic compounds on Europa is not yet available, but if the satellite's interior is anything like Earth's, the building blocks for organic chemistry, like methane, carbon dioxide, and hydrogen sulfide, may well be expelled from Europa's undersea rocky mantle. That leaves only energy as the missing requisite for life, and here the story is a little more complicated.

## 8.4 Energy for life on Europa

### 8.4.1 Light
Earlier we explained the advantages of light as an energy source for living systems. We know how it powers the vast majority of life on Earth, and probably energizes whatever life may float in the clouds of Venus. At the distance of Jupiter, light would have only about a twentieth of the intensity that it has on Earth, but this would probably be enough in principle to power surface-dwelling life on Europa. In all likelihood, though, the extreme cold and harsh radiation environment limit any life near the surface of Europa to microbial organisms embedded in hypersaline water inclusions within the ice, deep enough for some protection from radiation, but close enough to absorb the little bit of weak sunlight that filters through a few hundred millimeters (several inches) of overlying ice. Photosynthetic microbes could thus be the first life we encounter on arrival at Europa.

## 8.4.2 Radiation

Jupiter emits a powerful field of ionizing radiation, and particle radiation even from the distant Sun impinges on Europa without impediment. These highly energized particles could in principle deliver a powerful dose of energy to any surface or near-surface organisms capable of capturing it. Particle radiation, however, is paradoxically too strong, and probably too erratic to be used *directly* as an energy source. More likely is the possibility that the hydrogen generated from water split by collisions with the radiant particles could provide a source of metabolic fuel. If the splitting of water into hydrogen and oxygen occurs within the ice, perhaps a pool of hydrogen builds up sufficient to create a concentration gradient that favors diffusion of hydrogen into living compartments. If organisms have evolved the machinery for capturing energy as hydrogen ions diffuse through their outer membranes, usable energy could be harvested. Because they too would likely be encased in tiny water inclusions, these organisms would likewise be microbial.

For life in larger forms and greater variety, however, the ocean depths beneath the ice beckon. Even the thinnest models for the ice layer on Europa suggest that it has to be at least several kilometers thick. This renders the liquid ocean below, a realm of total darkness (though infrared radiation from some thermal vents can't be ruled out). If no light penetrates this marine habitat, other sources of energy must serve to sustain whatever ecosystems thrive there.

## 8.4.3 Chemistry

Looking first to forms of usable energy familiar to us, we know that carbon dioxide can be reduced to methane, if hydrogen is available, with a net release of energy. Both hydrogen and carbon dioxide could bubble up from the interior of the mantle through outgassing, or be produced by chemical reactions unknown to us. As we discussed in chapter 3, iron is a large atom that easily picks up and gives away electrons. So if iron at the ocean bottom is reduced (from $Fe^{3+}$ to $Fe^{2+}$) by sulfur or hydrogen vented from the interior, that too could provide energy. For these reactions to operate indefinitely, though, the oxidation-reduction cycle has to be completed. That is, the reduced compounds (methane or $Fe^{2+}$) would have to be reoxidized to carbon dioxide and $Fe^{3+}$, respectively. This could happen at the ocean's ceiling in the presence of oxygen that has worked its way down through the ice layer, or by chemical reactions within or at the bottom of the ocean that we don't know about yet. Organisms on Earth have evolved the capacity to carry out each of the reactions mentioned above inside their cells, where they capture the released energy in the form of chemical bonds that can be used to power all the living processes that the cell carries out. The same could obviously be true of life on Europa.

## 8.4.4 Fluid in motion

The convection of water currents beneath the ice provides another form of potential energy. While we don't know of any examples of organisms on Earth that harvest kinetic energy for supporting basic metabolism, we know it plays a role in certain living processes. Some coral-like organisms expose their flat fan-shaped surfaces perpendicular to the flow of the current in order to capture minute food particles. Hair cells in the inner ear of vertebrates are bent by vibrating pulses of fluid, transducing the mechanical energy of sound into nerve impulses. Air currents sustain the flight of birds, remove heat from animal bodies by evaporative cooling, and transport nutrients through circulatory systems in plants and animals alike. Perhaps cilia bent by currents of water add tensile energy to the shape of molecules, like winding a spring. When tension in the spring is released, energy for work is made available. Or maybe bending hair cells open semi-permeable pores that allow ions to flow down their concentration gradients. Flowing ions could energize the storage of chemical energy for later metabolic release. Given the inventive nature of evolution, any number of mechanisms can be envisioned.

### 8.4.5 Osmotic and ionic gradients

Another plausible source of power for living processes in the global ocean on Europa is the potential energy that can be harvested from ionic concentration gradients. As minerals from the mantle erupt into the water, or dissolve into it at the ocean floor, the salt concentration of the water is higher than it is at a distance above the ocean floor. By the same token, at the ocean's icy ceiling, the distinctly less salty composition of ice as it melts makes sea water near the ceiling more dilute than at greater depths. Living organisms could extract energy from these gradients merely by maintaining an internal ionic concentration, or tonicity, different from their surroundings (Figure 8.4). Near the ocean floor, for example, energy could be derived by allowing ions to diffuse from their area of higher external concentration into the more dilute (hypotonic) interior of a living cell. At the other extreme, organisms near the ice ceiling could draw energy from outwardly moving ions as they diffuse from their more concentrated (hypertonic) interior into their more dilute surroundings. The movement of water, or osmosis, is the mirror image of these phenomena; spontaneous osmosis of water into the cell near the ice ceiling, or out of the cell near the ocean floor, could be coupled to membrane machinery that alters the structure of chemical storage molecules, or drives chemical reactions that store energy for later use.

### 8.4.6 Heat

The kinetic motion of molecules, or heat, may be the most pervasive form of energy in the universe. Though heat is not a source of metabolic energy for life on Earth, probably because of its ready dispersal and its relative inefficiency compared to light, it could be a major source of energy in the eternal darkness of Europa's subsurface ocean (Figure 8.5). Heat could come from two sources: decay of radioactive minerals in Europa's metallic core (the source of heat that shuffles the continental plates on Earth to and fro), or the tidal tug on Europa caused by Jupiter's and Io's pull in one direction, while Ganymede and Callisto pull in the other. Some scientists think that the diapers, or upwellings in the surface ice, seen frequently on Europa result from thermal plumes emitted from the ocean

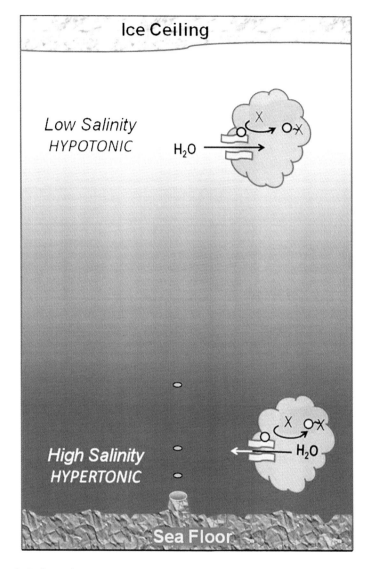

**Figure 8.4** Osmotic energy sources possible in the ocean of Europa. High-energy chemical bonds ($\sim$X) could be formed by reactions coupled to energy transferred by osmosis, or water moving down its concentration gradient. Ionic or osmotic gradients generated by salts dissolving from the ocean floor, and water melting from the ice ceiling, produce environments that are hypertonic and hypotonic, respectively, to living cells. Organisms could shuttle between areas of higher (hypertonic) and lower (hypotonic) salt surroundings, promoting the movement of ions or water in alternating directions. Salt concentration is indicated by the darkness of the shading. The color of the organism's interior is the same in both locations, showing how the gradients are reversed from the floor to the ceiling. The size of the organisms is vastly exaggerated for clarity (Art by Louis Irwin, adapted from original [10]).

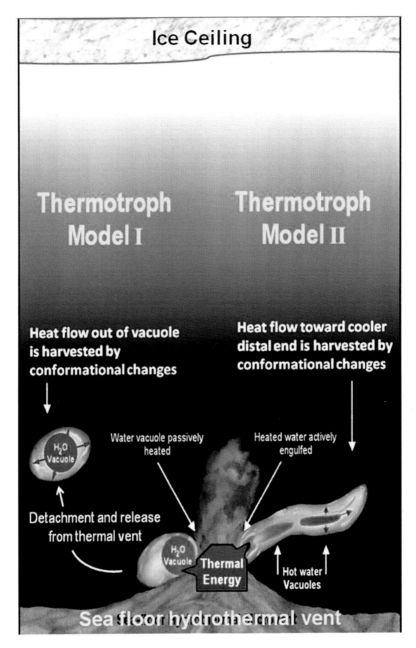

**Figure 8.5** Thermal energy sources possible in the ocean of Europa. Thermal gradients would be generated at points where hot water comes out of the ocean floor. Thermotrophs could harvest this energy either by floating up to release heat and sinking to absorb heat, or by conducting the heat through a long body extending from the heat source at the organism's anchor point to a heat sink at the end of the organism (Adapted from original art by Chris D'Arcy, Dragon Wine Illustrations [10]).

floor. And as previously noted, the tectonic rafting of ice islands, and the long curvilinear fault lines seen at the surface leave little doubt that forces capable of generating considerable heat contribute to Europa's dynamic nature. Whatever the source of the heat, once thermal gradients are established, the flow of heat down those gradients, or encapsulation of the heat from them, could serve as sources of energy.

### 8.4.7 Other long-shot possibilities

Jupiter also casts a huge magnetic field over all its major satellites, and the fluctuating pull of gravity itself represents a real if weak force. In principle, these forces too could be captured as energy for biological processes by an organism that has evolved appropriate transducing mechanisms for the task. In comparing the energy obtainable from these sources, however, we concluded in a careful analysis several years ago [2] that neither of them is likely to be powerful or efficient enough to fuel the energetic needs of a living system – at least not in comparison with the energy that can be harvested from any of the other sources discussed earlier. We think it unlikely, therefore, that they contribute to the putative biosphere of Europa; but we have mentioned them to honor the spirit of this book, which is to keep our minds open to any possibility that is not ruled out by the scientific facts or compelling logic.

## 8.5 Forms of life on Europa

### 8.5.1 Producers

Given that Europa has an abundance of water interfacing with solid surfaces both at the top and bottom of its global ocean, that it has an abundance of energy sources, and that it presents one of the most active geological surfaces in the Solar System, what forms of life could we expect to find there? The following speculations are based on the premise that organisms may have evolved into a considerable variety of forms, each making maximum advantage of a different type of enegy. Since the energy source that an organism adopts is typically rooted deeply in its ancestry, the mode of energy extraction can be used as a defining property for a whole class of living forms. Accordingly, we will group our putative organisms according to an energy-based taxonomy.

Those organisms that derive their energy from the abiotic resources of the environment are producers, the lowest level of the food chain. They will be considered first.

*Phototrophs.* Organisms that use sunlight for generating metabolic energy would be limited to the near-surface. Light at the distance of Jupiter is weak, temperatures at the surface are very cold, and the constraints that ice places on the size of such organisms, since they would likely inhabit only small to tiny inclusions, means that they would most likely be slow-growing microbes. We will call them ice algae.

*Chemoautotrophs.* The ability to extract energy from chemical reactions using substrates from the environment defines a chemoautotroph. Ice inclusions near the surface could harbor organisms sustained by chemical reactions based on reduction by hydrogen generated radiolytically from the breakdown of water. For the same reasons listed above, they would likely be slow-growing microbes as well. We will call them radiolytic chemotrophs.

We previously outlined several chemical reactions that could take place in the subsurface ocean of Europa. The number of different organisms that could thrive on these reactions is at least as numerous as the reactions themselves. Perhaps there are chemolithotrophs, (litho = rock) subsisting on the reduction of iron or other metals at the ocean floor. If the reducing agent is hydrogen erupted from the mantle, we'll call them hydrogen lithotrophs. If they derive their reductive capability from one of the compounds of sulfur, we could designate them sulfur lithotrophs. Or maybe a group of organisms similar to the methanogens on Earth make their living by reducing carbon dioxide to methane. We'll call them benthic methanogens. At the ocean's ceiling, where the availability of oxygen is greater, and therefore oxidation reactions are more likely, aerobic organisms could exist. We'll designate them upper level aerobes. All of these chemoautotrophs would be dependent on broadly dispersed chemical substrates, available in the immediate environment of the organisms. If there are fairly steep gradients of the substrates, as there might be for minerals and their reducing agents ebbing up from the mantle, or oxygen working its way down through the ice at the top of the ocean, the organisms would be expected to be restricted to the more concentrated regions of those substrates. Assuming a broad distribution of the substrates in horizontal layers, there would be no incentive for organisms to move about in search of their chemical nutrients. Rather, a horizontally flattened mat of sedentary organisms seems the most likely form for these organisms to take. The floor of the ocean could thus be covered with mats of benthic lithotrophs and methanogens, while the ceiling is home for a host of upper level aerobes.

*Kinetotrophs.* For the maximum capture of convective energy, kinetotrophs need to be anchored to a solid substrate so that the bulk of the organism holds still in relation to the moving current of its environment. At the ocean floor, long slender organisms like reeds could be anchored in the substrate. As they are bent by the passing currents, tensile potential energy is added to molecules specialized for the purpose that later transform their stored energy into metabolically useful reactions. We'll call them benthic reeds. Another means for capturing kinetic energy could be achieved by ciliated organisms. As their cilia bend with the passing current, tensile distortion at the base of the cilia is transduced into potential chemical energy. These would likely be pancake-shaped organisms adhering to either the ocean floor (benthic ciliates) or ceiling (ceiling ciliates).

*Osmotrophs and Ionotrophs.* For organisms to capture the potential energy provided by ionic and/or osmotic gradients, they would need to be mobile enough to move back and forth between areas of higher to lower solute concentration (Figure 8.4). This might not need to be a great distance, as the concentration gradients are probably fairly steep at both the ocean floor and ceiling. It is difficult to predict the size and shape of these organisms, since two antagonistic factors would serve as selective pressure during their evolution. On the one hand, a large surface area in relation to volume would provide a greater number of ionic or osmotic channels per organism. On the other hand, a larger organismic volume would be a better buffer against the changing concentrations that entering ions or water would induce. Maximizing surface area would be achieved by keeping the organism small and flattened. Maximizing volume would be achieved by making the organism large and round. Perhaps evolution would pursue both options, in which case Europa might harbor both microionotrophs and megaosmotrophs.

*Thermotrophs.* If thermal vents are found on the floor of Europa's ocean, they could host a biotic community as rich and diverse as the thermal vents on Earth do. Heat could be captured from these vents in at least two ways (Figure 8.5). A globular organism could engulf hot water at the vent opening, then float to a cooler region, where the outflowing heat adds potential energy to molecules through high-energy bonds or conformational changes in their structure. We'll designate this a bubble thermotroph. Alternatively, an elongated structure could convey hot water vacuoles from their origin at the basal end, to the cooler distal end of the structure, where heat flows outward into the cooler surroundings, again surrendering energy to appropriately configured molecules or chemical bonds. The latter could be called a cucumber thermotroph.

## 8.5.2 Consumers

All the producers listed above represent a source of food for consumers. The nature of the consumer depends to an extent on the nature of the food they consume.

The unicellular photoautotrophs could serve as food for microbial heterotrophs. Sedentary organisms, such as the benthic and ceiling chemotrophs and kinetotrophs, and the cucumber thermotroph, could provide a source of nutrients for parasites or symbionts that dwell on or inside their hosts. Many of the cohabitants would be small and stationary, just as algae in lichens and parasitic protists in animals remain embedded in their terran hosts. But if nutrients are abundant, they could become sizable, like tree fungi or intestinal worms. We might thus distinguish them on the basis of both size and host, such as megachemotrophic suckers or microthermotrophic scum.

The sedentary producers could also be food for small mobile organisms, like snails that scrape their food from the film of bacteria, protists, and smaller metazoans that live on the substrate. Assuming that organic films cover both the ceiling and the floor, we could distinguish between benthic scrapers and ceiling scrapers.

The mobile producers, like the benthic ionotroph, ceiling osmotroph, or bubble thermotroph, would be potential prey for both mobile and floating predators. Active swimmers like small fish and floating animals like jellyfish provide analogies from the oceans of the Earth. Let's call them mobile catchers and floatation catchers. We also might envision fan-like shapes, similar to the fan coral seen in marine environments on Earth. They could be attached at the floor of the ocean with their planar faces oriented perpendicular to the direction of tidal currents in the water. This orientation would optimize their ability to capture small organisms and detritus carried in the current. We'll call them floor fans.

Each of these primary consumers could in turn be preyed upon by larger but less numerous secondary consumers, just as large fish prey on smaller fish. Conceivably there could be as many predators as there are types of prey. For simplicity, we'll lump them into generic secondary or tertiary mobile predators, and secondary or tertiary floatation predators.

Every ecosystem contains decomposers, a category of consumer that lives off of dead and decaying organic material. As organisms die, they fall to the bottom of the ocean, where detrivores ranging in size from small bacteria to lobster-like animals consume them. Thus, detrivores tend to congregate on the ocean floor (so we'll call them benthic microscavengers or benthic megascavengers). Detrivores also operate at higher layers as well, however, so pelagic detrivores and ceiling detrivores could well be part of the Europan biota.

Clearly, the collection of producers and consumers enumerated above could constitute a sizeable biological community. Table 8.1 summarizes the types of organisms according to their level in the food chain. Note that each "organism" is actually a generic form of life. The variety of life that could exist on Europa is thus in theory much larger than the list in Table 8.1.

## 8.6 Possible evolutionary history for putative life on Europa

As on all planetary bodies, the evolutionary trajectory that life would have taken would have depended on the succession of geophysical stages through which it passed. On Europa, we distinguish between two major phases of planetary history: (1) an early stage of indeterminate but probably brief duration during which a solid silicate mantle had formed around a consolidated metallic core, with liquid water covering the globe; and (2) a later, longer stage that continues to the present, in which the surface has been frozen over, with a layer of liquid underlying the outer icy shell and overlying the solid mantle.

During the first stage, residual heat from the satellite's formation would have caused enough evaporation from the liquid ocean to generate a water-vapor atmosphere. The intense radiation from Jupiter and, to a lesser extent, the Sun, would likely have generated more oxygen by radiolytic degradation of water molecules than were found in the early atmosphere of Earth. Even later, after the ocean froze over, occasional large impacts may have melted the icy oceanic crust and instantly created a new water vapor atmosphere that dissipated as the ocean surface froze over again. With this planetary history in mind, we have sketched out a possible evolutionary history for putative life on Europa in Figure 8.6. The evolutionary steps referred to below reference the numbers in that figure.

**Table 8.1**   Summary classification of putative organisms on Europa

| TROPHIC LEVEL | ENERGY SOURCE | ORGANISM |
|---|---|---|
| Producers | Light | ice algae |
| | chemical reduction from radiolytic H | radiolytic chemotrophs |
| | reduction of metals by H from mantle | hydrogen lithotrophs |
| | reduction of metals by S from mantle | sulfur lithotrophs |
| | reduction of $CO_2$ by H from mantle | benthic methanogens |
| | organic oxidation | upper level aerobes |
| | current flow | benthic reeds |
| | current flow | benthic ciliates |
| | current flow | ceiling ciliates |
| | ionic gradients | microionotrophs |
| | osmotic gradients | megaosmotrophs |
| | heat | bubble thermotroph |
| | heat | cucumber thermotroph |
| Primary Consumers | lithotrophic producers | megachemotrophic suckers |
| | thermotrophic producers | microthermotrophic scum |
| | benthic producers | benthic scrapers |
| | ceiling producers | ceiling scrapers |
| | mobile producers | mobile catchers |
| | floatation producers | floatation catchers |
| | mobile or floatation producers | floor fans |
| Secondary Consumers | mobile or floatation consumers | floor fans |
| | mobile primary consumers | secondary mobile predators |
| | floatation primary consumers | secondary floatation predators |
| Tertiary Consumers | mobile secondary consumers | tertiary mobile predators |
| | floatation secondary consumers | tertiary floatation predators |
| Decomposers | benthic detritus | benthic microscavengers |
| | benthic detritus | benthic megascavengers |
| | pelagic detritus | pelagic detrivores |
| | ceiling detritus | ceiling detrivores |

There is no reason to doubt that life could have arisen during the first stage, when the global ocean was liquid, thermal energy from accretion was substantial, and organic molecules from the circumstellar disk and from meteorites were present in relative abundance. Soon after life arose as simple chemoheterotrophs metabolizing the available organic compounds as nutrients, other forms of energy metabolism could have evolved, like the reduction of $CO_2$ to methane *(step 1)*, or of iron by H or S *(step 2)*. At the surface of the ocean under an early water vapor atmosphere, sunlight should have been just bright enough to foster the evolution of simple organisms deriving their energy from photosynthesis *(step 3)*. Such a photoautotroph could then have evolved more

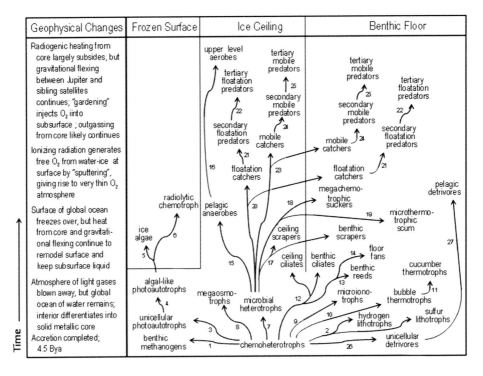

**Figure 8.6** Scenario for the evolution of life on Europa. Plausible steps in the major transitions that different generic forms of life could have undergone as Europa's global ocean cooled and froze over, culminating in three habitats: an icy surface under a thin oxygen atmosphere, an ocean ceiling beneath the ice, and the bottom of the ocean. Time, without units since the actual timing of the transitions is not known, is represented on the vertical axis.

complex cellular features similar to algae *(step 4)*. Osmotic and ionic gradients could have provided sources of energy for large osmotrophs *(step 8)* or microbial ionotrophs *(step 9)*, while heat emissions from the ocean floor may have supported various thermotrophs *(steps 10,11)*.

Once the initial producers had evolved, they could have served as nutrients for an ancestral primary consumer *(step 7)*. From these ancestral primary consumers, an evolutionary radiation of higher order consumers could have proliferated (Figure 8.6).

As soon as the shell of ice became permanent, two distinct interfaces with water would have been available for colonization. Beneath the ice at the ceiling of the ocean, a water-ice interface would provide a solid substrate, while the floor of the ocean would have provided a water-rock (or sand) interface. The vertical distance between the two is currently estimated to be tens if not hundreds of kilometers, so the two habitats probably would have evolved essentially as two distinct biospheres.

Once the ice shell formed, algal-like organisms could have persisted near enough to the surface to subsist on sunlight *(step 5)* or radiolytically dissociated H *(step 6)* as means of synthesizing high-energy compounds. Yet another form of producer may have evolved to harvest kinetic energy from water currents, either at the ceiling or the floor of the ocean *(step 12)*. Variations on this theme may have been reed-like *(step 13)* or filter-feeding fan-like *(step 14)* organisms. Another descendant of the ancestral heterotrophic producer could have been an anaerobe *(step 15)* that eventually gave rise to an aerobic form of life respiring oxygen near the surface *(step 16)*.

A third, minor biosphere may have resulted from the survival of some forms of life, either in liquid inclusions within the ice, or in cryptic form near the surface where occasional eruptions of water from below or bolide impacts from above could provide transient pools of liquid on top of the ice under a temporary water vapor atmosphere.

With a variety of organisms now thriving on the floor and ceiling of the ocean, consumers that could scrape them from their respective surfaces *(step 17)* and a variety of parasitic or commensal organisms *(steps 18,19)* could have evolved. Consumers adopting a floatation life style like jellyfish would likely have appeared both near the ceiling and floor *(step 20)*. Descendants of these passive feeders would probably have given rise to larger predatory forms *(steps 21,22)*. In a parallel manner, more mobile consumers could also have appeared *(step 23)*, likewise leading to higher trophic levels of predation *(step 24,25)*. Detrivores can be assumed to have arisen to feed on dead material that collected at the ocean bottom *(step 26)* or throughout the ocean depths up to the ceiling *(step 27)*. All forms of life suggested in Figure 8.6 are presumed to be capable of surviving to the present time.

## 8.7 Ecosystems on Europa

The trophic structure of an ecosystem describes the flow of energy through the system. Figure 8.7 represents a simple multilevel ecosystem, based solely on producers that harvest energy from the environment by reducing carbon dioxide with hydrogen, both of which are expelled from the mantle. The area of the polygons are proportional to the biomass of each trophic level. As explained in Chapter 3, each higher trophic level has less biomass than the level below it because the transfer of energy from each level to the next is far below 100% in efficiency. The efficiency of energy conversion among organisms on Europa, of course, is unknown, so Figure 8.7 has been constructed on the basis of known conversion efficiencies from a marine ecosystem on Earth, as we have previously calculated [4].

Figure 8.8 reveals a more complex potential ecosystem on Europa, consisting of communicating ecosystems at the floor of the ocean based on ionotrophic producers, and at the ceiling based on osmotrophic producers.

While the smallest size that a living organism can assume is not known, it's

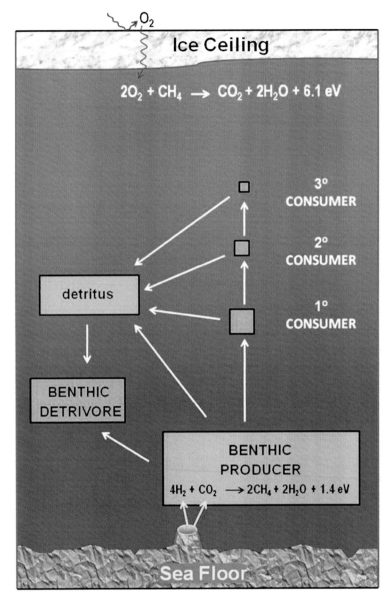

**Figure 8.7** Hypothetical Europan ecosystem based on chemical energy. Methanogens could harvest energy from the reduction of carbon dioxide to methane, generating high-energy chemical bonds ($\sim X$) in producer organisms that support a more elaborate ecosystem. An oxidant, such as $O_2$ gardened through the ice from the surface would be needed to complete the cycle. Arrows indicate the direction of energy flow. Rectangular areas are proportional to the approximate relative biomass at each trophic level in marine ecosystems on Earth (Art by Louis Irwin, adapted from Irwin and Schulze-Makuch [4]).

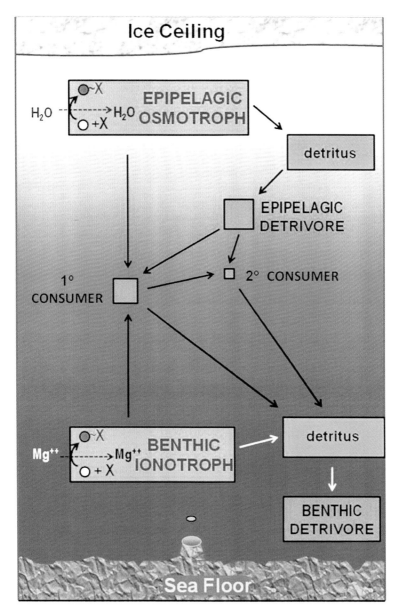

**Figure 8.8** Hypothetical Europan ecosystem based on energy from osmotic or ionic gradients. A different type of ecosystem might be based on coupling an ionotrophic producer at the floor and an osmotrophic producer at the ceiling of the ocean. An influx of $H_2O$ or $Mg^{++}$, for example, moving down their respective concentration gradients could power the synthesis of energy-rich chemical bonds ($\sim X$). Arrows indicate the direction of energy flow. Rectangular areas are proportional to the approximate relative biomass at each trophic level in marine ecosystems on Earth (Art by Louis Irwin, adapted from Irwin and Schulze-Makuch [4]).

safe to say that many if not most of the producers on Europa are microscopic. This is not true on Earth, where the abundance of sunlight makes possible the large size of plants. But even on Earth, the majority of total biological productivity is carried out by microbial-sized organisms (mainly in the upper layers of the ocean). It is conceivable that if some energy source on Europa is very abundant, and organisms have evolved to harvest that energy in a highly efficient manner, producers large in size could exist there. While we can't know with certainty, most sources of energy appear to be less amenable to efficient capture than sunlight, so it seems reasonable to predict that the total energy yield on Europa will be relatively modest. The total amount of energy available determines the total biomass that can be supported. Making reasonable assumptions about the sizes of organisms at each trophic level in a comparable marine ecosystem of Earth enables us to estimate the numbers and sizes of organisms at each trophic level in model ecosystems for Europa.

We and others have actually attempted to model what the energy availability on Europa might be, and what the resulting biomass could look like. The calculations are technical and full of assumptions that are little more than best guesses at our current state of knowledge. But the results of such models are instructive in guiding the nature and limits of our thinking about what life on Europa could be like.

Christopher Chyba and Cynthia Phillips [3] at the SETI Institute have proposed a model ecosystem based on radiolytic production of oxygen at the surface of Europa. Assuming that the oxygen works its way into the ocean beneath the ice, they have calculated that a biomass resulting in a tertiary consumer of about 1 gram (roughly the size of a tadpole), if evenly spread through the ocean of Europa, could occupy every 6.76 cubic meters of water volume (a cube with edges 2.6 meters long). We [4] have suggested a model ecosystem based on methanogenic producers that yields a biomass equivalent to the same sized tertiary consumer in every 32.5 cubic meters of ocean (a cube with edges 5.7 meters long). Mikhail Zolotov and Everett Shock [5] have proposed a third model ecosystem based on anaerobic reduction of sulfate as an energy source. Their results lead to a tertiary consumer in 729 cubic kilometers of ocean (a cube with edges 27 kilometers long). The differences in these projected densities for organisms at the top of the food chain in the ocean of Europa are due to different assumptions about the amount of energy harvestable by the producers.

We have no idea at this point which, if any of these models, approaches reality. They do, however, serve the purpose of enabling us to visualize what the possibilities might be, including the constraints on the support of any ecosystem. For instance, it seems unlikely that tadpole-sized organisms separated by an average distance of 27 kilometers, as in the model of Zolotov and Shock, represent a viable population, while the same organisms at an average distance of 5.7 meters (about the size of a residential swimming pool) could interact readily. Only real data can distinguish between these alternatives.

## 8.8 Biotic communities on Europa

With habitats and potential energy sources thus identified, and a plausible collection of organisms enumerated, we can begin to envision what biotic communities on Europa might look like. There are essentially three life zones. The first and most speculative is a community of organisms living near the outer surface of the ice. The other two, which are both likely to exist if life at all is found in the subsurface ocean, occupy the interface between water and ice at the ocean's ceiling, and between water and the mantle at the ocean's floor.

### 8.8.1 The near-surface community
The intense radiation and severe cold that characterize the surface of Europa provide formidable challenges to the existence of life unprotected on the surface. It remains possible, however, that within a few centimeters of the surface, a layer of phototrophic organisms could exist within small liquid inclusions in the ice. We have further suggested that other, non phototrophic organisms powered by hydrogen ion gradients generated by radiation sputtering at the surface, could share the same habitat. The tiny size of the inclusions would limit the size of the organisms to that of microbes, and access to their respective sources of energy would be limited to horizontal layers near but not at the top of the ice. This community would be seen in cross-sections of ice cores as horizontal bands, one of which would be colored (possibly even green), reflecting evolution of pigments adapted to absorb the wavelengths of light most efficient for driving photosynthesis.

### 8.8.2 The ice ceiling community
On the underside of the ice we could expect an expanse of mats and films consisting of ceiling ciliates and other organisms requiring attachment to a surface. Grazing across these planar populations would be a variety of consumers: snail like scrapers, crab-like detrivores, and numerous symbionts. Bobbing up and down near the ceiling might be osmotrophs, which could be both small and large, depending on their particular adaptations for moving up and down within the salinity gradient that would be diluted by melting ice.

### 8.8.3 The benthic community
The ocean floor could well be similar, with bottom-covering mats and films spread broadly across the horizontal extent of the substrate, consisting primarily of chemoautotrophs. A variety of symbionts, consumers, and decomposers could provide a considerable variety of organisms to this bottom-dwelling community. As at the top of the ocean, ionotrophs could be floating up and down, taking in ions as they dissolve from the floor, and releasing them in the more dilute waters above. Swimming about and feeding on them might be a variety of secondary consumers. Additional variety could be provided by thermal vents, which could have their own distinctive mix of producers – perhaps a variety of attached

thermotrophs and chemotrophs – with which a distinctive set of consumers would be entangled.

### 8.8.4 The pelagic community

More difficult to predict is the nature of the organisms that would be found in the open depths between the floor and the ceiling. To the extent that any organisms move into this region, there could be predators that would make use of them for food. The size and mobility of these consumers, and the number of trophic levels in the food chain, would ultimately depend on the total amount of energy harvested by the producers. Indeed, the amount of available energy is the single most vital piece of missing information about Europa.

## 8.9 Characteristics of Europan biota

The vast majority of biomass on Europa (all except for a thin layer of microbes near the surface) would be sequestered deep beneath the satellite's surface in perpetual darkness and cold. What would life under those conditions be like?

### 8.9.1 Metabolism

By comparison with Earth, life on Europa would likely appear to grow and move (if at all) in slow motion. The small size of Europa's rocky center probably means that radiogenic heating is limited, and any energy harvested from convection currents would probably required organisms to be anchored and stationary. While the potential for ionic and osmotic gradients is substantial, they would be harvested most likely either by elongated organisms attached to a surface, or by slowly ascending and descending floatation organisms. The combination of cold, relative scarcity of energy, and minimal need for autonomous movement would predict organisms with low metabolic rates.

### 8.9.2 Reproductive systems

If most of Europa's biomass is indeed sedentary, it would likely be more plant-like in its reproductive strategies. It is hard to predict how important sexual reproduction would be on Europa. On the one hand, Europa's planetary history has presumably been monotonously static since the global ocean froze over fairly soon after it formed. This would have placed a premium on stabilizing selection early in planetary history, leading to an optimization of the most simple and effective systems of reproduction. Typically, this means simple fission, budding, or other vegetative forms of non-sexual reproduction. On the other hand, parasitism could have been an important part of the biological dynamic on Europa. Since some theories of sexual reproduction on Earth propose that its purpose has mainly been to avoid the debilitating effects of parasitism, sexual reproduction could have a place on Europa. Even so, for the majority of the biomass that would likely be sedentary, even sexual reproduction would be plant-like, in consisting of the passive dispersal of reproductive cells and spores.

### 8.9.3 Motility

Few of the producer organisms in Europa's ocean would need to move of their own accord, as their sources of energy, whether chemical, osmotic, or kinetic, would be available pan-globally at the interface between ocean and substrate where they would mostly live. Even thermotrophs would have no need to move more than short distances, and passively if that, from their fixed sources of heat. Primary consumers, including detrivores, would need to move very little, since their food would not be moving. Only the secondary and higher level consumers would have need to move about, overtaking the minimal movements of the primary consumers, or defeating whatever evasive or clandestine strategies their prey may have evolved. It is possible that at the higher levels of the food chain, motility of both predator and prey would be driven to greater extremes by evolutionary competition, like the reactive selection for speed in both cheetah and elk. But this would likely involve only few forms of life, and would be limited by total energy availability, which as we've already pointed out, may not be great. Those few forms that would be motile, however, would require something akin to neural control systems to manage the motor apparatus that enables them to move about.

### 8.9.4 Sensory systems

Most forms of life on Europa would not be expected to have organized sensory systems. This is because the majority of organisms that can be envisioned for a deep, dark marine environment would be totally sedentary or passive floaters. They would have the ability to respond at a cellular level to their local environments, by opening or closing ion channels, depending on the surrounding salt concentration, for instance. They might also be capable of trophic responses, such as growth or ameba-like movement toward energy substrates, such as local pockets enriched in hydrogen ions. In this regard, they would most likely resemble plants and microbes on Earth. Consumers, on the other hand, would require some type of sensory capability in order to find and evaluate the nutrient value of the consumers on which they would be feeding. Clearly this sensory capability would not include vision, but the capacity to sense current flow, heat, and chemicals (odor) would be beneficial. The underwater habitats of Europa are likely to be silent as well as dark, since large animals needing to communicate by sound over long distances are unlikely to live there. However, the possibility cannot be ruled out that some consumers could use sonar to locate prey, or be capable of hearing bubbles of gas emanating from deep ocean vents. Sensitivity to pressure and heat, smell, and the possibility of hearing, would require something akin to a nervous system. These sensory systems, however, would most likely be found in only a few consumer organisms near the top of the food chain.

### 8.9.5 Cognition

Predicting whether organisms in the subsurface ocean of Europa would be cogent – that is, capable of evaluating information and making behavioral decisions – is not as difficult as predicting whether organisms will be mobile and have sensory abilities. The cognitive ability of animals on Earth is directly related to the complexity of their nervous systems, which is driven by the extent of their sensory capabilities and their need for motor coordination. Thus, if our speculation that Europa's biota would, for the most part, be neither very mobile nor capable of sophisticated sensory perception is correct, we can surmise that the cognitive ability of even the smartest organisms in Europa's ocean is not likely to be great. Europa may house alien life, but alien intelligence is not very likely to dwell there.

## 8.10 Properties of Europa *not* conducive for life

The possibility cannot be discounted that there is no life at all on Europa. The mantle and core of Europa are smaller than Earth's Moon, which appears to no longer be generating internal heat from radioactive decay, so Europa might have no internal source of heat left. It could be that there are no molecules, like carbon dioxide or sulfur, erupting from Europa's mantle to fuel chemotrophic metabolism. It's even possible that no liquid ocean exists beneath the ice covering, hence no currents or other forms of kinetic energy may be available. Finally, we may have assumed a planetary history for Europa that isn't the case; the period in which the water shell over the rocky interior was liquid may have lasted too briefly for life to have arisen. Any one of these possibilities, if true, would decrease the probability of life on Europa. The fact remains that its surface is clearly young and dynamic, reflecting at the very least substantial tidal flexing. That alone ensures that energy is present, whether thermal or kinetic or both, to be harvested by appropriately adapted organisms. A robotic mission to Europa, even if able only to sample the chemistry and measure the geological activity at the surface, would increase immensely our ability to determine which of the scenarios pictured in this chapter – a vibrant, multi-level ecosystem thriving beneath the icy covering, or a lifeless frozen snowball – is closer to the reality of Europa.

## 8.11 Enceladus: variations on a theme

We can't end our discussion of the potential for life on Europa without considering the same possibility on a similar moon, more distant from the Sun – Saturn's small icy satellite, Enceladus. This snowball moon defies our expectations in several respects, not the least of which is that a planetary body this small (with a diameter of only 500 kilometers) ought to be stone cold by 4.5 billion years after its formation. Contrary to this expectation, Enceladus shows clear evidence of recent resurfacing (Figure 8.9a) and is emitting a lot of heat energy from its south pole (Figure 8.9b). In November, 2009, the Cassini orbiter passed within 25 kilometers of Enceladus, and took dramatic pictures showing jet-like plumes emanating from its south pole (Figure 8.9c).

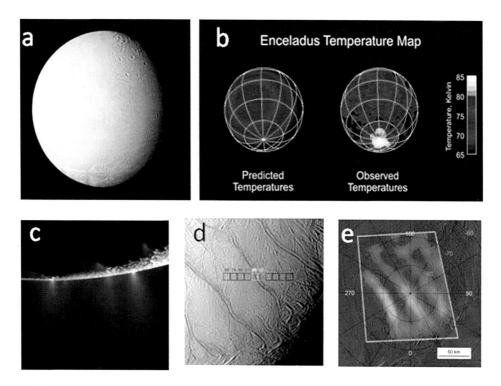

**Figure 8.9** Enceladus. Images and data from the Cassini orbiter have defied conventional wisdom about the geology of small icy satellites. (a) Enceladus shows features of old, well-cratered terrain (upper right) and smooth younger surfaces (equatorial regions). (b) Contrary to predictions, the south pole averages a warmer temperature than expected. (c) Dramatic plumes that turn out to be mostly water are seen jetting out of several points at the south pole. (d) Parallel trenches 150 km long and up to 2 km wide parallel to one another an average of 35 km apart at the south pole. (e) The parallel trenches, or "tiger stripes" emit significant amounts of heat from the interior. The hottest spots (in yellow) reach to less than 100°C below zero, compared with the surrounding surface at close to -190°C (JPL/NASA/Space Science Institute/ Goddard Space Flight Center/ Southwest Research Institute).

When better views of the south pole were obtained, the dramatic trenches now known as "tiger stripes" were revealed, along with evidence that the geysers and the heat are spewing out of distinct breaks along the stripes (Figure 8.9e), now thought to be slip faults in the rigid icy crust.

Explanations for the energy driving the geysers are varied [6], but the compositional data are pretty clear that the geyser contents are mostly made of water, with some methane and smaller amounts of organic compounds, along with nitrogen, carbon dioxide, and carbon monoxide. Experts now believe that Saturn's E-ring most likely is generated from these emissions, and sodium compounds found within the ring therefore most likely come from water beneath the icy crust in contact with a rocky core [7].

The possibility that liquid water may exist beneath the surface of Enceladus, and that energy in the form of heat and possibly organic chemistry is available, suggests a clear parallel with Europa [8,9]. Thus, predictions about what life, if any, could exist there, would follow the same pattern and logic that we discussed for Europa above.

One unresolved problem would be, again, the historical path to life beneath the ice of Enceladus. In the early age of their formation, Saturn's largest satellite, Titan, and Enceladus may have been similarly warmer and endowed with a reasonably rich repertoire of organic chemicals. Conceivably, incipient forms of life could have been exchanged between the two. Because of its small size and great distance from the Sun, Enceladus must have frozen over rather quickly, so any origin of life based on the presumption of liquid water at the surface would have had to happen relatively fast. But Enceladus defies conventional expectations already, so this possibility should not be precluded.

Assuming that life could have arisen, or been seeded, in the early ages of Enceladus, it seems most reasonable to assume that the biosphere there would have remained simple and quite static, once the surface froze over.

## 8.12 Significance of the potential for life on Europa or Enceladus

Europa occupies a privileged position in the thinking of astrobiologists for three reasons. First, if our Solar System is any guide, Europa may be much more typical of planetary bodies in the universe than either our inner rocky planets or the gas giants, since small icy planetoids are numerically the most common bodies in our region of space. Secondly, Europa is water rich and therefore likely to support forms of life that human explorers are more prepared to recognize. With a planetary history perhaps not unlike Earth's in its early stages, the early forms of life on Europa may have shared similarities with our own ancestral forms, albeit almost certainly independent from them in origin. Finally, the differentiated habitats that occur on a planetary body with an ice-covered ocean surrounding a differentiated interior gives rise, not only to a host of possible energy sources, but a number of multiple habitats, which the history of life on Earth suggests are important for generating biological diversity. On this consideration alone, with the possible exception of Titan and Triton, no place beyond Earth in our Solar System offers a greater prospect for biodiversity, in theory, than Europa.

Expectations for the existence of life on Enceladus would have to be rated less favorable than on Europa, because of the smaller size and greater remoteness of this small moon. But the unexpected geophysical activity evident there cautions us to be skeptical of conventional wisdom. Any evidence at all for life on Enceladus would have to be taken as a very strong signal that life is highly likely to exist on several other worlds in our Solar System.

## 8.13 Chapter summary

Europa is the most likely place in our Solar System other than Earth to be capable of supporting a reasonably complex biosphere consisting of forms of life more-or-less as we know them. This is because it holds a vast underground ocean of liquid water, with multiple sources of energy. While the chemical richness of the oceanic habitat is not known, there is no reason to doubt that the precursors of organic chemistry could be there. Further piquing our interest is the fact that a tenuous oxygen atmosphere enshrouds the ice shell of the ocean, and may feed into the subsurface habitat.

Except for a few ice-encrusted organisms near the surface, light is of no biological use on Europa. All other forms of life are encased by many kilometers of light-blocking ice. Beneath the icy shell, however, energy could be derived from many sources: a multitude of chemical reactions, thermal emissions from the interior, water currents, and ionic or osmotic gradients generated from the ocean floor and ice ceiling, respectively.

A variety of producers deriving their energy from any or all of these sources could lie at the base of a food chain that could achieve a fair degree of complexity. Plausible models have been proposed that envision ecosystems extending to the level of tertiary consumers consisting of organisms as large as the small to medium sized marine animals with which we are familiar on Earth.

Life on Europa is likely to be mostly sedentary, with any movements probably minimal due to the cold and (possibly) limited energy supply. This in turn would suggest simple sensory systems and limited, if any cognition. Reproductive strategies would also probably be simple, but sexual exchange of genetic material could have been favored to neutralize the incursion of parasites.

Europa is a particularly appealing target for astrobiological investigation because it (1) probably represents a large fraction if not a majority of the planetary bodies in the universe, (2) has the capacity for supporting water-borne life which could be somewhat familiar to explorers from Earth, and (3) consists of clear-cut interfaces between water, ice, atmosphere, and substrate, which generate the type of habitat fractionation needed for the development of complex biospheres. Indeed, the prospects for a considerable degree of biodiversity are as great on Europa as on any world we know beyond Earth.

Finally, the small snowball world of Enceladus deserves brief consideration because jets of water emissions from its south pole suggest a sizable subsurface oceanic reservoir, similar to Europa. If that can be confirmed, much of what is discussed in this chapter about Europa would apply to Enceladus as well.

## 8.14 References and further reading

1   Khurana K.K., Kivelson M.G., Stevenson D.J., et al. 1998. Induced magnetic fields as evidence for subsurface oceans in Europa and Callisto. *Nature* **395**: 777–780.

2    Schulze-Makuch, D. and Irwin, L. N. 2002. Energy cycling and hypothetical organisms in Europa's ocean. *Astrobiology* **2**: 105–121.

3    Chyba C.F.and Phillips C.B. 2001. Possible ecosystems and the search for life on Europa. *Proc. Nat. Acad. Sci. U.S.A.* **98**: 801–804.

4    Irwin L.N. and Schulze-Makuch D. 2003. Strategy for modeling putative multilevel ecosystems on Europa. *Astrobiology* **2**: 813–821.

5    Zolotov M.Y. and Shock E.L. 2003. Energy for biologic sulfate reduction in a hydrothermally formed ocean on Europa. *J. Geophys. Res.-Planets* **108**: art. no.-5022.

6    Gioia, G., Chakraborty, P., Marshak, S., et al. 2007. Unified model of tectonics and heat transport in a frigid Enceladus. *Proc Natl Acad Sci U.S.A* **104**: 13578–81.

7    Porco, C. 2008. The restless world of Enceladus. *Sci Am* **299**: 52–5, 58–63.

8    Parkinson, C. D., Liang, M. C., Yung, Y. L., et al. 2008. Habitability of enceladus: planetary conditions for life. *Orig Life Evol Biosph* **38**: 355–69.

9    McKay, C. P., Porco, C. C., Altheide, T., et al. 2008. The possible origin and persistence of life on Enceladus and detection of biomarkers in the plume. *Astrobiology* **8**: 909–19.

10.  Schulze-Makuch, D., and Irwin, L. N. 2008. Life in the Universe: Expectations and Constraints, 2nd ed., Springer-Verlag, Berlin.

http://photojournal.jpl.nasa.gov/targetFamily/Jupiter?subselect=Target%3A Europa%3A
Over 100 models and images of Europa, based primarily on the Voyager and Galileo missions.

http://galileo.jpl.nasa.gov
Home page for the Galileo mission to Jupiter and its moons.

http://www.jpl.nasa.gov/galileo/sepo
Background information and educational context for the images taken from the Galileo spacecraft.

http://solarsystem.nasa.gov/planets/profile.cfm?Object=Jup_Europa
Brief factual overview of Europa, with helpful interactive comparison of Europa with other Solar System bodies.

http://www.nineplanets.org/europa.html
Nice overview of Europa, with helpful links to other sites

http://opfm.jpl.nasa.gov/europajupitersystemmissionejsm
Home page for the Europa Jupiter System Mission (EJSM), scheduled to launch orbiters destined for Europa and Ganymede in 2020.

http://photojournal.jpl.nasa.gov/targetFamily/Saturn?subselect=Target%3AEn-celadus%3A Over 200 models and images of Enceladus, including the latest from the Cassini orbiter.

http://solarsystem.nasa.gov/planets/profile.cfm?Object=Sat_Enceladus
Brief fact sheet on Enceladus, including interactive comparison of Enceladus with other bodies in the Solar System.

# 9 Fire and Ice

## *Life at the interface of volcanic heat and frigid ground on a violent, sulfur-rich world like Io*

Linda Morabito was four years out of college with a degree in astronomy when she became a navigation engineer for the Voyager mission. On 5 March 1979, Voyager 1 made its closest approach to Io, taking pictures that revealed a world utterly unlike any yet seen in the Solar System (Figure 9.1a).

Linda's job was to locate the position of Voyager's target planets and moons as precisely as possible against the background of stars as seen from the spacecraft, to keep the craft on its proper path. On the morning of March 9, she was enhancing the contrast of a picture taken the previous day, to better see the faint stars behind Io, when a crescent-shaped object at the upper left margin of the moon came into view (Figure 9.1b). Knowing such a large object could not be another moon in that position, she began to realize she might be seeing something no one had ever seen before.

Recounting the discovery years later, the thrill had not abated. "It seemed unbelievable that something that big had not been visible before ... It was a moment that every astronomer, every planetary scientist lives for ... Those moments were the stuff of dreams."

For the next six hours, Linda and her colleagues went through every logical explanation for the anomalous image, until only one possibility remained: the crescent shaped object, as well as the bright spot at the terminator (boundary between light and dark) must be due to surface events on Io itself. Months earlier, Stanton Peale and his colleagues at the University of California, Santa Barbara, had published a prediction that Io should be volcanically active. It gradually became clear that what Linda had spotted was the misty fountain of a massive volcanic eruption.

That night at dinner with her parents, her father said, "Do you realize you may have discovered the first volcanic activity outside the Earth?" The full impact of her discovery hit home. "It was wonderful to hear him say that" Linda recalled.

## 9.1 Nature of Io

Within a few days, the discovery first enunciated by Linda Morabito's father had been confirmed, and was published in *Science* on 1 June 1979 [1]. For many

L.N. Irwin and D. Schulze-Makuch, *Cosmic Biology: How Life Could Evolve on Other Worlds*,
Springer Praxis Books, DOI 10.1007/978-1-4419-1647-1_9,
© Springer Science+Business Media, LLC 2011

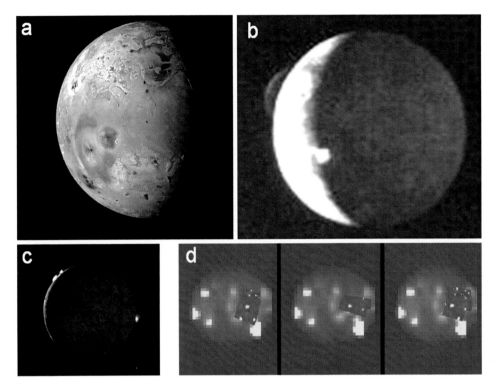

**Figure 9.1** Io. The nearest major moon to Jupiter is totally unique in the Solar System. (a) Io has a young and dynamic surface, speckled with multicolored compounds of sulfur emitted by persistent volcanic eruptions. (b) Evidence for volcanoes on Io, as first observed by Linda Morabito on 9 March 1979 in an image taken from Voyager 1 the previous day. (c) Further evidence for volcanic activity was seen in this image of flares at the upper left and lower right margins of the nighttime face of Io taken on 10 July 1979. d. These nighttime infrared images sensitive to heat reveal the multiplicity of volcanoes on Io (NASA/JPL).

planetary scientists, it represents the single most dramatic discovery of the Voyager mission. To this day, volcanism is *the* story on Io. Any process or phenomenon relating to that satellite, including its astrobiological potential, starts with the realization that Io is the most volcanically active planetary body known to humans in the universe.

### 9.1.1 Geology
Io is a victim of circumstance. Due to its accretion so close to Jupiter, and its placement between Jupiter and three sizable siblings, Io has been pummeled by an excess of meteorites, bathed by powerful magnetic fields, and wracked with gravitational stresses over its existence to a degree known by no other body yet observed. Io probably foreshadows what will ultimately be revealed about exoplanets orbiting close to their central stars. For now, its daily throes

are the ones we can observe at a range close enough to test the limits of life's versatility.

### 9.1.1.1 Interior
Io is the densest of all the major moons of the Solar System. Thus, of all the satellites, Io is most like a rocky planet. It has an iron or iron-silicate core that extends from 1/3 to 1/2 of its radius. Much but not all the remainder is made up of silicates like the rocky planets. Its density of 3.53 $g/cm^3$ is considerably less than Earth's 5.52 $g/cm^3$, though, indicating that a lot of liquid exists within or below the crust, estimated to be as much as 11 kilometers thick.

Three sources of energy are responsible for the heat that Io's interior gives off. First, the core is large enough that radioactive decay may still be generating thermal energy. Secondly, heat has accumulated from bolide impacts since Io's origin. Being the closest of the four major moons to Jupiter, it lies nearest the end of the "funnel" that draws infalling bodies toward a collision with Jupiter. Io has therefore likely intercepted more than its share of impactors. The highly cratered surface of Callisto, the most distant of the Jovian moons, indicates how much heavier the bombardment of Io must have been over its history.

By far the largest heat generator on Io is the gravitational flexing to which its position between giant Jupiter and its three siblings subject it (Figure 9.2). When Io is on the opposite side of Jupiter from the other moons, all the gravitational force is toward them. But when Io lines up with Jupiter on one side, and its siblings (especially Europa and Ganymede) on the other, it gets stretched in both directions. Expansions and contractions of the rigid crust have been estimated to be as great as 100 meters in amplitude. Since Io orbits Jupiter every 1.77 days, in a 2:1 resonance with Europa's orbital period of 3.55 days, this means the surface could be pumped up, then deflated, by crustal "tides" about every 43 hours (discounting the effect of Ganymede). The internal friction caused by these fluctuations keep the interior in a constant state of heated agitation.

### 9.1.1.2 Volcanism
Some evidence of volcanic activity is always present on Io. At any given time, one to two dozen active eruptions big enough to be seen from orbit appear to be underway (Figure 9.1d).

The dramatic plumes, like the one first detected by Linda Morabito (Figure 9.1b), are ejected with great force, at velocities of around a kilometer per second. This causes fountains of sulfur, sulfur dioxide ($SO_2$), and other volatile gases to rise as high as 300 kilometers before falling to the ground in concentric rings like the red circle around Pele seen in Figure 9.1a. These are thought to be generated by high subsurface pressures which keep $SO_2$ and other volatile gases in the liquid state until a rupture in the surface provides an outlet.

Plumes are also caused by the slow advance of lava across snow fields of $SO_2$. The frigid temperatures cause erupting lava to harden quickly at the surface, causing the liquid edge of the eruption to migrate slowly. Models suggest that the underground reservoirs of magma may be "rootless," or without a single

**Figure 9.2** Jupiter and its satellites. Io is the innermost of the four major Galilean moons. (a) When the other three moons (Europa, Ganymede, and Callisto, in that order) are on the far side of Jupiter, their gravity combines with Jupiter's to pull on Io in the same direction. (b) When all four moons are on the same side of Jupiter, Io is pulled by Jupiter's strong gravity field in one direction, but toward its sibling satellites in the other. Most of the time, Io is located at a position intermediate between these two rare alignments. The result is unremitting gravitational flexing. The moons are shown disproportionately large to Jupiter, but approximately in correct relative size to and distance from one another (NASA/JPL).

conduit to the surface. Rather, very hot lava melts the crust over which it flows, forming a 'boundary-layer slurry' that vaporizes as the erupting seam moves slowly across $SO_2$ snow fields, giving rise to "wandering" plumes [2]. The spreading lava eventually creates vast lava fields, crusted over lava lakes, and calderas resulting from sapping from the huge underground reservoirs. (Figure 9.3).

The high temperatures of close to 1700°C observed for these eruptions are consistent with magnesium-rich silicates, in addition to an abundance of sulfur compounds. The volume of erupted material is huge. The Amirani lava field, which stretches for over 300 kilometers (Figure 9.7), has been generated by lava disgorged at an estimated rate of 100 tons per second. The relative absence of large volcanic mountains on Io suggests that the lava is low in viscosity, consistent with a high liquid content.

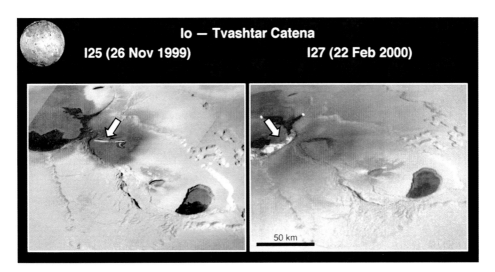

**Figure 9.3** Eruptions, calderas and sapping on Io. Volcanism on Io largely consists of ongoing emissions from seams at the edge of growing lava fields, gradually draining underground reservoirs of magma. This giant caldera is seven times larger than any found on Earth. At the lower right, a dark pool of crusted-over lava lies at the base of a broad mesa, whose collapsed margins give evidence of "sapping," or outflow of liquid from underground reservoirs. An erupting seam (white arrow) shows changes in position and extent over an 88-day period (NASA/JPL/University of Arizona).

### 9.1.1.3 Geochemistry

Io's atmosphere is very tenuous, consisting mainly of $SO_2$, with traces of sulfur monoxide (SO), hydrogen sulfide ($H_2S$), carbon monoxide (CO), and carbon dioxide ($CO_2$). It isn't known whether atmospheric $SO_2$ comes primarily from vaporization of surface $SO_2$ , or from volcanic emissions directly.

In Io's ionosphere are found analogues of those compounds, such as $O^{++}$, $O^+$ and $S^{++}$, $S^+$, and $SO_2^+$, probably generated by the ability of Jupiter's intense magnetic field to strip electrons away from their parent molecules.

Io's surface consists of a mixture of silicate and sulfur compounds, much of which is covered with a blanket of condensed $SO_2$. Io's technicolored surface is thought to be attributable to various compounds of sulfur. Much of the white coloring of the surface is due to condensed $SO_2$

The ices in the crust consist of frozen $SO_2$ primarily. Laboratory simulations that mimic infrared spectral data from the surface of Io are consistent with about 3% $H_2S$ and 0.1% $H_2O$ embedded as clusters in the frozen $SO_2$ [3]. Large reservoirs of $SO_2$ with some $H_2S$ must exist in liquid form under pressure beneath the surface.

### 9.1.1.4 Topology

Despite the unquestioned heavy bombardment that Io has been subjected to, its surface is perpetually young because it continues to be resurfaced by volcanic

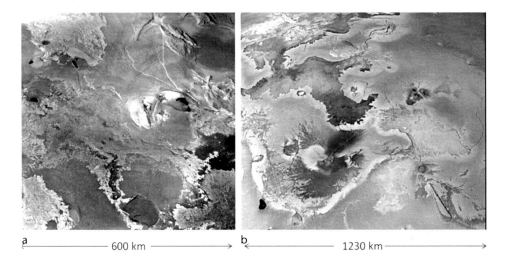

a ─────────── 600 km ───────────→   b ←─────────── 1230 km ───────────→

**Figure 9.4** Geological complexity of Io. From a distance, a mixture of complex geological features are revealed. (a) This image of the equatorial region shows mountains (center right), broad smooth plains, long relatively linear channels, and rough terrain indicative of erosion or other dynamic activity. (b) Vast lava plains reminiscent of basalt flood plains on Earth or the volcanic mare on the Moon cover much of Io. The lava lake to the upper left is about the size of Lake Superior (NASA/JPL/University of Arizona).

activity. No planetary body in the Solar System has had as many makeovers as frequently as Io.

Sharp boundaries occur everywhere. On a broad scale, wide flat plains are bisected by huge lava fields, interrupted by multi-terraced calderas, and cut by thin channels (Figure 9.4). At higher resolution, recently paved areas are superimposed on older, solidified flows, sometimes abutting against flat, rippled plains crumpled by unknown processes (Figure 9.5).

Except for a few shield volcanoes, the mountains on Io are not volcanic in origin. They result from uplifting along thrust fault lines due to compressional pressures in the interior. Jagged blocks of crust rise up from surrounding flat plains (Figure 9.6). Gorges slicing through the middle of these mountains are thought to provide added evidence for these fault lines, though erosional forces from flowing lave may also contribute.

Evidence of sapping is extensive (Figure 9.3). Long channels are also seen, created perhaps by sapping or by flowing liquid. Since no liquids would be stable at the temperature and pressure on Io's surface, any liquid flows would have to be primarily lava.

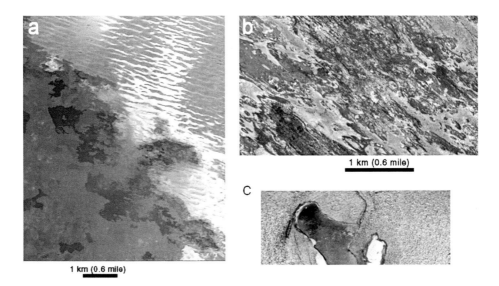

**Figure 9.5** Surface of Io at close range. Higher resolution images reveal surface complexity at a local level. (a) Part of the lava field produced by the Prometheus volcano. Younger (dark) areas probably represent more recently-emplaced lava, as older regions become covered with white sulfur dioxide fallout. The lava field abuts a rippled basin suggestive of liquid action on or beneath the surface, or compressive forces leading to crumpling. The ripples are coated on their lava-facing sides with what is thought to be condensed sulfur dioxide. (b) Elevated patches of ground coated with presumed sulfur dioxide condensates. At 5.2 meter (18 feet) per pixel, this is one of the highest-resolution images ever taken of the surface. (c) The bright white patch at the lower center of this image is probably a frozen-over lake of sulfur dioxide (NASA/JPL/ University of Arizona).

## 9.1.2 Thermal environment

The inherent surface temperature away from regions of volcanic activity on Io is around 90°K (-183°C). This very low temperature reflects Io's distance from the Sun and its lack of a substantial atmosphere.

Because of volcanic activity, however, hot spots abound on Io's surface, and temperatures rise in concentric gradients toward them (Figure 9.8a). At Io's most vigorous volcano, Loki Patera, temperatures reach almost 0°C, the melting point for water ice, at the perimeter of an oval 50 kilometers long (Figure 9.8b). Within a few kilometers of the hot spot, the surface temperature is up to 47°C, and at the point of magma eruption itself, the temperature may reach 2000°C.

The thermal gradients thus created at the surface may be fairly extensive, stretching out over a considerable distance; or they may be quite compressed. At Loki Patera, for example, the surface changes from 280°K (7°C) to 320°K (47°C) over a distance of about 32 kilometers (20 miles) to the left of the peak. The same transition occurs much more sharply, over a distance of about 1 kilometer, to the right of the peak.

a

b

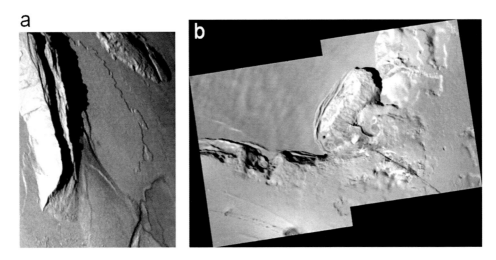

**Figure 9.6** Mountains on Io. Young uplifts along numerous fault lines course across the surface of Io. (a) Mongibello Mons rises as a jagged ridge to a height of 7 km above the plain. (b) A cliff 1–2 km high shows gravitational collapse along its western (left) edge (NASA/JPL/University of Arizona).

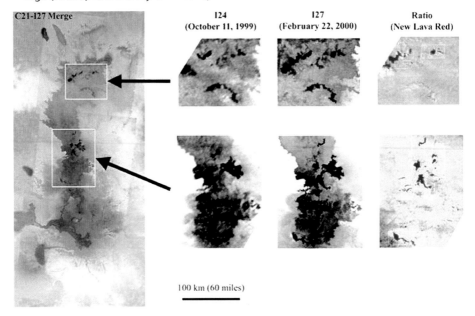

**Figure 9.7** Amirani lava field. The Amirani field is the largest currently active lava flow known in the Solar System, stretching for a distance of 300 km. Images taken 134 days apart reveal 24 areas with newly erupted lava, covering 620 square km. During a comparable period, Kilauea volcano on the Big Island of Hawaii covered 10 square km. The new flows (shown in red by contrasting the October and February images) required the eruption of about 100 tons of lava per second (NASA/JPL/University of Arizona).

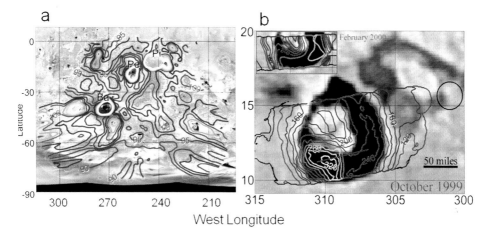

**Figure 9.8** Hot spots on Io. Volcanic activity creates thermal gradients. (a) This large scale view of 30% of the southern hemisphere plots isothermal temperatures based on infrared radiation. The surface away from hot spots, along the purple lines, registers 90°K (-183°C). Temperatures rise sharply in concentric circles toward the center of volcanoes Babbar (Ba), Pele (Pe), Pillan (Pi), and others. (b) At higher resolution around Loki Patera, the temperatures can be seen to reach 280°K (almost 0°C) at the perimeter of an oval 50 km long, and 320°K (47°C) within a few km of the hot spot, which itself exceeds 2000°C (NASA/JPL/University of Arizona).

Similar gradients must occur in the vertical direction underground, but they haven't been measured.

As lava that spills out of the ground at 1500°C or more and moves across a snow field of frozen $SO_2$ at -180°C, sharp thermal gradients will be generated. These movements have been measured to cover several centimeters (a few inches) a day. Thus the gradient creeps along. The lava fields may be quite long lasting. Once the surface hardens and provides some insulation for the molten lava that continues to spread out beneath it, the ground over which it flows will be heated, and the margins of the field will maintain a significant thermal gradient for some time. Both horizontal and vertical gradients will show an ever shifting profile.

### 9.1.3 Radiation environment

Io orbits within the magnetosphere of Jupiter. This creates, then strips ions from Io's ionosphere [4]. Jupiter's rapid rotation, with the correspondingly rapid rotation of its gravitational field, creates powerful currents. These have been estimated at 1,000 gigawatts, generating a potential electromotive force of 400 kilovolts.

The ions thus created, like $O^{++}$, $O^+ S^{++}$, and $S^+$, are strongly oxidizing and thus could contribute to oxidation-reduction cycles, in principle. However, most of the ions are probably found in the ionosphere, from which they are torn away by Jupiter's rotating magnetosphere at a high rate, so their contribution to oxidation-reduction cycles at the surface is questionable.

Electromagnetism and electricity represent potential energy sources, probably to a greater degree on Io than anywhere in the Solar System other than the atmospheres of the gas giant planets themselves.

## 9.2 Planetary history of Io

The accretion of the Jovian satellites from the protoplanetary disk that formed the Jovian system was recounted in chapter 8 (Figure 8.2). Because of its nearness to Jupiter, Io is the most distinctive of the four major moons.

Its mix of metals and silicates must have congealed from similar material that made up a much more massive core at the center of Jupiter. By virtue of its position so close to Jupiter and inside the orbits of the other, relatively close satellites, gravitational flexing would have dominated the character of Io from the beginning.

Io's core was too small to hold on to the hydrogen and helium that got drawn quickly into Jupiter's much larger gravitational field. Io's core would have been large enough to hold onto water, however, and it may well have had water briefly during the earliest periods of its existence [5]. Most likely, the water was lost rather soon due to sublimation, vaporization from constant volcanic heating, and heavy bombardment of bolides drawn into Jupiter's very close and dense gravity well.

Despite the constant outgassing of volatiles from the interior, an atmosphere would have been difficult to sustain from early in Io's existence, again because of the fierce magnetic and radiation environment, and high rate of bolide impacts. Indeed, today only a very thin atmosphere of $SO_2$, with trace amounts of other volatiles, is sustained. Lacking the thermal insulation of an atmosphere, Io grew very cold (about $-180°C$) at its surface early on. Roiled by the friction from gravitational flexing, however, the interior has continued to generate heat and conduct it to the surface through local and ever-changing outlets.

There is no reason to doubt that Io's technicolor splendor long predated the evolution of the human eye to perceive it. Indeed, Io has most likely been a volcanically convulsive body, remaking its surface constantly in the recurrent advance and retreat of the fiery heat of lava across an icy substrate of silicates and frozen $SO_2$ for four billion years or more.

## 9.3 Conditions for life on Io

The juxtaposition of fire and ice is more than a metaphor for Io. The very essence of what makes life conceivable there is the interplay between heat and cold – the clash of extremes that creates a boundary, ranging in thickness from millimeters to kilometers, between a frozen substrate too cold for liquid of any kind, and a blanket of spreading heat that makes retention of almost any liquid impossible. It is there, in the zone between fiery heat and frozen substrate that some form of

a

b

c

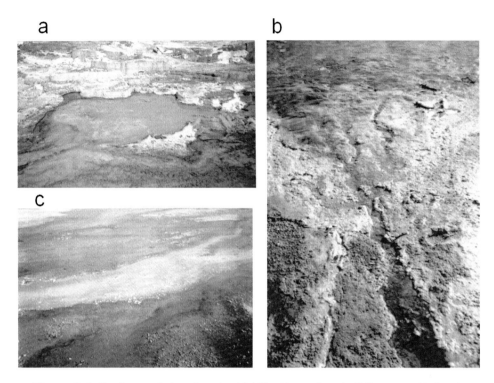

**Figure 9.9** Gradient variations in microbial life. Hot springs in Yellowstone National Park, Wyoming, USA, provide dramatic examples of microbial adaptation to narrowly constrained environmental conditions. (a) Travertine, a white deposit of calcium carbonate, hosts colonies of thermophyllic *Hydrogenobaculum* surrounding a hot pool paved on the bottom with a heat tolerant bacterium like *Thermus aquaticus*. (b) *Cyonidium*, an acid-tolerant thermophile lines a narrow channel draining a hot pool. An entirely different population of less heat-tolerant organisms grows above the water line. (c) *Sulfolobus*, an anaerobic extremophile, lines the bottom of a hot streambed, with maroon mats of *Zygogonium* growing at the cooler margins (Photographs by Louis Irwin, covering substrates about 4 m (a), 2m (b) and 1 m (c) in width).

life just might be able to thrive for the short span of time that a transient compromise between fire and ice is found.

Life thrives at boundaries: between surface and air, sea and sea shore, ocean and ocean bottom. And extreme life thrives at boundaries between extremes. Yellowstone National Park provides a high concentration of compelling examples, though similar sites are scattered across the Earth. The geysers, fumaroles, and heated pools found at Yellowstone in abundance provide one instance after another of water too hot, too acidic, or too alkaline for the survival of organisms used to more moderate conditions. Yet some extremophiles thrive in those very conditions – can't live outside them, in fact. Just a few centimeters away from the scalding stream or pool of water, extremophiles are replaced by other forms of life compatible with the more moderate conditions above or

**Figure 9.10** Surface and subsurface microbial gradients. Yellowstone National Park provides examples of gradients both above and below ground. (a) White mats of thermophiles line the center of a flowing stream of heated water, while acidophilic but non-thermophillic, maroon *Zygogonium* lines the cooler edges of the streambed. Further to the side, above the water line, different groups of grey-brown microbes thrive. (b) This vertical cut along an eroded cliff reveals vertical layering of microbes due to changing conditions at different depths beneath the surface. Stream bed (a) is about 2 m wide and cliff cut (b) is about 1 m high. (Photographs by Louis Irwin).

beyond the water line (Figure 9.9). Often, there are several gradations of life within a stretch of a meter or two (Figure 9.9c and 9.10a).

Anywhere that gradients can be established, a given form of life seems able to find its proper place within the gradient. This applies to vertical as well as horizontal transitions, as a cliff cut through successive layers of soil reveals (Figure 9.10b).

That life could exist on Io is not a widely-held view. The opinion of Bruce Jakosky, a planetary scientist at the University of Colorado, is typical. "The most reasonable conclusion is that . . . there is neither water nor life on Io," he wrote in 1998 [6].

There is now evidence for a small amount of water on Io, though admittedly it's a trace only, and hardly seems enough to support water-based life. Could any other solvent exist somewhere in the ever-shifting gradients found on Io? Can the other conditions for life be found there sufficient to question the conventional wisdom that life on Io is unlikely? Let's consider the possibilities.

### 9.3.1 Solvents for life on Io
Sulfur dioxide and hydrogen sulfide are both highly soluble in water. Thus, in principle, strong solutions of those gases (essentially dilute sulfurous and sulfuric acid) could lower the melting point of water sufficiently for it to be a liquid

beneath the ground at the margins of hot spots or lava flows. The very small amount of water detectable on the surface, however, suggests that water is an unlikely solvent for anything on Io.

Sulfur dioxide is a more promising candidate. It is abundant in the frozen state on the surface, but would be liquid beneath the ground at temperatures between -75°C and -10°C. This corresponds approximately to the 200°K to 260°K isobars in Figure 9.8b, encompassing hundreds of square miles – essentially all of the black ground in the image – on the right hand slope of Loki Patera. While the low vapor pressure at the surface would result in immediate vaporization, underground $SO_2$ would remain liquid under pressure in stable reservoirs.

Hydrogen sulfide is also theoretically possible as a solvent, with a temperature range for liquidity of -85°C to -60°C [10]. So it too could be a liquid underground at hot spot margins. Its abundance on the surface, however, is so much lower than that of $SO_2$ that the latter seems more likely to be a widespread solvent.

Io could present a case in which the solvent in the ambient environment differs from the liquid medium in which metabolic processes take place. $SO_2$ is not an ideal solvent for hosting biochemical interactions because it doesn't form hydrogen bonds readily and isn't that conducive for the easy changes in structural conformation that macromolecules need to undergo to carry out their functions. $H_2S$ does have those properties, but is much less abundant. However, its range of liquidity (-60°C to -85°C) overlaps with that of $SO_2$. Therefore, living cells conceivably could actively absorb $H_2S$ and make use of it as their intracellular solvent, while subsisting in an environment of primarily liquid $SO_2$.

### 9.3.2 Chemical building blocks for life on Io

It would appear, then, that whatever could serve as a building block for life on Io would have to be able to form complex biomolecules that could sustain metabolic interactions in liquid $SO_2$ or $H_2S$ in the vicinity of –50° to –70°C. The biochemistry of carbon-based polymers under conditions so far removed from the water-based biochemistry at higher temperatures with which we are familiar is not known well enough to evaluate the possibility that life on Io could be built from compounds with a carbon backbone. There is as yet no evidence for organic compounds, and relatively little carbon of any kind other than trace amounts of CO and $CO_2$, in the atmosphere and in the frozen substrate of Io.

What is very abundant, of course, is sulfur. One of sulfur's most attractive features as a potential building block for biomolecules is the large number of oxidation states it can assume, including some that are even fractional: +7, +5, +4, +3⅓, +2½, +2, –0.4, –½, –⅔, –1. This gives sulfur the capability of forming a great variety of compounds. Sulfur forms a variety of ring structures, some accounting for the vivid coloration of Io's surface. Perhaps in the frigid but liquid solvents of $SO_2$ and $H_2S$ beneath the surface, chemical complexity based on sulfur rings or polymers could arise.

Mixed polymers of carbon, sulfur, nitrogen, oxygen, and phosphorus could form biomolecules, as they do in living cells on Earth. We have no evidence that they do, however, and little empirical information on what the possibilities would even be under Ionian conditions. As in many aspects of astrobiology, empirical research in chemical laboratories on Earth would greatly help narrow the possibilities.

### 9.3.2 Energy for life on Io

Heat would appear to overshadow all other forms of energy on Io, partly because its flux is so visible. Not only is it available in abundance, the frigid ambient temperatures with which it contrasts provide a readily available sink, ensuring that heat will flow from its source. And as pointed out in Chapter 2, energy flowing through a system tends to organize the system. Though heat is easily dissipated as increased entropy, and therefore is very inefficient, it is so readily available on Io that life could make use of it there nonetheless.

Another obvious source of energy is oxidation-reduction chemistry. The most visible and obvious element on Io, sulfur, exists in an oxidized form as $SO_2$, in a neutral form as S and $S_x$, and in a reduced form as $H_2S$. The potential for oxidation reduction cycling thus exists, in theory. The thermodynamic reality – whether oxidation and reduction reactions can actually take place in a way that supplies energy – and whether mechanisms have evolved that can be coupled with those reactions to capture that energy in a biologically usable form, may be another matter. Again, we fall back on the inventiveness of evolution: organisms evolve under the circumstances available to them, and the history of life on Earth shows that evolution is at least as inventive as our imaginations.

A third source of energy needs to be considered seriously for the first time in the case of Io. That is electromagnetism. As noted earlier, Io circumvents Jupiter within that giant planet's magnetosphere. While the amount of energy obtainable from electromagnetism is generally non-competitive with chemical or light energy on Earth [7], the magnetic field at Jupiter is at least 12 times greater than on Earth, and sunlight is 25 times dimmer, so magnetism could power an appropriately configured biological process on Io. Electricity generated from the constant rotation of Jupiter's magnetosphere through Io's orbital path could also be tapped. Organisms have evolved on Earth to detect both magnetic fields and electric currents. Since sensation is simply the detection of energy fluctuations in a biologically meaningful way, a logical corollary is that energy fluctuations can affect biological processes. To the extent that energy can be coupled to a mechanism for storing it, that energy can in principle be used to power living systems.

### 9.4 Origin of life on Io

Unlike the other planetary bodies we've considered – Mars, Venus, Europa – where the origin of life was easier to envision than its persistence, Io presents us with a formidable challenge. It's not that hard to see how a contemporary microbe, for instance, could lie in the frozen substrate in cryptobiotic hibernation till a flow of lava overhead liquefies its subsurface surroundings and brings it to life. It is, however, rather harder to see how that microbe could have evolved in the first place.

The problem is that we know almost nothing about the earliest stages of Io's existence. What we do know leads us to doubt that it could have held much water for very long, or had an atmosphere for more than a brief period, before the incessant pounding of incoming bolides and the huge distortions of gravitational flexing came to dominate its nature. We are tempted to think that it must have become too dry, too cold, too fast for the slow clockwork we associate with life's origins in water to play out. Of course, we don't really know how long it takes for life to get going in an abstract sense, and we don't know whether the conditions that would have made life's origin possible existed for long enough on Io.

So what if water was short-lived, or maybe never even there? Maybe life sprung forth in cold pools of $SO_2$ that could have been scattered over the surface under a tenuous atmosphere very early in Io's existence. We can't deny the possibility, but we have neither evidence nor convincing models that support it. Since we don't even know how life originated on Earth, perhaps the admission that we don't know how it could have started on Io should not be viewed as a fatal flaw. And since this book is not about what *was*, but about what *could have been*, we will content ourselves for now with the assumption that life may indeed have formed on Io, despite the chaos and violence that brought the satellite itself into being.

## 9.5 Habitats for life on Io

Any surface habitats that would have been suitable for life on Io were almost surely short-lived. Since Io and Europa are both assumed to have formed at a temperature of -23°C or higher [8], the surface could have held pools of liquid $SO_2$ and $H_2S$, long after any water froze completely, assuming a sufficient atmosphere was present to slow evaporation. Those pools may have lasted just long enough for life to form in them, perhaps on the substrate at the bottom, or in small puddles where evaporation concentrated critical reactants. As soon as the atmosphere was stripped by Jupiter's powerful magnetosphere or pounded away by early bombardment, however, the pools would have evaporated or frozen as Io settled toward its global average temperature on the surface of about -180°C.

Beneath the surface, where the temperature is higher and the pressure greater, certain strata would persist at a temperature in the -10°C to -75°C range, conducive for keeping $SO_2$ in the liquid state. In the overlying crust, silicates mixed with frozen $SO_2$ and other volatiles created a hard, cold ground. Periodically, moving sheets of lava would flow across the frozen ground, melting it temporarily and turning it into a slushy mix of minerals and solvents. Once the lava solidified and its heat was dissipated – a process that could take from minutes to centuries, depending on the volume of lava overhead – the ground would revert to its frozen state.

Io is basically a dual-habitat world, then: one consisting of more-or-less permanent, underground pools of mostly $SO_2$; the other composed of transient slurries of soil and liquid, melted at the interface of extreme heat and cold anywhere on the globe where hot spots or spreading lava abut frozen ground, but destined sooner or later to revert to frozen solidity. We don't know how long these two habitats have been the only ones there, but we think it most likely that they have been the only ones realistically capable of hosting life on Io almost from the beginning. Our speculative trajectory for evolution on Io therefore assumes a single geological state, invariant over time from the origin of life on that planetary body.

## 9.6 A possible evolutionary history for life on Io

If life has ever existed in Io, it could have been transported there from another point of origin – most likely Europa, since that is the closest and most probable body in the Jovian system to have spawned any form of life. The chances of that don't seem good. Some type of organism conceivably could have been imbedded in the icy crust of Europa, then blasted off the satellite by a meteorite. But even if the chunk of ice were big enough to avoid being melted by the blast that sent it flying (which seems unlikely), the Europan organism upon arrival at Io would find an environment so alien from its point of departure, that survival in the land of fire and ice seems highly improbable.

For the sake of this speculative history, we will therefore assume that life arose *in situ* on Io, in or beneath a pool of liquid $SO_2$, perhaps in the presence of enough $H_2S$ to facilitate increasingly complex, sequestered chemical interactions to form the protocellular origins of life. These primitive steps toward the living state would most likely have made use of whatever compounds they found concentrated in ice inclusions or adhering to mineral substrates, for the building blocks and energy they would need to form boundaries and make the larger molecules necessary for becoming fully alive. They would, in other words, have been ancestral chemoheterotrophs. From them, the rest of life on Io could have evolved, in something like the following steps, as diagramed in Figure 9.11.

It would be a reasonable assumption that fully formed and reliably replicating chemotrophs could have emerged both on substrates beneath the liquid *(step 1)*, and within microinclusions in the overlying or surrounding $SO_2$ ice *(step 2)*. Given the abundance of thermal energy on Io, it would only be prudent to suppose that this abundant energy source would likewise have been tapped in the evolution of an ancestral substrate thermotroph *(step 3)*. Free-floating chemotrophs *(step 4)* would probably have evolved from ancestors encased in the ice inclusions.

Though we have noted several times previously that heat is not an efficient basis for bioenergetics, its abundant availability at the margins of heat sources like spreading lava fields makes us seriously consider a branch of life on Io that could have evolved to depend on it. Free-floating thermotrophs *(step 5)* would

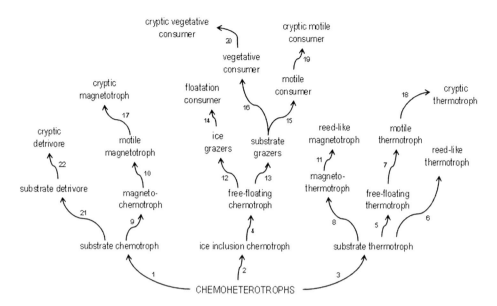

**Figure 9.11** Possible evolutionary trajectory for life on Io. Life is assumed to have originated as protocellular chemoheterotrophs that evolved in surface or underground pools of liquid sulfur dioxide ($SO_2$) into (1) chemotrophs on mineral substrates or (2) chemotrophs within ice inclusions. (3) Simple substrate thermotrophs may also have been early descendants of protocellular life. For subsequent evolutionary transitions, see text.

have been able to absorb heat from a source, like the substrate, then harvest its energy by floating to a cooler level of liquid where the outflow of heat could be coupled to chemical reactions that captured the energy. An alternative mechanism can be visualized in which a long reed-like thermotroph *(step 6)* harvests energy from the flow of a heat source at its anchor point in the substrate to the sink at the cooler end of the elongated structure. Eventually, the free-floating thermotroph could have evolved the ability to propel itself and become a motile thermotroph *(step 7)*.

The presence of the strongest magnetosphere outside the Sun in the Solar System may have presented a compelling source of energy as well. The earliest organisms to evolve the ability to use magnetic energy were surely transitional, like a magneto-thermotroph *(step 8)* and a magneto-chemotroph *(step 9)*. These, in turn, could have evolved into fully magnetotrophic forms. A motile magnetroph *(step 10)*, harvesting energy through cellular specializations that separated charges to create energy-yielding gradients, like the thylakoid membranes in plant chloroplasts that sequester electrons split from water with energy from light. A variation on this theme would have been reed-like magnetotrophs *(step 11)*, powered by electrical currents induced by the Lorentz force over the length of their elongated structures.

With a variety of producers now available for food, consumers could have evolved to feed on each type. Ice grazers *(step 12)* and substrate grazers *(step 13)*, feeding on producers growing on the ice and substrate, respectively, would probably have been among the earliest in this category.

Another step up the food chain could then have been taken by the emergence of floatation consumers *(step 14)*, then more active motile consumers *(step 15)* feeding on whatever they could catch, and vegetative consumers *(step 16)*, devoted to a diet of whatever reed- or plant-like organisms had evolved by then.

On the dynamic world that Io has always been, life would have been able to thrive in those underground lakes of liquid, and surge into the transient slurries of substrate, only so long as the temperatures remained warm enough to prevent them from freezing. Inevitably the freezing would come, though. Hence, from an early stage in the history of life on Io, selective pressure would have favored the evolution of cryptic forms, able to hibernate in suspended animation for long periods of time, till the next wave of lava or a new nearby fissure in the crust marked the arrival of rejuvenating heat. Hence cryptic versions of magneto-trophs *(step 17)*, thermotrophs *(step 18)*, motile consumers *(step 19)*, and vegetative consumers *(step 20)* are very likely to have evolved.

As in all ecosystems, detrivores *(step 21)* would have made an early appearance, and cryptic detrivores *(step 22)* would also have evolved with all the other cryptobiotic forms.

## 9.7 Ecosystem possibilities for life on Io

Our absence of knowledge about the extent and longevity of underground lakes of liquid, either as pools in lava tube caves or as analogues to groundwater on Earth, where space and some habitat fractionation would enable the development of a moderately complex ecosystem, makes it difficult to speculate about the nature of that ecosystem. If liquid-filled underground strata are extensive and long-lasting, a multilevel trophic structure could exist. Figure 9.11 postulates 11 producers, five consumers (three of which could be at the secondary or higher trophic level), and two detrivores living together, as a minimal degree of biodiversity that could be found on Io. It could, of course, be much more extensive than that. Or, if energy is simply not extractable or biochemistry really won't work in that environment, biodiversity and ecosystem complexity could be overestimated even by the simplifying scheme in Figure 9.11.

The alternative habitat, consisting of cryptic forms embedded in substrate that comes to life only for transient periods, would be much simpler, like all ecosystems restricted to life in the interstitial spaces between granules of sand and soil. A few microbial producers and a smaller number of microbial consumers would be about the extent of life at the slushy margins of lava fields and hot spots.

As for all ecosystems, the total biomass would be a function of the amount of biologically useful energy that could be harvested from the environment. Energy is one thing that Io has plenty of, but we know so little about how effectively that energy can be used (if it is at all), that estimating biomass and inferring ecosystem complexity on Io is nothing more than a guess.

## 9.8 Characteristics of life on Io

Life on Io in underground pools of liquid $SO_2$ would have some of the characteristics of life in the subsurface ocean of Europa, where it is dark and most of the action occurs on the floor and at the ceiling of liquid habitats. But there the similarities end. The solvent is $SO_2$ (with $H_2S$ possibly playing an intracellular role) rather than $H_2O$; therefore the macromolecular structures and biochemistry are bound to be very different from the water-based systems considered so far. The temperature is much colder. And the energy mix is probably quite different. But above all, there is an *urgency* to life on Io that is totally absent on Europa, in that life only has a narrow thermal window in which to flourish, and that window opens and closes very episodically and unpredictably.

### 9.8.1 Radiation resistance

Particle emissions from Io's surface interact with Jupiter's powerful magneto-sphere to form a plasma radiation environment that would be deadly to the most radiation-resistant organism on Earth. Therefore, for any form of life based on macromolecular integrity, a robust resistance to damaging radiation would need to exist. Some insight into how this could be achieved can perhaps be gained by looking at radiation-resistant organisms on Earth.

One of the best-known radiation-resistant microbes is *Deinococcus radiodurans*. It has apparently evolved a resistance to extreme radioactivity and ultraviolet radiation, genotoxic chemicals, heat, and extreme kinetic forces as a side-effect to natural selection for desiccation resistance. *D. radiodurans* has an unusual tightly packed and laterally ordered DNA toroid morphology, which might contribute to its radioresistance and protection against dehydration. The DNA toroid is composed of four compartments, each containing one copy of DNA, restricted within the ordered DNA toroid. Further, the DNA is organized in a unique ring that prevents broken pieces of DNA generated by exposure to radiation from floating freely within the cytoplasm. Therefore, unlike other microorganisms, *D. radiodurans* is able to retain loose genetic information by keeping the severed DNA fragments tightly locked in the ring. The fragments, held close, eventually come back together in the correct, original order, reconstructing the DNA strands [9].

Whether organisms, if they exist on Io, contain DNA or anything like it is unknown. The point provided by this Earth analog of radiation resistance is simply that certain intracellular architectures can help protect against the molecular fragmentation that would be the greatest hazard from Io's powerful radiation environment.

### 9.8.2 Metabolism

Metabolic rates will be slow by any standard familiar to us because they can only occur at temperatures well below the freezing point of water. In fact, if $H_2S$ plays a prominent role in the intracellular life of Ionian organisms, metabolism is effectively restricted to the overlapping range of temperatures, -60°C to -75°C, at which $SO_2$ and $H_2S$ are both liquids. Narrow thermal ranges of activity are not a problem in principle – all the metabolism of birds and mammals is restricted to a narrower range than that. It does mean, though, that reaction rates are expected to be slower in general.

### 9.8.3 Growth and reproduction

Io presents the classic case for what ecologists call *r*-selected populations. That notation refers to a mathematical model for "opportunistic" population growth, in which the window for reproduction and growth is typically brief and mortality is high. Populations subjected to *r*-selective pressure reproduce quickly and grow rapidly as soon as favorable conditions arise. This is precisely the situation that would pertain when cryptic organisms are brought to life by a wave of warmth that suddenly liquefies their surroundings. While less obviously characteristic of longer-lasting underground pools, the erratic geological activity on Io could cause an entire reservoir to erupt and vaporize at any time. Thus, *r*-selective pressures probably operate in both habitat types.

Reproduction in *r*-selected populations is fast; the number of offspring is high, and their average size is small. Growth rates are rapid, a trait favored by natural selection in habitats that don't remain conducive for growth for very long.

These are theoretical predictions. The actual reproductive and growth patterns of real populations depend on additional factors, like space, available food supply, degree of competition, and ecosystem complexity – none of which can be known at this time for possible life on Io. It seems reasonable to expect that most of the life that might be found there is microbial, and therefore characterized by simple, rapid, non-sexual reproduction, but even that can't be said with confidence.

### 9.8.4 Motility

For habitats that are sizeable and persistent, like the larger underground reservoirs of liquid might be, the evolution of some motile forms at the secondary consumer or higher level is a reasonable prediction. A conservative view of energy actually extractable by organisms on Io is probably prudent. This would translate into a fairly small number of motile forms, for which the low temperatures at which they would operate would predict a fairly low level of motility. Most life on Io, like subsurface life everywhere except in deep liquid reservoirs, is more likely than not to be small and sedentary.

### 9.8.5 Sensory Systems

Sedentary organisms are typically keen on sensing the environment with which they are in direct contact. Hence touch, smell, and taste are important. For life on Earth, these sensory abilities are properties of individual cells, including single-celled microbes; so the simplest organisms on Io would likely have these capabilities. The senses for distance information, like sight and sound, are unimportant for an organism that isn't seeking anything and can't go anywhere. Motile organisms, on the other hand, inevitably evolve the ability to detect information about features or objects in their environment at a distance. Photoreception in the subsurface of Io would be useless, but sound (or vibration) and electromagnetic perception could be quite important for the few forms of life expected to be mobile.

### 9.8.6 Cognition

For a planetary body on which most scientists doubt that life exists at all, it may seem presumptuous to suggest the possibility that some organisms could actually be equipped with a modicum of intelligence. Certainly, the vast majority of the biomass that could reasonably be expected to exist on Io is bound to be more analogous to microbes or fungi than to anything with the slightest degree of behavior, much less cognitive capacity. But – making the best case scenario to emphasize the point – if semi-permanent pools of liquid host ecosystems dense and complex enough to include secondary or tertiary consumers big and mobile enough to require an integrative apparatus like a nervous system, a degree of cognitive ability cannot be ruled out. It might be well to remember that the largest brained animals on Earth today, and the most intelligent of all animals other than hominids, are marine mammals with poor eyesight but exquisite hearing.

### 9.9 Chapter summary

Io is the fourth largest moon in the Solar System, with a density just slightly less than that of Mars. But it looks more like a pizza than a planet, because of the rich variety of multicolored compounds of sulfur it emits from its unending volcanic convulsions. Orbiting between the giant pull of Jupiter and three sizable companions, it undergoes gravitational flexing to a degree matched nowhere else in our part of the universe.

This extraordinary volcanic character is the feature that dominates Io. Fountains of superheated sulfur compounds spew to a height of hundreds of kilometers above the surface, then fall back to it as snow. A surface temperature of $-180°C$ keeps even $SO_2$ and $H_2S$ frozen solid, except where heat flowing out from the interior melts them into subsurface pools of liquid. At any given time, a dozen or more volcanic hot spots are disgorging lava that vaporizes the frozen substrate at its advancing margins, generating a second, transient habitat of liquid $SO_2$ mixed with the mineralized crust.

These two unpredictable and ephemeral reservoirs of liquid $SO_2$ represent the most likely habitats for life, if it exists, on Io. $H_2S$ might be more likely as an intracellular solvent, and larger molecules would probably make greater use of sulfur, nitrogen, phosphorus, and other elements in constructing their back-bones. It would be a form of life consisting of chemistry, governed by metabolism, and displaying structures and functions quite alien to our experience. It would all be underground and able to thrive only when temperatures reached an appropriate narrow range. But that would happen, for at least a brief period, in local pools or over short stretches of ground, every time a hot spot erupted or a sheet of lava advanced.

Life on Io would be very much an *edge* effect – as it is everywhere, really. It's just that nowhere else in the Solar System are the edges more evident and ever changing. This places a premium on biological features that are opportunistic, like rapid reproduction and fast growth, coupled with the ability to survive in a cryptic state of suspended animation for long periods of time. Life on Io may well represent the pinnacle of evolution toward cryptobiotic forms in our Solar System.

Io is not a high priority on anyone's list of targets for further astrobiological investigation. It would be a mistake to disregard it completely, however. Science and logic support the possibility of life there, even if the prospects don't look good from our terracentric point of view. If life indeed is found there some day, it will enlarge our view of the nature of life considerably. With the possible exception of the next two worlds we will consider, no other place holds as much promise for surprising and delighting us, just as the first images of this pizza-colored world did for Linda Morabito and the rest of us in the Spring of 1979.

## 9.10 References and further reading

1   Morabito, L. A., Synnott, S. P., Kupferman, P. N., et al. 1979. Discovery of currently active extraterrestrial volcanism. *Science* **204**: 972.

2   Kieffer, S. W., Lopes-Gautier, R., McEwen, A., et al. 2000. Prometheus: Io's wandering plume. *Science* **288**: 1204–8.

3   Salama, F., Allamandola, L. J., Witteborn, F. C., et al. 1990. The 2.5–5.0 micrometers spectra of Io: evidence for $H_2S$ and $H_2O$ frozen in $SO_2$. *Icarus* **83**: 66–82.

4   Geissler, P.E., McEwen, A.S., Ip, W., Belton, M.J., Johnson, T.V., Smyth, W.H., and Ingersoll, A.P. (1999) Galileo imaging at atmospheric emissions from Io. *Science* **285**: 870–874.

5   Fanale, F. P., Johnson, T. V. and Matson, D. L. 1974. Io: a surface evaporite deposit? *Science* **186**: 922–925.

6   Jakosky, B. 1998. *The Search for Life on Other Planets*. Cambridge: Cambridge Univ. Press.

7   Schulze-Makuch, D. and Irwin, L. N. 2008. *Life in the Universe: Expectations and Constraints*. Berlin: Springer-Verlag, 2nd ed.

8   Consolmagno, G. J. and Lewis, J. 1976.Structural and thermal models of icy Galilean satellites. In: T. Gehrels (ed). *Jupiter*. Tucson: Univ. of Arizona Press; pp. 1035–1051.

9   Zahradka, K., Slade, D., Bailone, A., et al. 2006. Reassembly of shattered chromosomes in *Deinococcus radiodurans*. *Nature* **443**: 569–573.

10  Schulze-Makuch, D. 2010. Io: Is life possible between fire and ice? *J. Cosmol.* **5**: 912–919.

http://photojournal.jpl.nasa.gov/targetFamily/Jupiter?subselect=Target%3AIo%3A
Over 200 models and images of Io, based on numerous robotic missions to and flybys of the Solar System's most volcanically active world.

http://www.lpi.usra.edu/resources/outerp/io.html
Information about and images of Io, including a motion picture tour of its surface.

http://www.nineplanets.org/io.html
Nice overview of Io, with helpful links to other sites.

http://galileo.jpl.nasa.gov
Home page for the Galileo mission to Jupiter and its moons.

http://www.jpl.nasa.gov/galileo/sepo
Background information and educational context for the images taken from the Galileo spacecraft.

# 10 Petrolakes

*Life in pools of organic liquids as on Titan*

Saturn was the furthest planet from Earth in the known Solar System in 1655, and the crude telescopes available at the time revealed an enigma. At times, the planet looked like it had "handles" on either side, but at other times the handle-like bulges disappeared. Christiaan Huygens, born into a wealthy Dutch family of high status, was given an allowance by his father to do nothing but pursue his interest is mathematics, mechanics, optics, and astronomy (What a dad!). With a telescope much improved over earlier versions, he deciphered enough detail about far away Saturn to propose that the handles were actually rings encircling the planet, seen sometimes at an angle facing Earth, and at other times viewed edge-on from Earth to the point of becoming invisible.

In the course of gazing at Saturn's bulges over many evenings in succession, Huygens noticed that what he first took to be a bright star was actually moving ahead of, then in back of the planet, more or less in the same plane as its rings (Figure 10.1). He deduced correctly that what he was seeing was the first moon of a planet other than Earth and Jupiter. The moon, which he called *Luna Saturni* but which would later be named Titan, became the sixth known satellite of planets in the Solar System, after Earth's moon and the four moons of Jupiter discovered by Galileo.

Gerard Kuiper, another Dutch astronomer, was working on radar counter-measures in the United States during World War II, when he decided to take a break in the late winter of 1943 to work at the newly-built McDonald Observatory in the Davis Mountains of West Texas, not far from the small town of Marfa which would later become famous for occasional nighttime lights of mysterious origin. By luck one evening, several of the major satellites in the outer Solar System came into alignment, so he trained his spectroscope on each of them. Quite by surprise, an unmistakable signature for the gas methane showed up in the spectrum from Titan, the first evidence of an atmosphere on a moon. Later, the atmosphere of Titan was shown to consist mostly of nitrogen, making Titan the only other body in the Solar System with a nitrogen-rich atmosphere like Earth's. The added presence of methane, an organic compound and possible biosignature for life, heightened interest in Titan, and inspired the most complicated and expensive robotic expedition in the history of space exploration – the Cassini-Huygens mission to Saturn and Titan.

L.N. Irwin and D. Schulze-Makuch, *Cosmic Biology: How Life Could Evolve on Other Worlds*, Springer Praxis Books, DOI 10.1007/978-1-4419-1647-1_10, © Springer Science+Business Media, LLC 2011

**Figure 10.1** Saturn and satellites. This is probably a better view than that which Christiaan Huygens (inset) saw through his telescope in 1655, since the smaller moons, Dione and Rhea, would not be discovered for another couple of decades. (Photograph from Lee Jenkins [1], with permission).

## 10.1 Nature of Titan

Before the Cassini-Huygens mission took flight, Titan was visited by Voyager 1, which took a close-up photo during its passage through the Saturnian system in November, 1980. To say that this much-anticipated glimpse at Titan was a disappointment is an understatement. Even at the closest range, Titan was nothing but a featureless orange ball (Figure 10.2a), immersed in a haze so thick that its surface was totally obscured. A blue haze in the highest reaches of the atmosphere (Figure 10.2c) was consistent with the known presence of methane, while the smog below, it was assumed, must be a hydrocarbon haze of undetermined composition.

### 10.1.1 Atmosphere and climate
Titan has an atmosphere composed mostly of nitrogen, with smaller amounts of methane, ethane, argon, and carbon monoxide, and trace amounts of numerous hydrocarbons, hydrocyanic acid, and nitriles (organic molecules containing nitrogen). It towers to an altitude of well over a hundred kilometers, producing a column of air that weighs 1.5 times the weight of Earth's atmosphere, despite having only a tenth of Earth's gravity.

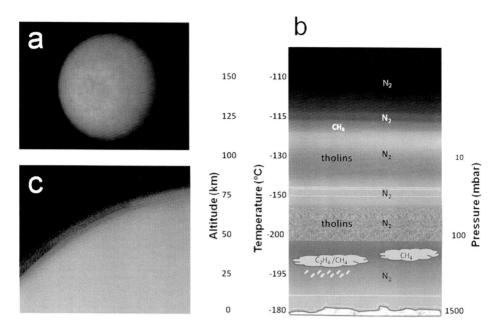

**Figure 10.2** Atmosphere of Titan. An opaque atmosphere envelopes the organic-rich icy satellite Titan. (a) The view from Voyager 1 on 4 November 1980 at a distance of 12 million km. (b) A dense atmosphere consisting mostly of nitrogen towers to a height well into space. At high altitudes, methane is dissociated by ultraviolet light into reactive species that form a complex set of organic compounds called tholins. They collect in two layers, a relatively warmer one at about 100 km up, and a cooler and presumably denser one at about 50 km. Clouds of methane ($CH_4$) and ethane ($C_2H_6$) move about the planet at lower levels, precipitating under appropriate conditions to form lakes and seas with a mixture of methane and ethane in local regions. (c) Lighter methane under lower pressure rises as a gas into higher altitudes where it reflects blue wavelengths above the orange haze. This was the view of Voyager 1 from 470,000 km (over a quarter of a million miles) out on 12 November 1980. (Art by Louis Irwin (b); Images for panels (a) and (c) NASA/JPL).

The heavy isotope of nitrogen ($^{15}N_2$) is found in greater abundance relative to that of the lighter isotope ($^{14}N_2$) than the solar abundance would predict. Since lighter isotopes preferentially escape to space over time, this means the original atmosphere of Titan was probably denser and more massive than it is today [2].

At high altitudes, ultraviolet light and ionizing radiation cause the methane and nitrogen to form reactive molecules that lead to synthesis of larger organic compounds, including benzene and polycyclic aromatic hydrocarbons (PAHs). Large, ionized organic compounds are also generated, which can react with benzene and PAHs to form the complex hydrocarbon-nitrile complexes called tholins [3]. These eventually float down to lower, cooler regions of the atmosphere, where they condense into layers of smog-like haze. At even lower altitudes, somewhere in the 30 to 50 kilometer range, ethane and methane form

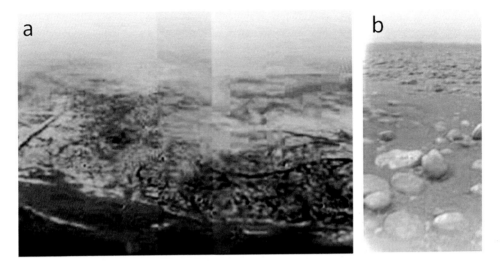

**Figure 10.3** Titan's surface. The first clear views of Titan's surface were obtained by NASA's Descent Imager aboard the Huygens lander on 14 January 2004. (a) At an altitude of 8 km (5 miles), a large uplifted land mass is bordered by a smooth basin toward which the Lander is descending. (b) Shortly after touchdown, this panorama was acquired, revealing pebble-sized rocks of water-ice, possibly containing solid carbon dioxide and hydrocarbons as well. The light flat rock just below the center of the image is about 15 cm (6 inches) long. The substrate had the consistency of wet sand, and gave off puffs of methane when heated by instruments aboard the Lander (NASA/JPL/ESA/University of Arizona).

clouds that become locally abundant, perhaps where methane is being outgassed from the interior (Figure 10.2b).

The temperature at the surface is about $-180°C$ ($292°F$ below zero). While a major part of the mass of Titan is made up of water-ice, at that temperature ice is as hard as rock – and looks it (Figure 10.3b). However, methane and ethane can persist as liquids at that temperature and pressure on the surface. Once they fall to the lower, colder parts of the atmosphere, they condense into drops of organic liquids and rain down from the clouds, forming drainage channels and collecting into basins (Figure 10.4). This effect is more pronounced in polar regions, so not surprisingly, this is where lakes of ethane have been found to be concentrated (Figure 10.5).

Time lapse photos have shown cloud movements over periods of hours to days. Dune formations show evidence of surface winds that are locally variable as well. The polar lakes of methane and ethane are prevalent at the pole in winter darkness, and less prevalent at the pole in summer sunlight, suggesting seasonal variations. Thus, the weather on Titan includes wind, rain, and seasonal changes.

## 10.1.2 Topography

For decades we've known that Titan holds the only nitrogen-rich atmosphere

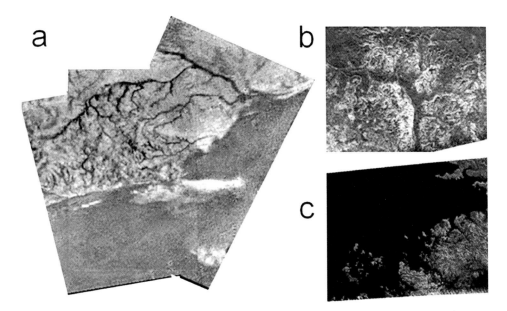

**Figure 10.4** Upland channels and seas on Titan. (a) Dendritic network of channels emptying into the broad basin, as photographed by the Huygens Lander shortly before it settled into the basin in the picture. (b) Complex hilly area scoured by runoff channels and a broad, curving canyon. (c) The coastline of a liquid sea, most likely of ethane and methane, about 70 degrees north latitude as imaged by the Cassini orbiter. Bays, inlets, and offshore islands are evident in the image, which covers an area of 160 by 270 km (100 by 170 miles). The absence of any light areas beyond the landforms suggests a liquid with a depth of tens of meters or more (NASA/JPL/ESA/University of Arizona).

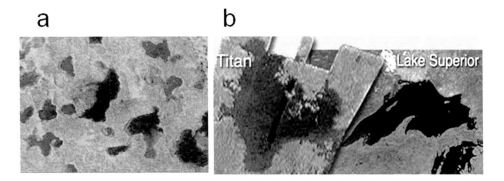

**Figure 10.5** Northern lakes on Titan. The Cassini orbiter's radar confirms the existence of widespread lakes of ethane in the far north. (a) Multiple lakes are scattered across an area of about 100,000 square kilometers, larger than all the Great Lakes between the United States and Canada. While these are likely too large to be ephemeral, smaller lakes and ponds may dry up, then refill on a seasonal cycle. (b) One of the bodies of liquid ethane is as large as Lake Superior (NASA/JPL).

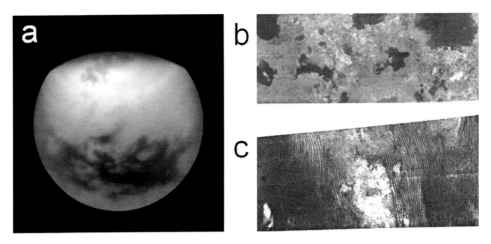

**Figure 10.6** Topogaphy of Titan. Earth-like features dominate the landscape of Titan. (a) This image created by superimposing photographs at different wavelengths taken by the camera aboard the Cassini spacecraft at a distance of 1.5 million km, reveals a continental-sized area of darkness reminiscent of an inland sea, while smaller darkened areas are evident at the north pole. (b) Cassini's radar confirms the existence of a scattered network of lakes at the north pole. (c) The inland "seas" in equatorial regions now appear to be relatively smooth but liquid-free basins streaked with dunes surrounding radar-bright "islands" of rougher and probably elevated terrain (NASA/JPL).

other than Earth in the Solar System. Once the flyby of Voyager 1 in 1980 determined Titan's density to be similar to that of the other icy satellites, and measured its surface temperature to be cold enough to liquefy hydrocarbons, it was assumed that Titan would turn out to have a surface something like Ganymede's, another icy satellite just barely larger than Titan, with a relatively homogeneous cratered icy surface, possibly holding lakes or seas, or even a global ocean, of ethane or methane. Thus, when radar first penetrated Titan's orange haze and began to reflect the outline of continental-sized features quite unlike that seen on the other icy satellites (Figure 10.6a), but quite reminiscent of features seen of Earth, interest in Titan was dialed up another notch.

Titan resembles neither Earth nor the other icy satellites in the composition of its surface. Ice and granular organic compounds, rather than silicates as on Earth, or mainly water-ice as on Ganymede, make up the crust of Titan. But in its topology, Titan looks a lot like Earth. It has coarse and fragmented land masses surrounding low-lying basins, filled at the poles with liquids and at the mid-latitudes with sand-dunes (Figure 10.4–6). It features elevated terrain cut by dendritic channels like converging streams, or sometimes thick conduits for large volumes of liquid that empty into smoother low-lying plains (Figure 10.4).

As radar imaging of the surface has become more extensive, it appears that the lakes or even seas of liquid hydrocarbons, which have long been hypothesized, do exist, but mainly at the poles in the wintertime. Recent images of the north pole of Titan, taken when that pole was tilted away from the Sun, include one such body as large as Lake Superior, on the border of Canada and the USA (Figure 10.5). Basins large and small are found at many latitudes, however, suggesting that bodies of liquid may occasionally flood low-lying areas. When the Huygens lander settled into such a basin, heat from the craft caused methane to sublimate. Such basins may be saturated with hydrocarbons left over from periodic inundations, which tend to evaporate quickly. Some of the lakes near the poles are very non-reflective of radar, however, suggesting lakes of considerable depths which therefore may be longer lasting (Figure 10.4c).

It could be that some of the channels cutting through uplands, or some of the basins, have resulted from sapping, or the collapse of overlying ground due to outflow of underground liquid reservoirs. Thus, rain from above and liquid displacement underground may both contribute to the topology of Titan's surface.

Rifts, discontinuities, and other evidence of geological activity suggest that dynamic forces have shaped the face of Titan. Volcanic features can be seen on Titan, though the outflow from such activities must mainly be in the form of a slush of ammonia water. Impact craters are few and far between. Because of the dense atmosphere, only large meteorites ever reach the surface, and fewer than a dozen impact craters have been identified. By comparison, hundreds cover the surface of Callisto, Jupiter's moon of comparable size. Thus, the history of Titan must be characterized by extensive weathering and resurfacing.

### 10.1.3 Interior
By piecing together various lines of evidence, including data on its density and the composition and evolution of its atmosphere, a credible model of Titan's interior can be proposed. While details among competing models vary, they can all be generalized as shown in Figure 10.7.

A solid core consisting of silicate and/or heavier metals differentiated from the water and other volatiles, including large amounts of ammonia and methane, and settled to the center of Titan in the early stages of its formation. Gravitational collapse and radioactive decay continue to emit heat that energizes ongoing geological activity.

Heat flux from the core drives softer ice with dissolved ammonia and methane toward the surface. The volatiles substantially lower the melting point of water. At some distance below the surface, the outflowing heat melts the ice, creating a global layer of a highly alkaline solution of water, ammonia, and other organic constituents. Nearer the surface, the ammonia-water solution freezes solid, at temperatures cold enough to keep even methane entrapped as a solid. When episodic local fluctuations in heat flow or other conditions melt the water, a cryovolcanic eruption of ammonia-water results, which includes the outgassing of methane.

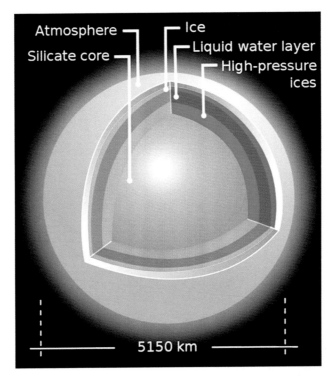

**Figure 10.7** Internal structure of Titan. Current models of Titan's interior envision a solid core surrounded by layers of liquid water and ammonia under high pressure, with an outer shell of water-ice and solidified tholins with the consistency of sand. Substantial stores of methane are thought to be held in the interior, trapped within the mineral structure of ice (NASA/JPL).

## 10.2 Planetary history of Titan

Saturn's largest satellite, Titan, is a planetesimal without precedent in the Solar System. In its atmosphere, it most resembles that of early Earth with its major gases of nitrogen and methane, yet is much smaller and much further from the Sun than Earth. In its size and density, it most resembles Ganymede, but is totally different from Ganymede in atmosphere and surface characteristics. The only way to understand the uniqueness of Titan is by reconstructing its history – a challenge, at best, for such a remote and enigmatic body. What follows is our simplified version of the work and models set forth by experts in the planetology of Titan [2, 4–6].

1. In the distant suburbs of the Solar System four-and-a-half billion years ago, the protoplanetary disk that would become Saturn and its satellites began to coalesce. Whether the distance from the Sun, or the very low temperature that far away, or other factors were responsible is unclear,

but an unusually high concentration of chemically reduced (hydrogen-bearing) compounds were part of the mix. Methane ($CH_4$) and ammonia ($NH_3$) were the most conspicuous among them.

2. Saturn itself got the bulk of the mass in the disk, compressing the heavy elements into a metallic core massive enough to hold on to hydrogen and helium, forming a gas giant planet exceeded in size only by Jupiter in our Solar System.

3. Over 90% of the remaining mass accreted at a distance of 1.22 million kilometers (about 758,000 miles) into a mixture of minerals and ice large enough to hold on to the volatile methane and ammonia, but not to the lighter hydrogen. This became Titan, the largest satellite of any planet beyond the Jovian system – second in size, just barely, to Ganymede.

4. The atmosphere of methane and ammonia appears to have been massive – perhaps the thickest, densest atmosphere of any planetary body other than the gas giants in the earliest eons of the Solar System. Titan's location relatively distant from Saturn's gravity well – further out than Ganymede is from Jupiter – spared it from the level of meteorite bombardment that appears to have driven off the atmospheres from the comparably large Ganymede and Callisto.

5. Like all the outer, icy satellites, Titan had an abundance of water; but at -180°C, it was frozen solid at the surface. Below the crust, however, aqueous solutions of ammonia heated from the core most likely formed.

6. Meanwhile, methane was reacting high in the atmosphere with ultraviolet and ionizing radiation to create more complex organic molecules that formed a global smog of tholins (Figure 10.2).

7. The lower atmosphere was so cold that the air couldn't even hold methane and ethane in gaseous form, so they began to rain down, along with the heavier tholins. Ethane pools, possibly of oceanic size at one time, flooded the surface and remain today in ponds, lakes, and seas near the poles (Figures 10.4 and 10.5).

8. The same radiation that was turning methane into a more complicated aerial broth of organic compounds was also breaking up ammonia into its constituent nitrogen and hydrogen atoms. Titan was too small to hold on to the hydrogen, so it escaped to space, leaving behind an atmosphere increasingly dominated by molecular nitrogen ($N_2$), with an elevated $^{15}N_2/^{14}N_2$ ratio.

9. As Titan grew older it lost a lot of its atmosphere, but curiously held on to methane. All the calculations show that methane should have been long gone by now, unless it were being renewed. This is true of Earth's atmosphere as well, where the unexpectedly large content of methane is attributed mainly to biogenic sources. While this could be the case on Titan also, it appears equally plausible that outgassing of methane from the interior, along with the cryovolcanic eruptions for which there is now strong evidence, could account for the bulk of methane in Titan's atmosphere.

10. Occasionally, large meteorites have crashed into Titan, but they've had to be large in the first place to make it all the way through the thick atmosphere. The craters created by these rare catastrophes have apparently not lasted long, due to weathering and cryovolcanism. Hence, the surface of Titan remains forever young.

## 10.3 Conditions (good and bad) for life on Titan

In some ways, conditions for life on Titan are great. Carbon-based building blocks for biomolecules are abundant, liquids are found above and below ground, and a dense protective atmosphere enshrouds the satellite. In other ways, from the perspective of life as we know it, conditions for life are terrible. The surface temperature is far too cold for water-based organisms, the liquids are highly non-polar, and the prospects for chemical cycling appear to be limited. The question is, can life as either we do or do not know it, exist in some quarter of this exotic world?

Every press release on the latest observation of Titan either starts or ends with the statement that scientists view Titan as a possible model for conditions on the early Earth. This is true with regard to the ready availability of organic chemicals. But that's pretty much it. Titan's composition, distance from the Sun, orbital characteristics, size, temperature, and planetary neighbors are so different from those of Earth that the history of the two bodies has to have been distinct almost from the start. If Titan turns out to harbor life, it seems more likely that what we'll learn is how different life can be on other worlds, rather than gaining much insight into how life arose and evolved on Earth.

### 10.3.1 Chemistry

*10.3.1.1 Chemistry for building blocks*
Titan's atmosphere is a veritable factory for organic compounds (Table 10.1). Starting with methane and nitrogen in the upper atmosphere, a great variety of hydrocarbons, nitriles, and PAHs are produced. Many of the smaller organic compounds are ionized, leading to more favorable reaction kinetics at the cold temperatures of Titan's atmosphere. Even some negatively-charged compounds at least as large as several amino acids have been detected [3]

For living organisms as we have defined them to be a reality on Titan, at least two types of building blocks have to be present. First, there has to be the capacity to form boundaries like the membranes that enclose metabolic systems in Earth-like organisms. In the aqueous-ammonia solvents that we strongly suspect fill underground reservoirs on Titan, lipophilic membranes of the type with which we are familiar can serve as we know they do in any aqueous solvent. For the surface pools of ethane and methane, however, membrane chemistry is bound to be more exotic. The tholins that rain down as solid precipitates are not very soluble in ethane. They may crosslink to form some type of sheath or film that is

**Table 10.1**  Organic compounds identified or inferred on Titan

Ultraviolet light and ionizing radiation in the upper atmosphere cause extensive ionization. This lists only a partial sample of the compounds detected [7].

| Hydrocarbons | | Oxidized Carbons and Nitriles | |
|---|---|---|---|
| Molecule | Formula | Molecule | Formula |
| methane | $CH_4$ | carbon monoxide | $CO$ |
| ethane | $CH_3CH_3$ | carbon dioxide | $CO_2$ |
| ethylene | $CH_2CH_2$ | hydrogen cyanide | $HCN$ |
| acetylene | $C_2H_2$ | cyanogen | $C_2N_2$ |
| benzene | $C_6H_6$ | acetonitrile | $CH_3CN$ |
| Ionic hydrocarbons | $CH_5^+$, $CH_2CH_3^+$ | Ionic HCN | $HCNH^+$ |

structurally stable in hydrophobic solvents. Silicon, as silane or polysilane, conceivably could be part of some type of structural barrier, since silanes are stable at very low temperatures under reducing conditions. The possibilities for constructing some type of semi-permeable boundary stable in organic solvents seems reasonable, given the range of glasses, plastics, and other materials manufactured on Earth to hold hydrophobic liquids. Such possibilities remain speculative only, however, at this stage of our ignorance about the surface chemistry on Titan.

The second need for chemical building blocks is for the construction of informational molecules – i.e. large molecules of variable structure. This should be easy on Titan, since carbon polymers form the basis *par excellence* for macromolecules with high information content. One simple route to the construction of a biopolymer using compounds known to exist on Titan is shown in Figure 10.8. Using hydrocyanic acid and other nitriles, plausible reactions can be envisioned that lead to a long-chain polymer with a carbon-nitrogen backbone (-N=C-C-N=C- C-) in which a variable chemical group, like the four different nitrogen bases in DNA (Figure 2.7) would be attached to every other carbon atom. The sequence of the variable groups would then constitute information. Assuming the existence of a template mechanism for directing the order of side group addition, a process for replicating information can readily be seen. While no such polymer is known yet to exist on Titan, the point is, that it or something like it, could.

### 10.3.1.2 Chemistry for energy

You would think that on a world where something akin to gasoline rains down from the sky, chemical energy would not be a problem. On Earth, hydrocarbons from methane to octane, and carbohydrates like glucose and sucrose, are harvested for energy by oxidizing them to water – by using oxygen to break apart the C-C and C-H bonds and capturing that energy in other molecules. Free oxygen, however, is found in only trace amounts in Titan's atmosphere.

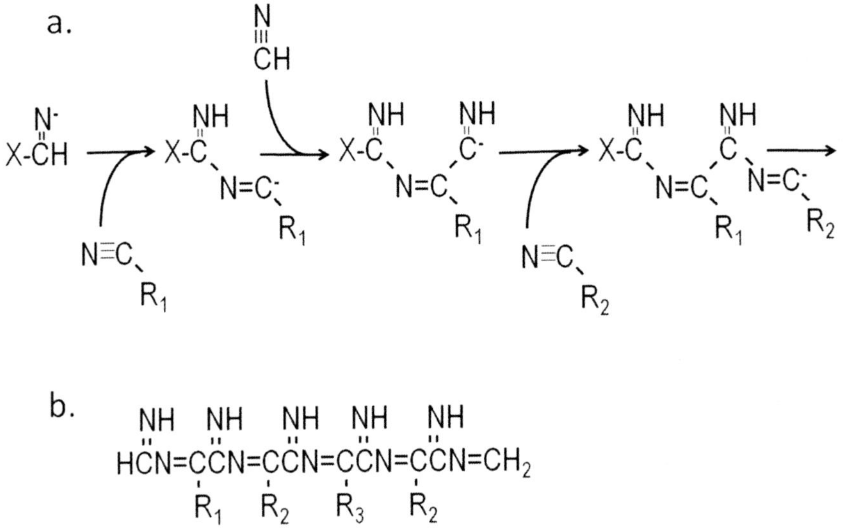

a.

b.

$$HCN=CCN=CCN=CCN=CH_2$$

**Figure 10.8** Theoretical pathway for creation of information-containing polymer. (a) A sequence of reactions in which nitriles with different side groups (R) add to a growing chain could result in a long molecule with a unique sequence of R groups. Adapted from Shaw [7] (b) The number of structural variants would be given by $x^n$, where x is the number of different side group variants (e.g., x=4 for the four bases in DNA), and n is the number of R groups in the complete chain. This molecule, for instance, would be one of 81 possibilities from a pool of 3 side group variants ($R_1$, $R_2$, or $R_3$) assembled with a chain length of 4 total side groups (n=4), since $3^4 = 81$.

While water is a plentiful potential source of oxygen on Titan, it is completely locked up in solid ice and therefore isn't available for chemical reaction. Carbon dioxide is likewise frozen solid. Larger amounts of carbon monoxide are found in the atmosphere, and recent studies suggest that a slow influx of oxygen ions from space, and possibly of water from neighboring satellites like Enceladus, could provide enough oxygen to account for the observed amounts of atmospheric carbon dioxide and carbon monoxide [8]. If so, other molecules may have captured the energy from these reactions, in a way that would make it available for biological uses. It needs to be pointed out, however, that the amount of oxygen available for this purpose, is way below the amount of oxygen available for life on Earth.

Another form of energy on Titan may make more sense. The photochemical production of larger organic molecules in the upper atmosphere of Titan generates a large amount of potential chemical energy. Acetylene ($HC\equiv CH$) is particularly attractive in this regard, since its $C\equiv C$ triple bond contains a lot of energy. When acetylene reacts with hydrogen to form methane, that triple bond is broken and energy is released [9, 10]. This is a particularly attractive possibility on Titan, since it can take place in an organic solvent like methane and ethane. Again, though, the question is whether there is enough reactive hydrogen available to make this a realistic possibility.

*10.3.1.3 Chemistry for solvents*

Liquids have now been confirmed to exist on the surface of Titan (Figures 10.4, 10.5 and 10.6). As theory would predict, they consist primarily of liquid ethane, no doubt with dissolved methane and other organic compounds in solution. Water cannot exist on the surface permanently as liquid because of the extremely low temperature. Beneath the surface, however, the temperature rises to a point where methane and ethane cannot remain liquefied (though methane is probably held as a clathrate, or "caged" molecule within a lattice of water under high pressure). At this higher temperature, ammonia (which depresses the freezing point of water) most likely exists in aqueous solution, which must have a strongly basic pH.

## 10.3.2 Temperature

Titan is very, very cold. At the surface, a constant temperature of about $-180°C$ ($290°F$ below zero) means that water is totally solid. This temperature is slightly above the melting point for methane and ethane, though, so they can exist on the surface as a liquid. At higher levels of the atmosphere, sufficient solar heat is absorbed to raise the temperature enough for methane and ethane to vaporize, which is why they float as gases in the atmosphere until they sink to the lower colder regions closer to the ground and condense out as rain (Figure 10.2b).

Calculations have shown that the impact of large bolides would release enough heat to liquefy water-ammonia mixtures, which could exist as pools on the surface temporarily. Though they would freeze over quickly on top, an impact crater at least 15 kilometers in diameter could hold liquid ammonia-water solutions for hundreds to thousands of years beneath a frozen crust [11]. There is no evidence at this time that bodies of water *currently* exist on the surface of Titan, though.

Beneath the ground, temperatures rise toward Titan's core. With depth, the heat is great enough to exceed the melting point for aquatic solutions of ammonia. Two lines of evidence support the probability of substantial liquid stores beneath the surface. First, the rotation of the surface is dissociated enough from Titan's rotational period to suggest that a global layer of subsurface water decouples the mantle from the interior of the satellite. Secondly, radar imaging reveals underground depots of liquid that are quite extensive. Heat generated both radiogenically and from gravitational contraction apparently drives cryovolcanic eruptions, giving further evidence of a warmer interior.

While Titan's extreme cold would put a damper on aquatic-based metabolism of the type with which we are familiar, it could have the advantage of moderating the activity of highly reactive chemical constituents like free radicals. Such reactions could turn out to be critical for an entirely different form of energy metabolism that, though unworkable on Earth, would serve perfectly well in the thermal and solvent environments on Titan.

## 10.3.3 Habitats

Nowhere else in the Solar System do we find a body as large as Titan with two radically different liquid habitats (Figure 10.9).

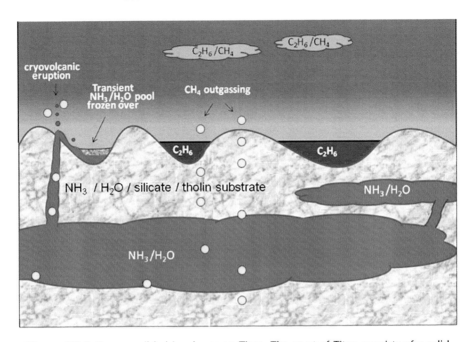

**Figure 10.9** Two possible biospheres on Titan. The crust of Titan consists of a solid substrate of frozen ammonia and water, intermixed with tholins, probably silicates, and liquid hydrocarbons – mostly ethane ($C_2H_6$) and methane ($CH_4$) – in pore spaces. Circumstantial evidence suggests the existence of underground reservoirs of liquid ammonia ($NH_3$) and $H_2O$ constituting a hydrosphere, while the surface holds pools of liquid $C_2H_6$ and $CH_4$, making up a liposphere. The two putative biotic habitats are separate and distinct, though occasional cryovolcanic eruptions disgorge the $NH_3/H_2O$ onto the surface where it quickly freezes over. Atmospheric $CH_4$ is destroyed photolytically at upper levels of the troposphere, but is regenerated by outgassing from the interior. Conceivably, biogenic production of methane could also contribute to this cycle (Art by Louis Irwin).

Well beneath the surface, an aqueous solution of ammonia and organic compounds provides a potential habitat for water-based life. This habitat may be global in extent, and therefore not unlike the subsurface ocean on Europa, except for its more alkaline character.

Above ground, ponds, lakes, and seas of ethane – also rich in larger, more complex organic compounds – wax and wane through seasonal cycles of abundance and scarcity. Occasional downpours scour the surface and recharge the basins with liquid ethane and methane (Figure 10.10). While the average precipitation on Titan is only a few centimeters per year, it appears that organic rain on Titan is heavy, very similar to downpours in the desert southwest of the United States, but with perhaps even more violence. The downpours are likely to follow a pattern based on atmospheric circulation, thus monsoon seasons on Titan similar to those on Earth seem to be a definite possibility. However, how frequent the monsoons would occur in any location on Titan is unclear.

**Figure 10.10** Stormy weather on Titan. "Gully washers" of ethane rain, pouring from the sky and channeling down into lake-forming basins, as envisioned by the art of Michael Carroll (NASA/JPL).

Where liquids flow or exist for a time, they appear to leave behind layers of tholins. This liposphere (from "lipids") of non-polar habitats resembles those desert habitats on Earth that are dry much of the time, but are recharged by occasional downpours which make the creeks run and the basins fill with water that brings them to life – the difference being that the liquid on Titan is ethane and methane rather than water.

Through occasional geological events, like cryovolcanic eruptions, the two habitats probably come in contact briefly, but overall they remain two worlds apart.

## 10.4 A possible evolutionary history for putative life on Titan

Among the most enticing aspects of Titan from an astrobiological perspective is the very high likelihood that any life that might be found there almost surely began there. This is because Titan is so remote from any other body where life is likely to have existed, that immigration by panspermia of life alien to Titan is a very remote possibility. Above and beyond this consideration, however, is the further possibility that life may even have originated independently in two different habitats on Titan. At the very least, whatever and wherever life originated there, two very different scenarios need to be envisioned for the evolution of life since its inception on Titan (Figure 10.11).

### 10.4.1 An aqueous origin and evolution for life on Titan

In Titan's youth, when the satellite was likely covered with a widespread if not global ocean, the origin and early evolution of life should have followed a trajectory similar to that under similar conditions on Europa. One difference would have been the greater abundance of organic chemicals, which if anything, may have accelerated the origin of self-organizing systems into living organisms faster than anywhere in the Solar System.

Another contrast between Titan and Europa would have been a considerably lower intensity of sunlight. At Titan, sunlight is less than 2% as bright as it is on Earth. While photosynthesis at such dim intensities cannot be ruled out, we suspect that other forms of energy would have been relatively more competitive. Obviously, organic chemical energy was available in abundance, hence a primordial chemoheterotroph seems the most likely candidate for the ancestor of life in the early aquatic environment on Titan. However, heat and kinetic energy may have provided alternative bioenergetic sources as well, hence thermotrophs and kinetotrophs could have evolved from this ancestral form fairly soon. As discussed in section 10.3.1.2 above, methanogenesis based on the reduction of acetylene is such an attractive possibility that we think both benthic and lithotrophic (subsurface) methanogens were among the earliest forms of life on Titan.

Once the biochemical pathways for processing certain organic chemicals readily available in the early ocean had evolved, a microbe specialized for utilizing this abundant source of high-energy fuel would have become the ancestral organochemotroph deeply rooted in the tree of aquatic life on Titan. From this ancestor could have sprung a variety of descendant forms, including microbial biofilms, first on the ocean floor, then at the ice ceiling, once the ocean froze over. Floatation filter-feeders, something like jellyfish on Earth, and other types of mobile consumers could also have evolved, and they in turn could have provided nutrients for higher order floatation and mobile consumers.

Meanwhile, the ancestral kinetotrophs might have given rise to both benthic and ceiling ciliates, as well as benthic reeds harvesting energy from water currents. Consumers such as ceiling scrapers could then have evolved to feed on these producers. Other consumers, in the form of microbial detrivores and more complex fungal-like detrivores are equally likely.

From the beginning, the water would have been very alkaline because of the high dissolved ammonia concentration, so all forms of life in this habitat would have had to adapt to that condition. An early descendant of the ancestral organochemotrophs may have evolved another specialization as well – a tolerance for hydrocarbons. This adaptation would have been particularly suitable for those occasional bodies of water on or near the surface where

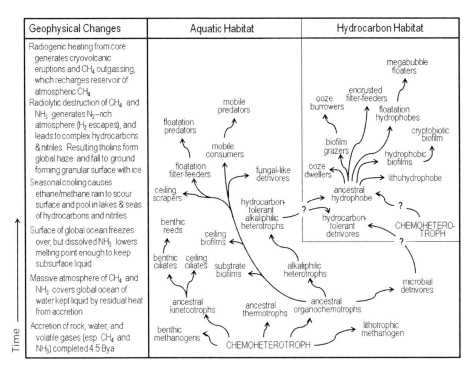

**Figure 10.11** Possible evolutionary trajectory for life on Titan. Life is presumed to have first originated when a liquid ocean covered much if not all of Titan. After the surface froze over and atmospheric hydrocarbons were cooled enough to precipitate, a surface liposphere of liquid hydrocarbons and layered tholins provided a distinct biosphere for hydrophobic-based life, while a separate biosphere of aquatic-based life persisted in an underground system of liquid ammonia-water reservoirs.

hydrocarbons would have been found in higher concentration. After the surface froze over, sizable meteorite impacts that created transient ammonia-water filled craters for hundreds to thousands of years would have represented a habitat in which a hydrocarbon-tolerant alkaliphilic heterotroph could have thrived, perhaps regressing to a cryptic form once the aqueous solution completely froze over.

## 10.4.2 A hydrocarbon habitat for the origin and evolution of life on Titan
By the time it became cold enough for ethane and methane to rain from the sky, the surface of Titan was frozen solid, encasing whatever liquid remained of the primordial ocean beneath a thick layer of granular ice and organic material. Basins formed within the rugged terrain, where ethane collected in pools large enough to form seas (Figure 10.5). Two different "worlds" now existed on the same planetary body – an underground "hydrosphere" of ammonia-water, and a surface "liposphere" speckled with ponds, lakes, and seas of hydrocarbons (Figure 10.9).

Life is most likely to have appeared first in the older, now subterranean hydrosphere, kept liquid by Titan's internal heat and the freezing-point depression of ammonia and other solutes. The question therefore arises as to whether life in the liposphere above could have descended from an aquatic precursor, as opposed to originating independently in a separate, isolated hydrocarbon habitat on the surface. On the one hand, life in ethane would need to be so different that an independent origin seems very plausible. On the other hand, Titan's transition from a liquid water world to a frozen cradle for lakes of ethane and methane would not have been instantaneous. As tholins built up in the atmosphere, then fell to the surface, and methane became encased in the water as it turned to ice, the substrate would have taken on the nature of petroleum-rich sludge gradually, providing pressure but also time for directional selection of increasingly hydrocarbon-tolerant organisms.

An ancestral hydrophobe is presumed to have been a common ancestor for all the forms of life in the liposphere. (Given the rich inventory of organic compounds on Titan, multiple origins for proto-biological systems cannot be ruled out, but a single common ancestor is our simplifying assumption.) Figure 10.11 considers the possibility that the ancestral hydrophobe could have originated in two ways. Assuming an origin independent from the aqueous biosphere, chemoheterotrophs arising *de novo* as self-organizing systems in liquid ethane/methane are envisioned as one pathway to life in hydrocarbons. Alternatively, a hydrocarbon-tolerant aquatic ancestor could have become sufficiently preadapted to the demands of life in liquid ethane that it gave rise to the ancestral hydrophobe at the root of life in the liposphere.

Regardless of how it originated, the last common ancestor and all subsequent forms of life in liquid hydrocarbons would have to have had properties very different from any we have yet encountered or discussed. These properties include (1) a structural boundary stable in a hydrophobic solvent, (2) an internal solvent remaining liquid at temperatures precluding liquid water, and (3) the capacity for maintaining metabolism under thermal conditions much colder than any that have ever been studied or even seriously modeled.

As exotic as these properties sound, they are not without known precedents on Earth. A strain of the fungus, *Fusarium alkanophyllum*, has been shown to grow in and degrade saturated hydrocarbons with little or no oxygen and a minimum of water [12]. The membranes of this organism are enriched in fatty acids, and depleted of more hydrophobic lipids, thereby establishing a stable boundary between the interior of the cells and their hydrophobic surroundings. Other radical departures from common cellular architecture include the absence of internal membranes. Also, abundant microbial life has been found in a liquid asphalt lake in Trinidad, an analogue site used to study life in hydrocarbons [13]

Imagining an appropriate solvent for life at -180° C is something of a challenge, but two good candidates are theoretical possibilities. The first is an ammonia-water mixture, which can stay liquid down to -98° C at the surface pressure on Titan – lower even than pure ammonia – and could serve as a polar solvent in cells warmed above their surroundings by metabolic processes. In this case, a polar solvent-based biochemistry something like the one we already understand, would not be implausible. The other is non-polar methane, with a range of liquidity between -182° (melting) and -162° C (boiling) that matches well the actual temperatures encountered in the liquid ethane-methane habitat. Biochemistry in such a solvent could proceed at temperatures close to that of the surroundings, but it would be a biochemistry totally unfamiliar to us.

The metabolic challenges are formidable as well, beginning with the need to metabolize highly-reduced organic molecules in an environment that isn't very oxidizing. A natural strain of the bacterium *Bacillus cereus* has been found to degrade n-hexadecane, a long hydrocarbon, in an oilfield in China [14]. A special

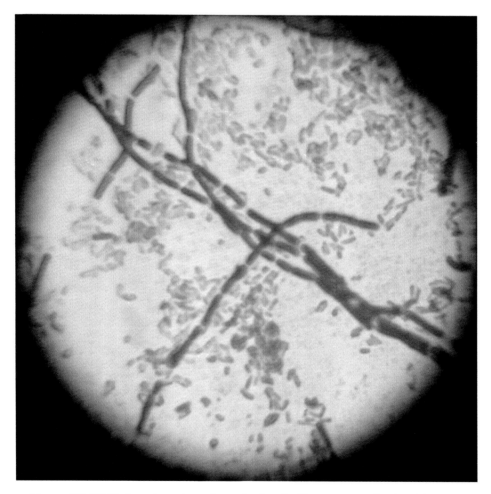

**Figure 10.12** Hydrocarbon-degrading microbe. *Bacillus cereus*, a microbe able to degrade long-chain hydrocarbons, is stained blue-green in a culture which includes *Escherichia coli*, stained in pink (Microscopic photo by Bibliomaniac).

set of enzymes on the bacterium's inner membrane has apparently evolved to degrade the unusual carbon source. A variety of anaerobic bacteria have been shown capable of living in and biodegrading petroleum sludge. Indeed, a typical strategy for bioremediation of hydrocarbons is to artificially select for microbial strains that progressively shift their reliance for nutrition from conventional carbon sources to their hydrocarbon substrates over a succession of generations. What humans accomplish in an instant (geologically speaking) by artificial selection should be an easy task for natural selection acting over vastly longer periods of time.

None of the Earth analogies comes close to matching the actual conditions on Titan. What they do illustrate, along with all the other examples of "extremophilic" life on Earth, is that natural selection can favor the evolution of exotic properties in the face of exotic selective pressures. With many millions of years to evolve on a world becoming progressively colder and increasingly hydrophobic, there is no reason that organisms highly exotic by Earth's standards could not come into being.

Whatever the pathway to its origin, we think it likely that ancestral hydrophobes would have given rise early on to lithohydrophobes to occupy niches beneath the substratum. Since heat from the Huygens lander kicked up a puff of methane from the substrate, the ground on Titan could be saturated with hydrocarbons. If that's the case, microbial hydrophobes dwelling in the saturated substrate might be widespread, whether buried beneath pools of hydrocarbon or not. If such organisms happen to be methanogens and they exist in massive numbers, the methane they produce could be part of the reason that atmospheric methane appears to be continually recharged.

Another early likely descendant of the common ancestor for the liposphere would have been a biofilm covering the bottom of the ponds, lakes, and seas of ethane. These hydrophobic biofilms may have consisted of aggregates of organisms, or conceivably have taken the form of an extended expanse of a unitary organism, something like the way that slime molds lose their cellular identity and merge into a megaorganism under certain environmental circumstances. Such biofilms would probably have spread across the floor of hydrocarbon pools, protected from desiccation by the overlying liquid. When the pools evaporated away, however, cryptobiotic biofilms may have survived to be regenerated by the next downpour of ethane.

Wherever biofilms could grow, biofilm grazers evolved to feed on them would likely have been around. Floatation hydrophobes and encrusted filter-feeders are two other, among many, descendant forms that could have evolved. They, in turn, would have provided the opportunity for evolution of higher order consumers, such as a megabubble floater large enough to feed on smaller residents of the liposphere.

The Huygens lander and Cassini orbiter have relayed pictures back to Earth that make the channels that scour Titan's surface appear as though they are coated with a dark material, reminiscent of tar. Many scientists think this may be a layer of tholins expected on theoretical grounds to coat much of Titan's surface. We will simply call it an "ooze" of some sort. Whatever it consists of, it could well harbor organisms peculiar to its particular composition. We will simply speculate generically that some type of ooze dwellers occupy this niche. It seems reasonable to assume that if biofilm grazers have evolved to feed on biofilms, they might well have given rise to ooze burrowers that dig into the ooze and feed on the ooze dwellers.

## 10.5 Ecosystem possibilities for life on Titan

Since two putative biospheres are found on Titan, two distinct ecosystems would have to exist. The subsurface hydrosphere would likely look like that of Europa, with the distinction that Titan's subsurface hydrosphere is most likely more alkaline. As on that body, the hydrosphere would be subdivided into two principle habitats, dominated by microbial and immobile producers, that would hover on the floor and at the ceiling, respectively, of the underground ammonia-water reservoirs. Floating forms might occupy the volumes in between, but at a lower density with increasing distance from the producers.

The liposphere would not look like anything we know on Earth, and may not exist anywhere else in our Solar System with the possible exception of Triton (Chapter 11). We envision two principle subdivisions of this biosphere: one lying at the floor of liquid pools of hydrocarbons; the other lining the channels where a liquid ethane-methane mixture flows toward lower elevations, in the form of a tar-like ooze of tholins.

We suspect that the trophic structures of both major ecosystems on Titan are shallow, meaning food chains would not be very extensive. Our reasoning is based on the assumption that (1) the extremely cold temperature severely restricts mobility, and therefore the efficiency of a predatory lifestyle; and (2) the abundance of organic nutrients renders consumers relatively less likely. Consumers are certainly a theoretical possibility, but their inclusion in Figure 10.11 is highly speculative.

## 10.6 Characteristics of life on Titan

Extreme cold is the dominant factor for any life that may exist on Titan. A close second is the abundance of organic compounds for potential energy and building blocks. Assuming that a means of harvesting this abundant chemical energy evolved during the earliest stages of evolution, the characteristics of life on Titan would be defined by the success of metabolic innovations driven by the special chemistry of that world, and constrained by thermal limitations on what that chemistry could accomplish.

### 10.6.1 Metabolism
Given the pervasiveness of organic compounds in the air, on the surface, in the

lakes and seas of petrochemicals, and under the ground on Titan, it seems highly likely that self-organizing metabolic systems capable of harvesting the available chemical energy would have emerged early in evolution. The two concerns that could dampen enthusiasm for the readiness with which metabolism could have quickly stabilized in this environment are (1) the relative lack of free compounds for sustaining robust oxidation-reduction cycles, and (2) the limitations that the cold would impose on reaction rates.

The first problem may be an impediment in our own terracentric thinking, which is accustomed to relying on oxygen-containing compounds for oxidation of organic molecules. While some oxygen does appear to make its way into the atmosphere of Titan, it would appear to be a very sparse supply as a global basis for bioenergetics. And while the apparent abundance of unsaturated (double and triple) carbon bonds provides a great potential energy source, as in the reduction of acetylene to methane described in section 10.3.1.2, whether sufficient reactive hydrogen is available for that purpose is not yet known. Perhaps nitrogen could take the place of oxygen in completing oxidation-reduction cycles on Titan [15]. Of course there's no end to the carbon-bonded hydrogen in a sea of hydrocarbons, provided that thermodynamically favorable reactions can be invoked to transfer them to molecules like acetylene in order to release energy. We need to remember that evolution is very innovative, though. There is no reason to doubt that novel metabolic pathways as yet unknown to us could have evolved under Titanian conditions to provide the energy required for whatever life may live there.

Likewise, "cold" is a relative term. The evolution of extremophiles on Earth shows that adaptations appear to whatever the conditions demand. Any chemical reaction is going to proceed more slowly at lower temperatures, but natural selection would require that reactions be fast enough at the temperature at which they must take place, to accomplish their necessary function. In other words, metabolic rates will evolve to be as fast as they need to be. From our perspective, metabolic rates on Titan are probably going to be slow; but even here we need to be cautious, not knowing at all what the nature of those metabolic transformations will look like.

## 10.6.2 Growth and reproduction

Continuing the theme of abundant building blocks and energy on a cold, cold world, we would expect growth to proceed at a leisurely pace and reproduction to be simple. In homogeneous and stable environments on Earth, reproductive rates are low and growth is slow. This is because (to oversimplify a bit) evolution has tended to optimize adaptations for success in relatively unchanging environments, thereby minimizing mortality rates. Hence organisms can conserve energy by lowering the number of their offspring, and they can afford to grow slowly toward generally larger terminal sizes.

While there is some evidence that weather on Titan is dynamic, and a degree of habitat fractionation is evident, the underground hydrosphere, as well as the subliquid or ooze-embedded liposphere probably enjoy long-term stability. So the general principal of low rates of growth and reproduction should apply on Titan.

Sexual reproduction would not be expected. Finding mates and exchanging genetic material takes energy and entails some danger, so biologists generally believe that the advantages of genetic scrambling have to be great enough to counterbalance those negatives. The advantages are generally assumed to be the generation of an ever-changing genetic repertoire in order to pre-adapt the organism for new challenges and to evade the ravages of parasitic degradation and disease. If no new challenges arise over long periods of time, and the abundance of nutrients makes chemoautotrophy a simple and straightforward life style readily available to all, reproductive simplicity – like growth by mere expansion, or asexual budding – is more often likely to be the case. On the other hand, if parasitism should turn out to be a prevalent feature of life on Titan, the evolution of sexual reproduction might be favored.

We previously mentioned the possibility of periodic monsoons of precipitation, perhaps creating rivers, streams, and lakes in seasonal cycles. Like organisms on Earth whose life cycles are closely tied to significant seasonal variations in the nature of their environment, putative life on Titan might similarly be tied to seasonal oscillations. For instance, particularly in the desert dunes of the mid-latitudes of Titan, organisms might be hunkered down in an inactive state until the ethane-methane monsoon season arrives, very much like plants in the desert southwest of the United States await the occasional downpour to bloom. Other organisms living at or in the ethane-methane lakes closer to the poles may also wait for the monsoon to wash organic nutrients out of the atmosphere and down from the hills. Our lack of solid information about seasonal fluctuations on Titan makes even informed speculation difficult. However, assuming that any life on Titan would be adapted to thrive in organic hydrocarbons, it would be astonishing if their life cycle would not be intrinsically linked to the seasonally regulated downpour and enrichment of organic compounds.

### 10.6.3 Motility
If searching for mates is not necessary, and organic nutrients are everywhere, why go anywhere? This is the likely logic by which natural selection on Titan would have no basis for favoring the evolution of motile forms of life. They could exist at one or more consumer levels (Figure 10.11), but would probably be slow and not very numerous, in keeping with the relatively flat ecosystem structure mentioned in section 10.5 above. For the most part, life on Titan is likely to resemble a moss-covered forest floor, with a bare minimum of animal-like forms moving about.

### 10.6.4 Sensory Systems
Sensory systems evolved on Earth as a navigational aid for motility – telling organisms where they are in space, whether they are approaching friend or foe, and how their muscles need to move in order for the organism to behave in a

manner appropriate to its situation. If the organism is not moving, its need for sensory information is fairly minimal. In a rich chemical environment, however, something like olfaction would probably be well developed. Sensitivity to touch would be important as well, since the social life of microbes in general and biofilms in particular depends on mutual recognition or antagonism. The lack of much light in any of Titan's habitats would render vision unimportant. Sound detection is important for predator-prey interactions, and therefore not likely to be of much need where predators are few.

### 10.6.5 Cognition

Both touch and olfaction are ancient sensory capabilities of life on Earth, developed in the most ancestral microbes without benefit of specialized neural tissues. The more sophisticated senses of vision and audition are the ones that require nerve cells, but if these sensations are absent, the need for a nervous system is not evident. Since cognition is an emergent property of nervous systems, cognition is not likely to be a property of organisms that have no need for a nervous system. Again, the metaphor for life on Titan is the grass-covered plain or the algae encrusted pond. The biomass might be considerable, but not at all intelligent.

### 10.7 Note of caution

For those who *want* there to be exotic forms of life on another world in our region of the universe, Titan is an incredibly attractive candidate. It combines the components that we know life can use, like carbon-based chemistry, with totally alien features, like liquids of organic compounds and awesomely frigid temperatures. Indeed, an analysis [16] by the National Academy of Sciences concluded that "… if life is an intrinsic property of chemical reactivity, life should exist on Titan." The temptation to let our imaginations roam wild is compelling. For that very reason, we need to be cautious in looking too hard for what we want to find.

In total keeping with the spirit of this book, we have aggressively plumbed the range of possibilities for life on Titan. We have offered nothing that isn't scientifically plausible, within the limits of our knowledge at the present time. But that knowledge is very limited, and the possibilities are extremely speculative. We offer no apology for reasonable speculation, but that same appeal to objectivity warrants a corresponding note of caution.

Titan is very unlike the habitats for life as we know it to exist with certainty on Earth. We simply don't know enough about the generic limits of life to know if it can exist under such conditions as unearthly as those found on Titan. Our premise is that life most likely can exist under conditions beyond our imagination. But we could be wrong. If life turns out to be possible only in liquid water, for example, it cannot exist anywhere near the surface of Titan, and perhaps not anywhere on that tantalizing world.

## 10.8 Chapter summary

From the night of its discovery over 350 years ago, Titan has been the delectably different satellite in our Solar System. Only the sixth known moon of any planet at the time of its discovery, it was long thought to be the largest, and is the furthest of any large moon from the planet it orbits. Discovery of a nitrogen-rich, organic-containing atmosphere added to Titan's allure, but that same atmosphere shrouded it in mystery until the Cassini orbiter and Huygens lander penetrated its opaque organic haze to reveal a surface with surprisingly terrestrial features.

Rugged uplifts cut by dendritic channels draining into basins suggest a surface subjected to cycles of precipitation and erosion. Cryovolcanic eruptions and tectonic-like features suggest ongoing geological activity. Impact craters are rare and large where they are found, indicating infrequent impacts, vigorous erosional activity, or both. The recent discovery of substantial pools of ethane at the poles confirms theoretical expectations that Titan's frigid temperature should be cold enough at -180°C for ethane, and some methane, to rain from the atmosphere and last on the surface for extended periods of time.

Titan's atmosphere appears to be derived from a much more massive primordial envelope of gases that consisted mostly of ammonia. As photolytic destruction converted the ammonia to nitrogen (with hydrogen escaping to space), destruction of methane proceeded as well, generating a rich reservoir of hydrocarbon compounds and ions that turned the atmosphere into a cauldron of organic chemicals. Some of those compounds formed tholins, and layers of those found their way to the surface. Mysteriously, methane remained undepleted, suggesting ongoing outgassing from an immense subterranean reservoir, or perhaps even biogenic production by methanogens.

The prospect for life on Titan lies in two distinct habitats. Circumstantial evidence strongly indicates extensive subsurface reservoirs of liquids, possibly global in extent, that could be primarily aqueous solutions of ammonia. Such a hydrosphere might resemble the putative biosphere in Europa's subsurface ocean, hosting a complete aquatic ecosystem. A separate and distinct habitat is potentially provided by the surface ponding of liquid ethane, and the tar-like layers of tholins that cover the bottom of channels and perhaps basins where liquid hydrocarbons flow or stand. This liposphere would have to host a very different biosphere of organisms, evolved to flourish in organic solvents at temperatures far below the possibility of liquid water.

On theoretical grounds, it seems most likely that if life exists on Titan, its biomass could be considerable but its trophic structure would be simple, its different forms mostly sedentary, and its metabolism totally unique to our experience. The quest for life on Titan is of such great value precisely because of the way it will expand our understanding of life in general, if it is found there. But scientific caution requires us to admit the possibility that it might not be there at all.

## 10.9 References and further reading

1   Jenkins, L. 2008. Saturn, Dione, Rhea, and Titan. http://eltiriel.wordpress.-com

2   Lunine, J. I., Yung, Y. L. and Lorenz, R. D. 1999. On the volatile inventory of Titan from isotopic abundances in nitrogen and methane. *Planetary Space Sci* **47**: 1291–1303.

3   Waite, J. H., Jr., Young, D. T., Cravens, T. E., et al. 2007. The process of tholin formation in Titan's upper atmosphere. *Science* **316**: 870–5.

4   Grasset, O., Sotin, C. and Deschamps, F. 2000. On the internal structure and dynamics of Titan. *Planetary Space Sci* **48**: 617–636.

5   Niemann, H. B., Atreya, S. K., Bauer, S. J., et al. 2005. The abundances of constituents of Titan's atmosphere from the GCMS instrument on the Huygens probe. *Nature* **438**: 779–84.

6   Tobie, G., Grasset, O., Lunine, J. I., et al. 2005. Titan's internal structure inferred from a coupled thermal-orbital model. *Icarus* **175**: 496–502.

7   Shaw, A. M. 2006. *Astrochemistry: From Astronomy to Astrobiology.* West Sussex: John Wiley.

8   Hörst, S. M., Vuitton, V. and Yelle, R. V. 2008. Origin of oxygen species in Titan's atmosphere. *J. Geophys. Res.* **113**: E10006.

9   McKay, C. P. and Smith, H. D. 2005. Possibilities for methanogenic life in liquid methane on the surface of Titan. *Icarus* **178**: 274–276.

10  Schulze-Makuch, D. and Grinspoon, D. H. 2005. Biologically enhanced energy and carbon cycling on Titan? *Astrobiology* **5**: 560–567.

11  O'Brien, D. P., Lorenz, R. D. and Lunine, J. I. 2005. Numerical calculations of the longevity of impact oases on Titan. *Icarus* **173**: 243–253.

12  Marcano, V., Benitez, P. and Palacios-Pru, E. 2002. Growth of a lower eukaryote in non-aromatic hydrocarbon media >= C-12 and its exobiological significance. *Planet. Space Sci.* **50**: 693–709.

13  Schulze-Makuch, D., Haque, S, Resendes de Sousa Antonio, M., Ali, D., et al. 2010. Microbial life in a liquid asphalt desert. *Astrobiology*, in review.

14  Wang, H.-Q., Chen, Y.-J. and Qin, B.-Y. 2009. Degradability of n-hexadecane by Bacillus cereus DQ01 isolated from oil contaminated soil from Daqing oil field, China. *Int. J. Environ. Pollution* **38**: 100–115.

15  Raulin, F. and Owen, T. 2002. Organic chemistry and exobiology on Titan. *Space Sci Rev* **104**: 377–394.

16  Baross, J. A., Benner, S. A., Cody, G. D., et al. 2007. *The limits of organic life in planetary systems.* Washington, D.C.: National Academies Press.

http://www.jpl.nasa.gov/missions/missiondetails.cfm?mission=Cassini
Home page for the Cassini-Huygens mission to Saturn and Titan.


http://photojournal.jpl.nasa.gov/targetFamily/Saturn?subselect=Target%3A Titan%3A Repository for hundreds of images and models of Titan.

http://www.esa.int/SPECIALS/Cassini-Huygens/index.html
European Space Agency home page for the Cassini-Huygens mission to Saturn and Titan.

http://solarsystem.nasa.gov/planets/profile.cfm?Object=Titan
Brief fact sheet on Titan, with interactive comparison of Titan with other Solar System bodies.

http://www.lesia.obspm.fr/cosmicvision/tandem
Home page for the TANDEM mission, scheduled to send exploratory probes to Titan and Enceladus in about 2030.

# 11 Exotic Cocktails

*Life in liquid mixtures of organic solvents, solutes, and water beneath the surface on unimaginably cold worlds like Triton and Pluto*

Every 175 years, the outer gas giants line up in a way that makes the shortest and most energetically efficient pathway from Earth to each of them possible. In the mid 1960s, when it was realized that this infrequent window would open up between 1976 and 1978, a grand tour of the outer planets and their satellites was planned for the robotic exploration that came to be known as the Voyager Mission.

By the time the two Voyager spacecraft were launched in the summer of 1977, funding had forced the grand tour to be scaled back to an exploration of the Jovian and Saturnian systems only. Nonetheless, Voyager 2 was launched on a trajectory that would make travel to Uranus and Neptune possible. When Voyager 1 and 2 succeeded spectacularly in revealing the wonders of Jupiter and its four exotic moons, and Saturn with its marvelous rings and enigmatic giant moon, Titan, funding was found to send Voyager 2 toward the outer gas giants, restoring the original plan of the grand tour. After its closest approach to Saturn on 25 August 1981, the trajectory of the spacecraft was adjusted to set it on a course for Uranus and Neptune. Eight years to the day later, Voyager 2 made its closest approach to Neptune and its largest satellite, Triton. The vast majority of what we know about those two bodies is based on that one encounter.

Pluto and its co-rotating sibling, Charon, have never been visited by any spacecraft, but ground-based observations have given us enough information to whet our appetites for their first encounter with a robotic visitor, the New Horizons spacecraft, on target to arrive in 2015. In the meantime, what we do know leads us to believe that the Pluto-Charon pair, as well as Triton, are intruders from a birthplace beyond the orbit of Neptune. Once classified as a planetary pair because they orbit the Sun, Pluto and Charon were reclassified as dwarf planets in 2006, after even larger asteroids were found to be orbiting the Sun. Triton most likely originated from a similar population of dwarf planets, though classified technically as a moon, once it was captured by its adoptive parent, Neptune.

L.N. Irwin and D. Schulze-Makuch, *Cosmic Biology: How Life Could Evolve on Other Worlds*,
Springer Praxis Books, DOI 10.1007/978-1-4419-1647-1_11,
© Springer Science+Business Media, LLC 2011

## 11.1 Nature of dwarf planets

The creation of the Solar System was an incomplete event. At this point, 4.5 billion years into its life-span, the Solar System has accreted into a Sun and eight planets, but remnants of the earliest formative stages remain unassimilated. Those earliest stages of accretion saw the gradual growth of dust and gas into particles, then boulders, then bodies large enough for us to call them asteroids today. The largest asteroids grew larger by sucking smaller ones into their gravitational fields, and the largest of these became planetesimals. In the region of the inner Solar System, four planetesimals would give rise to our four rocky planets – Mercury, Venus, Earth, and Mars. In the region of the outer Solar System, four larger rocky planetesimals would attract enough gases left over from the Sun's formation to become the four gas giant planets – Jupiter, Saturn, Uranus, and Neptune.

As the eight survivors grew larger, they swept up most of the smaller asteroids in the vicinity of their orbits. Thus, the inner Solar System was cleared of most of its asteroids by the relatively closely spaced rocky planets. Though much further apart, the gas giants were massive enough to effectively clear out the protostellar disk all the way from inside Jupiter to a little beyond Neptune. A number of the asteroids and planetesimals that became moons must have formed along with the parental planets that they encircled, since they orbit in the same direction as their central planets rotate, and they lie roughly in the ecliptic – the flat plane in which all eight planets move.

The distance between Mars and Jupiter was apparently too great for either of those planets to clear asteroids from the space between them, so a belt of unassimilated asteroids still lies within that interval – perhaps as an incompletely accreted planet. (An alternative model poses the possibility that catastrophic collisions between fully formed planets created the asteroid debris there secondarily.) Lacking the formation of a large enough planet beyond the orbit of Neptune to assimilate them, another vast expanse of asteroids stretches for six billion kilometers beyond Neptune to the edge of the Solar System. This outer band of asteroids is named the Kuiper Belt, after Gerard Kuiper who suggested their existence in 1951, though a similar idea had been advanced by the Irish astronomer, Kenneth Edgeworth, eight years earlier.

Asteroids large enough to be several hundred kilometers in diameter have enough gravitational force to shape their masses into spheres. The process of accretion for bodies approaching a thousand kilometers in diameter involves enough thermal energy to melt their interiors and cause differentiation into a core and mantle. In the outer Solar System where temperatures are cold enough to enable water to stick to mineral substrates, the mantle of asteroids includes varying amounts of water-ice, which is why the sizable moons of the gas giants, and the larger asteroids of the Kuiper Belt, are mostly snow-ball worlds with densities roughly between 1.5 and 3.0 (the density of pure water being 1.0). Those that circle the Sun are dwarf planets. Those that orbit other planets are moons of those planets. Based on formation theory, they are generically the same.

It now appears that asteroids, like stars, commonly form in pairs. Pluto and Charon are examples. Triton may have been part of such a binary at one time, as discussed below. All three are clearly snow-ball worlds of the type with which the Kuiper Belt abounds.

## 11.2 Outline of the history of dwarf planets like Pluto and Triton

Triton is a moon because it orbits Neptune, and Pluto and Charon are classified as dwarf planets because they orbit the Sun, but all three probably have a common origin in the Kuiper Belt of asteroids and planetesimals that circle the Sun beyond the orbit of Neptune. While their origins cannot be known with certainty, the following outline attempts to trace a plausible history for where they came from and how they ended up in their current positions.

1. The protostellar cloud of gas and dust that formed our Solar System contained enough heavy elements to be "mineral rich." Therefore, dust particles, rocks, and millions of boulder-sized accretions formed as the bulk of the cloud collapsed toward its gravitational center.

2. The vast majority (over 99.9%) of the mass of the protostellar cloud collapsed into the Sun, while the remainder spun out into the protostellar disk. Either by luck or by rules we don't yet understand, eight planetesimals grew larger than the others, sucking in to their gravitational fields all the smaller objects in their orbital paths. These became the eight full-sized planets, with the four outer ones distant enough from the Sun to retain large volumes of hydrogen, helium, and other gases.

3. Of the residual material from the protostellar disk, millions of smaller accretions failed to consolidate into full-sized planets. Some became satellites orbiting the planets, like the Jovian moons and Titan. Others became asteroids, orbiting the Sun in a belt between Mars and Jupiter, or in a much broader plane, the Kuiper Belt, extending for about 6 billion kilometers beyond the orbit of Neptune.

4. A few asteroids grew massive enough to produce sufficient heat from radiogenic decay, which, in addition to the kinetic energy from ongoing impacts with smaller bodies, caused their interiors to melt and differentiate. With this metallic core and a mantle of minerals and water, spherical planetesimals with diameters of about 1000 kilometers or more formed what are now classified as dwarf planets. Many of them formed binaries (planetary pairs) which rotate around a center of gravity between the two.

5. The largest dwarf planet known in the Kuiper Belt is Eris, with an uncertain diameter of 2400 to 3000 kilometers. With similar sizes, Triton (2700 km) and Pluto (2300 km) presumably formed in the Kuiper Belt as well. At such a far distance from the Sun, their surface temperatures were cold enough to condense methane and even nitrogen.

6. Kuiper Belt objects are periodically perturbed by the passage of massive Neptune within their vicinities. Some are thrust into more eccentric and inclined orbits around the Sun. This appears to have happened to Pluto and its smaller companion, Charon, which revolve around one another in gravitationally-locked synchrony, and now together orbit the Sun in an elliptical path that brings them closer to the Sun than Neptune for 20 out of their 248.6 year orbital period.

7. Other Kuiper Belt objects were captured by the larger planets. This appears to have been the fate of Triton, which may have been the surviving member of a dwarf planetary pair. The angle and complexity of the encounter thrust Triton into a retrograde, oblique, and originally eccentric orbit around Neptune.

8. Triton's encounter with Neptune probably was energetic enough to melt water in the mantle, and the eccentricity of its original orbit continued to provide tidal heating that could have kept water liquid for up to a billion years, though the surface would likely have frozen over quickly.

9. Viscoelastic tidal forces due to Triton's retrograde orbit, along with friction from debris left over from the encounter with Neptune, eventually regularized Triton's orbit to its current near-perfect circle. However, radioactive decay in its core could still be generating energy internally.

10. The circumstantial evidence in favor of a family resemblance between Pluto, Charon, and Triton includes the fact that their bulk compositions are similar. With a density just over 2.0, they consist of about 2/3 rocky mineral and 1/3 water-ice and other frozen gases. Both have tenuous atmospheres of mainly nitrogen, volatilized from polar regions that alternate in facing toward and away from the Sun because both are severely tilted toward the plane of the ecliptic.

11. Because of its retrograde orbit around Neptune, gravitational drag is drawing Triton ever closer to the larger planet. Eventually, Triton will get close enough to Neptune to be pulled apart or crash into its adoptive parent.

## 11.3 Nature of Pluto and Charon

Pluto and its half-sized sibling, Charon, are a pair – probably born together of the same process. They orbit one another in synchrony, each revolving around the other, and on their own axes of rotation, every 6.38 Earth days (Figure 11.1a). Thus they keep the same face to one another at all times.

Having never been visited by any spacecraft from Earth, the nature of Pluto and Charon has been deduced entirely from ground based observations. These are barely enough to discern a little bit of surface detail on Pluto (Figure 11.1b,c). Charon is simply too small to make out even that much. For simplicity, we will speak of Pluto only instead of Pluto-Charon for the remainder of this chapter,

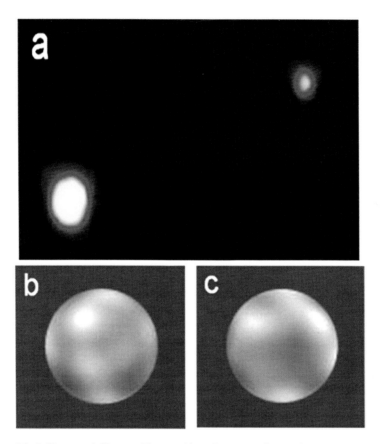

**Figure 11.1** Pluto and Charon. Pluto, with a diameter of 2274 km, and its half-sized companion (at 1206 km) form a binary system, like many asteroids in the Kuiper Belt. (a) Pluto and Charon orbit one another at a distance of 19,640 km, as seen in this image acquired by the European Space Agency's (ESA) Faint Object Camera aboard the Hubble Space Telescope, at a distance of 4.4 billion km (nearly 30 AU) from Earth – equivalent to photographing a baseball 65 km (40 miles) away. (b,c) Two faces of Pluto, viewed 3.2 Earth days apart to visualize opposite hemispheres. White polar caps, and striking contrasts suggesting major topographical variations can be discerned (Images from Dr. R. Albrecht, ESA/ESO Space Telescope European Coordinating Facility, and NASA (a) and Alan Stern (Southwest Research Institute), Marc Buie (Lowell Observatory), ESA, and NASA (b)).

recognizing that what we say about Pluto may or may not apply as well to Charon.

Pluto's origin as a Kuiper Belt object is indicated not only by the fact that most of its orbital period is spent in that region, but by its highly elliptical orbit and its inclination of more than 17 degrees from the plane in which the other planets orbit the Sun – characteristic of short period comets also presumed to come from the Kuiper Belt. Pluto orbits in a 3:2 resonance with Neptune, and has a radical

(122°) tilt to its axis, resulting in rotation that is retrograde to Neptune and most of the other planets.

Pluto's density of 2.03 g/cm$^3$ suggests a dwarf planet composed of roughly 70% rock and 30% water-ice. Its surface has areas that are very bright, suggesting the high reflectivity of frost or ice, and other regions of darkness, indicating quite a bit of topographical heterogeneity. In fact, Pluto shows the second greatest degree of surface contrast (after Saturn's moon, Iapetus) in the Solar System. In the absence of higher resolution images of the surface, the state of geological activity on Pluto is impossible to gauge.

Spectroscopic evidence indicates that Pluto has a very tenuous atmosphere composed mostly of nitrogen, with small amounts of carbon monoxide and methane as well. Its surface temperature is about -230°C. Therefore, the surface almost surely consists of nitrogen ice, with traces of frozen carbon monoxide and methane, as well as water ice.

## 11.4 Nature of Triton

Triton is by far the largest moon of Neptune, and the seventh largest satellite in the Solar System. It follows a nearly circular orbit around Neptune in the retrograde direction, moving opposite to the direction of Neptune's rotation (Figure 11.2). Triton's orbital plane is highly inclined toward the ecliptic at the present time, and as Neptune orbits the Sun, Triton's poles take turns facing the Sun with the result of extreme seasonal changes in the polar areas. These are orbital characteristics that would suggest an origin independent of Neptune's accretion. Triton's density of 2.08 g/cm$^3$ and diameter of about 2700 kilometers are slightly greater than those of Pluto, and typical of large Kuiper Belt objects. Most astronomers believe that Triton was captured from the Kuiper Belt by Neptune. The possibility that Triton originated as part of a dwarf planet pair, then encountered Neptune in a vigorous interaction that tore its partner away, has been proposed as a plausible model [1].

### 11.4.1 Composition and chemistry

Triton's size and density suggest a composition of about 2/3 to 3/4 rocky material, with the remainder as water ice and other frozen gases. Subtle yellow to peach-colored hues on its surface suggest the presence of organic compounds, perhaps reaction products from methane [2]. The blue-green areas closer to the equator may reflect frosts or ices of nitrogen and nitriles (Figure 11.3).

Triton's surface is the coldest yet measured in the Solar System: –235°C. At this temperature, even molecular nitrogen ($N_2$) condenses, as do hydrocarbons and nitriles. The surface is thought to consist of frozen nitrogen, water, and carbon dioxide, in that order of abundance, with traces of methane, carbon monoxide, and possibly ammonia.

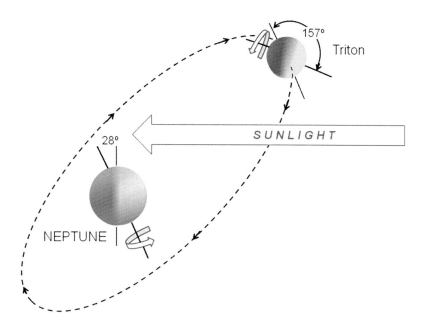

**Figure 11.2** Triton's orbit and orientation. Neptune orbits the Sun in a flat plane with the other planets (the ecliptic), tilted at an angle of 28° from a line perpendicular to the path of sunlight. It rotates on this tilted axis from west to east. Triton orbits Neptune in a retrograde, or opposite, direction to that of Neptune's rotation; and is tilted on its axis 157° from Neptune's axis. It rotates from east to west, tidally locked to Neptune so that one complete rotation on its axis takes the same amount of time, 5.9 Earth days, as one complete orbit. In this orientation, its south pole is pointed almost directly into the path of sunlight during its southern summer, which lasts over 41 Earth years. This was its orientation at the time of its encounter with Voyager 2 in August, 1989. Triton is shown disproportionately large relative to Neptune for clarity (Art by Louis Irwin).

### 11.4.2 Atmosphere

Triton has a tenuous atmosphere composed mostly of nitrogen and methane. A very thin haze extends 5–10 kilometers above the surface (Figure 11.4a), and evidence of air flow can be seen in streaks of dark material ejected from geysers (Figure 11.4b). At upper levels of the atmosphere, organic compounds are generated from methane and nitrogen by high energy particles from Neptune's magnetosphere [3]. Some must fall to the surface, but their life spans are thought to be brief [2]

### 11.4.3 Topography

Triton has a relatively smooth but varied topography, with some features seen on other worlds, and others seen nowhere else.

Because of Triton's extreme tilt, its poles point nearly directly toward or away from the Sun, during the summer and winter, respectively (Figure 11.2). At the time of its encounter with Voyager, Triton's south pole was pointed toward the

**Figure 11.3** Triton. This composite image was synthesized from pictures taken by Voyager 2 as it passed by Triton's southern hemisphere. A global surface of frozen nitrogen is differentiated into a complex polar region with pink-colored products of methane and white frost streaked with grey material released from the interior. "Cantaloupe" terrain crossed by lines and ridges occupies the equatorial region (NASA/JPL/USGS).

Sun, and the stark contrast between the polar and equatorial regions may have been a consequence of that seasonal timing.

The south polar region features a pink- to peach-colored substrate with a mottled white overlay. Grey streaks extend uniformly toward the east (Figure 11.4b). These are thought to be debris expelled from the interior by geysers, which presumably erupt when constant sunlight warms colored material beneath the transparent layer of frozen nitrogen on the surface. Nitrogen and particulate matter spew out of fumarole-like openings up to a height of 8 kilometers, then are blown downwind as far as 150 kilometers.

At the northeastern edge of the south polar region, three conspicuous splotches can be seen near the eastern horizon in Fig 11.3. A closer view of this area (Figure 11.4c) reveals them to be uniformly dark features, perhaps basins, rimmed with white material, presumably nitrogen frost. The largest on the lower right is roughly 200 by 100 kilometers, about the size of New Jersey. They

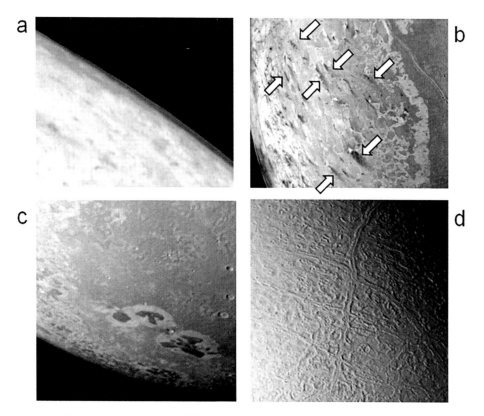

**Figure 11.4** Surface features of Triton. Numerous contrasting images are seen on Neptune's largest moon. (a) A very thin haze reflecting Triton's tenuous atmosphere can barely be discerned at the horizon. (b) Wispy streaks of gray material (arrows) are ejected from the interior and carried downwind in a uniform direction. The white mottled features are probably nitrogen frost. North is to the right. (c) Dark splotches of unknown origin apparently ringed with nitrogen frost near the transition from southern polar to equatorial latitudes. The center splotch is 130 km across the top with a 50 km stem. Several sizable impact craters are visible to the northeast. (d) "Cantaloupe" terrain grooved with ridges of various ages. Very few impact craters are evident (NASA/JPL).

somewhat resemble lag deposits from desiccated lakes on Earth, but otherwise are unique. Whether their presence at the transition from polar to equatorial zones is suggestive or just coincidental is unknown.

A truly unique topographical feature seen nowhere else in the Solar System is the "cantaloupe" terrain featured in the western equatorial region in Figure 11.3, and closer up in Figure 11.4d. The terrain derives its name from its resemblance to the skin of a cantaloupe. The depressions measure 30–40 kilometers in diameter with walls several hundred meters high. Their spacing and size are too regular for them to be volcanic calderas or impact craters. The favored hypothesis for their formation is diapirism – the upward thrust of less dense material through a denser crust. While the cantaloupe nature of the terrain is unique to Triton, grooves of troughs and ridges course across the terrain, closely resembling those seen abundantly on Europa and many of the other icy satellites. The cantaloupe terrain would thus appear to combine features unique to Triton with processes known to occur elsewhere.

Much of Triton resembles broad volcanic plains, similar to the mare on the Moon. At higher resolution, these regions are seen to be quite complex, like the line of fumaroles or calderas, closely cropped ridge lines, and depressions possibly resulting from faulting and/or sapping seen in Figure 11.5a. Broad, flat depressions look like the remnants of extensive flooding (Figure 11.5b,c) of the type that could be caused by the spread of lava from cryovolcanic eruptions. Consistent with this view is the fact that the surface appears quite young, based on crater counts which are four to five times lower than on the much smaller moons of Uranus.

### 11.4.4 Geological activity

The apparent youth of Triton's surface, estimated at 5–50 million years, is a primary reason to assume that it is now or recently has been geologically active. The superimposition of newer surface features over older ones indicates that resurfacing has been ongoing on Triton for some time. Furthermore, contemporary evidence of geyser eruptions during the southern summer indicate dynamic activity on a seasonal cycle.

The large-scale resurfacing is somewhat mysterious, in that Triton's near perfectly circular orbit would cast doubt that tidal irregularities with Neptune could be pumping energy into Triton. Perhaps interactions with other Kuiper Belt bodies, or ongoing radiogenic decay, provide the energy for these eruptions.

### 11.5 Conditions conducive for life on Triton and Pluto

We simply don't know about Pluto. Readers to this point will have learned we don't shrink from informed speculation, but when information is too scant for plausible ideas, we have to decline to offer any. What we *can* say is that Triton and Pluto, as best we can tell, are virtual twins in size, probably origin, and even the inclination of their axes of rotation. Therefore, what we will do is analyze the conditions conducive for life on Triton, recognizing it could be a proxy for whether life could exist on Pluto as well. Soon after New Horizons arrives at Pluto in 2015, we'll have the information we need to start making educated guesses about that world. For the remainder of this chapter, we will restrict our consideration just to what we know about Triton.

### 11.5.1 Energy for life on Triton

At a distance of 30 astronomical units from the Sun, Triton receives 1/900th of the intensity of sunlight that strikes the Earth. Even if means had evolved to capture such a faint amount of light for biological purposes, it seems doubtful

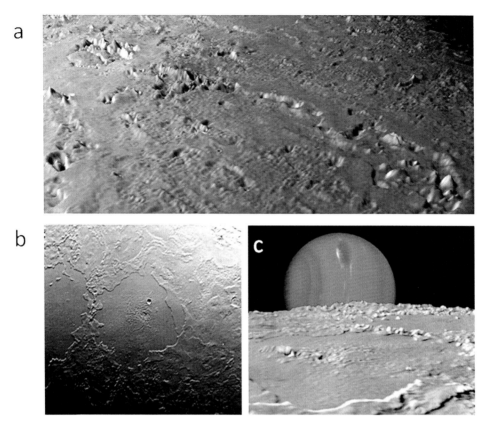

**Figure 11.5** Volcanic Plains of Triton. Broad areas of Triton resemble smooth volcanic plains, like the mare on the Moon. (a) A line of pits, depressions, and calderas stretching for about 500 km across an otherwise smooth plain. The pits are typically 10 km across, while the large depressions with central peaks at the far left and right are 50 and 80 km across, respectively. (b) Remnants of two broad plains, suggestive of flooding by cryovolcanic lava are seen from above. (c) Computer reconstruction of what the plains in panel (b) would look like from a height of 45 km, looking toward Neptune on the horizon. Vertical relief is exaggerated by 25× in panel (a) and 30× in panel (c) to aid visualization (NASA/JPL/USGS/Universities Space Research Association/Lunar & Planetary Institute).

that any liquid could exist other than transiently, near enough the surface of such a cold world to make biological use of it. Therefore, sources of energy other than light for putative life on Triton need to be sought.

The potential for organic chemistry does appear to be present. Methane and some of its reaction products clearly exist near the surface. Conversion of methane and nitrogen into other organic compounds apparently occurs in the upper atmosphere, though certainly to a lesser degree than on Titan. We don't know whether sufficient organics are present, along with the other chemical reactants needed to support energy-yielding reactions like oxidation-reduction cycles, on Triton. To the extent that color is an enticing indicator of some degree of chemical complexity, Triton does display that. And the brown material discharged and carried downwind from geysers is reminiscent of tholins seen on Titan. It could, of course, be only dirty water. The bottom line about chemical sources of energy for putative life on Triton is that they probably are there.

From several indications, geothermal energy is still in play on Triton. The youth of the surface, the flood plains that seem best explained as evidence for extensive cryovolcanism, the abundant pits and occasional calderas that betray more than occasional eruptions – all these support the view that internal heating continues to drive geological activity. So heat with bioenergetic potential could be available from the interior.

Neptune's magnetosphere is very weak, and magnetic energy is very inefficient to begin with, so this is probably not a viable option. Other forms of energy, like osmotic and ionic gradients, or kinetic energy could be present if extensive ocean-like reservoirs of liquid exist beneath the surface, which appears to be a distinct possibility.

### 11.5.2 Building blocks for life on Triton
Carbon is so superior as a building block for life, under so many of the conditions encountered on nearly all worlds other than stars (see section 2.2 in chapter 2), that it has to be considered the leading candidate for the building blocks that make up any biomolecules on Triton. And organic carbon is clearly available on Triton.

If there is any place in our Solar System, however, where silicon could conceivably be a building block for life, Triton is the leading candidate. Silanes are silicon compounds structurally analogous to hydrocarbons, with the carbon simply replaced by silicon (Figure 11.6). Conditions that favor the existence of polysilanes include the following: extremely cold temperatures, absence of liquid water, lack of reactive oxygen species, restricted abundance of carbon, solvents like methane or methanol compatible with silanes, and high pressure [4]. Triton meets most of these requirements. Its subsurface down to an unknown depth is too cold for pure water to be liquid but is probably under high pressure and mixed with other components such as ammonia which would raise its melting point. Further, temperatures are cold enough to liquefy methane and methanol. Oxygen is restricted to frozen reservoirs of carbon monoxide and dioxide, and the other forms of carbon, though organic, do not appear to be in high abundance. Furthermore, Triton provides conditions under which the formation of silanes is favored: a cryogenic environment, bolide impacts on cold planetary bodies, and ultraviolet radiation of ice-coated silicate grains [5].

An obvious catch to the possibility that silanes could ever be prevalent is the fact that if methane is abundant enough to be a solvent, it may provide enough carbon to be competitive with silicon. Thus, silanes as building blocks for putative live on Triton may be a long shot, but we don't know enough about the behavior of silanes in unusual chemical cocktails under high pressure at extremely low temperatures to rule them out as building blocks for life on

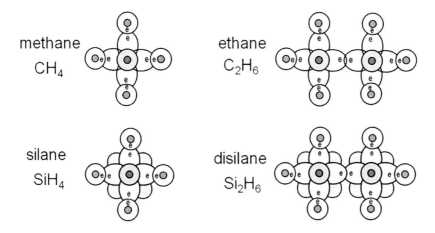

**Figure 11.6** Similarity of silanes and hydrocarbons. The two simplest hydrocarbons differ from the two simplest silanes only by the substitution of silicon for carbon. Refer to Fact Box 2.2 for structural conventions.

worlds at the margin of our Solar System. Silanes are rarely found in meteorites or at other interstellar locations, but the protostellar disk W33A appears to be an exception. Thus, a rare set of circumstances may be found where silanes can form into biomolecules. The possibility that Triton could be such a world cannot be excluded.

### 11.5.3 Solvents for life on Triton

Two questions need to be answered in considering whether solvents could exist for the support of life on Triton. The first is what their chemical nature would be. The second is whether they could be liquid in any putative habitat on Triton.

Water makes up ¼ or more of the bulk composition of Triton, but if it exists there persistently in liquid form, it must be well beneath the surface. The abundant evidence for cryovolcanism indicates that a subsurface cocktail of liquids which includes water as a major component must be heated enough on occasion to erupt to the surface, though just how much of cryovolcanic lava is made up of water rather than methane, nitrogen, or ammonia, is uncertain.

As on Titan, underground reservoirs of liquid water-ammonia are possible, but they require considerably higher temperatures ($\geqslant$ –80°C) than are found anywhere near Triton's surface. Liquid methane is a better bet – the question there being whether methane can be concentrated enough to form substantial solvent pools.

Nitrogen ($N_2$) poses an intriguing possibility. Triton's surface is covered with frozen $N_2$ at a temperature of about -235°C. $N_2$ melts at –210°C, then vaporizes at –196°C. Thus it stays liquid only in a very narrow thermal range, but one likely to be found at some depth beneath Triton's surface. That might not be far down, either, since the transparent layer of frozen $N_2$ on the surface appears to be thin enough to warm underlying colored chemicals through a greenhouse effect. If under the summer sun, enough energy can be gathered to generate geysers, then there may be enough to form narrow pockets of liquid $N_2$ – perhaps a film of liquid between overlying frozen $N_2$ and underlying frozen $H_2O$; or maybe liquid $N_2$ inclusions within the layers of frozen $N_2$. As a liquid, nitrogen's low dielectric constant and zero dipole moment would mean it would act like a non-polar solvent – not the ideal for carbon-based biochemistry. That might be just the solvent, though, in which polysilanes could achieve a measure of complexity and dynamic interaction.

### 11.5.4 Habitats for life on Triton
The surface of Triton is too cold, and its atmospheric pressure too low, for any liquid to remain stable. Below the surface, two microhabitats and a third major one are conceivable.

The first microhabitat would be tiny water-ammonia-saline ice inclusions within the bulk of water ice that makes up much of the mantle. Though probably too cold for such inclusions near the surface, at greater depth, internal heating, high pressure, and the freezing point depression provided by dissolved ammonia and solutes could result in stable liquid inclusions capable of supporting halophilic (salt-tolerant), water-borne, carbon-based life.

The second microhabitat, alluded to above, would consist of thin films of liquid $N_2$ between overlying frozen $N_2$ and an underlying solid substrate like frozen $H_2O$, or liquid $N_2$ inclusions embedded in frozen $N_2$. The seasonally driven geysers that appear to result from solar heating of underlying volatiles show that liquefaction of frozen $N_2$ happens. Thin layers of liquid $N_2$, or liquid $N_2$ inclusions, could thus exist.

The third habitat would be the major one for life beneath the surface of Triton. This would be whatever is left of the global ocean that likely enshrouded the satellite for some time after its capture by Neptune. Circumstantial evidence that liquid water still exists underground on Triton includes the compelling case for cryovolcanism which is probably still occurring, and the cantaloupe terrain, which is easiest to explain by the dynamic action of subsurface liquid reservoirs. To be sure, the cantaloupe terrain is like no other seen in the Solar System, so whatever gives rise to it might be quite an exotic cocktail of subsurface liquids. But given the amount of Triton's bulk composition estimated to be water, it must be a water-based cocktail to a substantial degree. With a negligible amount of tidal heating from Triton's circular orbit, internal temperatures are presumed to be kept high enough to liquefy water and drive cryovolcanic eruption by radiogenic decay in the core.

## 11.6 Scenarios for the possible evolution of life on Triton

With Triton, we encounter our first example of a world on which life probably did not originate early in its planetary history. This view is qualified by the fact that we know very little about the early eons of a Kuiper Belt body's existence. But consider the challenges facing the origin of life on a dwarf planet more than 30 AU from the Sun. Clearly, water and some organic compounds would be available, but energy from sunlight and magnetic fields would be negligible. Cataclysmic collisions would have been frequent, as they were for the larger planets; but on dwarf planets, cooling would have ensued much faster after their version of the Great Bombardment subsided – especially on those as relatively small as Triton and Pluto. While it isn't impossible, so far as we know, for life to get going pretty fast, the alternative, more conservative view is that life on Triton had a better chance to form after its capture by Neptune rather than before. That's the scenario we propose for the possible evolution of life on Triton.

In the beginning, the only one of our three putative habitats for life on Triton would have been the global ocean, liquefied by the capture encounter with Neptune [6]. Even if the surface froze over fairly quickly, the mass of water and the thermal energy available would have kept it in liquid form for hundreds of millions of years, according to current models. Access to the surface would have

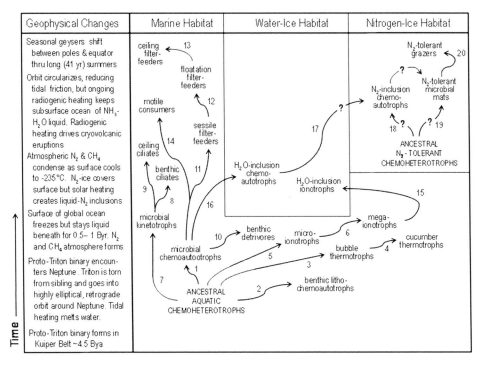

**Figure 11.7** Possible trajectory for the putative evolution of life on Triton. Geophysical changes over the planetary history of Triton are shown in the left-hand column. Time is represented on the vertical axis, without dimensions since the actual timing of each stage is not known. Life is presumed to have arisen only after Triton's encounter with Neptune melted its global ocean.

been of no use, since sunlight for phototrophy would have been negligible. The possibility that UV photodissociation of $H_2O$ could have produced some $O_2$ for use in oxidation-reduction reactions cannot be discounted, but chemoautotrophy in or at the bottom of the ocean seems more likely.

As Triton settled into its initially eccentric orbit around Neptune, then, it may have looked a lot like an early Europa. Indeed, the evolutionary trajectories for life on the two worlds may well have paralleled one another quite closely. Our version for Triton is shown in Figure 11.7, to which the steps in the following narrative refer.

We envision that water-borne, carbon-based life likely began with bottom-dwelling heterotrophic proto-cells subsisting on organic chemicals from their environment. From these ancestral chemoheterotrophs, microbial chemoautotrophs would have arisen (*step 1*). They could have been methanogens, or microorganisms similar to our sulfur-iron bacteria – we don't have any way of knowing what, so will refer to them in generic terms only. Some form of lithoautotroph probably arose soon as well (*step 2*). Other early descendants of the last common ancestor for life on Triton could have included a succession of thermotrophs (*steps 3 & 4*), ionotrophs (*steps 5 & 6*), or kinetotrophs (*steps 7–9*), as well as the usual detrivores (*step 10*).

Hundreds of millions of years would have presumably been long enough for complexity to evolve, at least to the point of macroorganismic consumers capable of filter feeding, first from a fixed position on the ocean floor (*step 11*), then as floaters like jellyfish (*step 12*), and finally by attachment to the ice ceiling at the top of the water column (*step 13*). With an abundance of producers available for food, consumers would inevitably have evolved (*step 14*).

The ice ceiling would also have created a new habitat in the form of liquid inclusions within the ice. Iono- or osmotrophs ought to have been pre-adapted to take advantage of the high saline content of these aqueous microhabitats (*step 15*). Equally plausible would be the sequestration of psychrophilic (cold-tolerant) chemoautotrophs (*step 16*). The evolution of these ice-dwelling organisms would not have represented a radical transition from their open ocean ancestors.

As Triton's orbit gradually became less eccentric, tidal heating would have diminished, though enough remained along with radiogenic heating from the core, to keep an unknown amount of the ocean liquid. Outside, however, the temperature would be plunging to levels not seen on Europa, or eventually, on any other world in the Solar System. Frigid Titan would have been the closest analogue, and indeed, at this point in Triton's history, it would have begun to look a little more like Titan, a little less like Europa. Especially, with its exotic mix of organic compounds and their breakdown products, hydrophobicity would have begun to look a little more like a respectable life style. But how to get from hydrophilic forms of life to something so different?

If we're going to be brave enough to propose the possibility of life on Triton, we may as well go all the way and suppose that life in liquid nitrogen might be the ticket to survival at temperatures under 40 degrees above absolute zero. For that matter, silanes might make more appropriate backbones for biomolecules in an oxygen-poor hydrophobic environmnent like liquid nitrogen. So the idea is that nitrogen-borne, silane-based life just might have taken hold in the tiny cracks, fissures, and inclusions that inevitably would open up inside frozen nitrogen.

Titan, appropriately enough, provides us with a way of thinking about this problem. As on Titan, where the challenge was in evolving life from water-based to ethane-based forms, it could have happened on Triton in two ways. First, life may have arisen *de novo* within the liquid nitrogen fissures or inclusions, giving rise to an endemic $N_2$-tolerant chemoheterotroph. Or, with $H_2O$-ice in contact with $N_2$-ice, transitional forms that gradually shifted from hydrocarbon-based to silane-based building blocks may have evolved over time (*step 17*). The change from $H_2O$ to $N_2$ as a solvent may have been aided by transitional forms through, first, a mixture of $H_2O$ and ammonia ($NH_3$), then later, a mixture of $NH_3$ and $N_2$. Whether the barriers between the different melting points for $H_2O$, $NH_3$, and $N_2$ could have been surmounted, and whether biochemistry is even possible in all these solvents is, of course, questionable. It may be a long shot, but we think it worth considering.

Assuming that a $N_2$-tolerant chemotroph of some sort eventually found itself living in liquid nitrogen, evolution of fully autonomous forms could have evolved to occupy either inclusions (*step 18*) or fissures (*step 19*), or both. Biofilm mat-like organisms spreading across frozen surfaces in liquid nitrogen may even have given rise to some type of consumer (*step 20*), thereby providing at least two trophic levels of silane-based life in liquid nitrogen – arguably the most exotic ecosystem that might exist this side of the Oort Cloud.

## 11.7 Chapter summary

Triton and Pluto are dwarf planets by birth, and immigrants into the family of major planets in our Solar System. Pluto still dwells most of the time in the region where both were born, the Kuiper Belt of asteroids beyond Neptune. Every 260 years or so, its orbit dips closer to the Sun than Neptune's for a 20 year visit before retiring with its binary companion, Charon, to a point reaching 40 astronomical units from the Sun. Triton is now a permanent resident, having been snatched by Neptune from the Kuiper Belt to serve as its largest satellite.

Both bodies display the characteristics of asteroids born far from the Sun. Two-thirds to three-fourths of their mass is rock; the rest is mainly water. Among the coldest bodies ever measured, the ice is frozen as hard as rock, at least on the outside. Organic compounds in the form at least of methane and some of its reaction products including nitriles are among its constituents, but nitrogen is the other major volatile found on them. Enough nitrogen coats the surface, in fact, to generate a very thin atmosphere of $N_2$, with trace amounts of the other usual gases: methane, carbon monoxide, and carbon dioxide.

Triton is the strangest looking body in the Solar System. The combination of extreme cold, colored chemicals, severe tilt that causes winter to alternate between the equator and the poles, and likely underground reservoirs of liquid give rise to a complicated surface topography and appearance that is both familiar and unique. Exotic by every standard that applies to Earth, the prospects for life there appear daunting from our perspective, but far from impossible.

The history of Triton's formation and capture suggest that a global ocean covered the dwarf planet-turned-satellite after it went into orbit around Neptune. Enough time should have elapsed for life to emerge, with the organic chemicals and a choice of several energy sources that were available. Though probably encased in an icy crust fairly soon, a Europa-like trajectory is certainly plausible.

What makes Triton truly unique is its transition from a Europa-styled satellite to something more like Titan, where organic chemistry is greater in evidence. But then Triton pushed past Titan in cooling to the point where even nitrogen froze. This event provided a new microhabitat – reservoirs of liquid nitrogen – to join the two other possible habitats: aqueous inclusions in water-ice, and extensive subsurface pools, if not an ocean, of water-based solutions.

Assuming the origin of life on Triton in water, the major line of evolutionary descent would have looked like that on Europa for the most part. Emergence of the liquid nitrogen microhabitat, however, opened the possibility to a more exotic trajectory – one in which nitrogen-tolerant organisms, possibly relying on silane building blocks in whole or in part, came to occupy the fissures or inclusions of liquid nitrogen. Getting "there" from a Europa-like "here" may have been a challenge, but nature has a tendency to show more creativity than the human mind can imagine.

## 11.8 References and further reading

1   Agnor, C. B. and Hamilton, D. P. 2006. Neptune's capture of its moon Triton in a binary-planet gravitational encounter. *Nature* **441**: 192–4.
2   Thompson, W. R. and Sagan, C. 1990. Color and chemistry on Triton. *Science* **250**: 415–8.
3   Thompson, W. R., Singh, S. K., Khare, B. N., et al. 1989. Triton: stratospheric molecules and organic sediments. *Geophys Res Lett* **16**: 981–4.
4   Schulze-Makuch, D. and Irwin, L. N. 2008. *Life in the Universe: Expectations and Constraints*. Berlin: Springer-Verlag, 2nd ed.
5   Bains, W. 2004. Many chemistries could be used to build living systems. *Astrobiology* **4**: 137–167.
6   Shock, E. L. and McKinnon, W. B. 1993. Hydrothermal processing of cometary volatiles–applications to Triton. *Icarus* **106**: 464–77.

http://nineplanets.org
Good starting point for basic facts about Triton, Pluto, and Charon.

http://photojournal.jpl.nasa.gov/targetFamily/Neptune?subselect=Target%3A-
   Triton%3A
Excellent images of Triton from the Voyager mission.

http://solarsystem.nasa.gov/planets/profile.cfm?Object=Neptune&Display=-
   Moons
Brief overview of Triton.

http://solarsystem.nasa.gov/planets/profile.cfm?Object=Dwarf&Display=Over-
   view
Good explanation of the definitions of planets, dwarf planets, and plutoids
   (dwarf planets in orbits beyond Neptune).

http://pluto.jhuapl.edu
Home page for New Horizons mission, launched in January 2006 for encounter
   with Pluto and Charon in July 2015.

# 12  Biocomplexity in the Cosmos

## Factors favoring the evolution of intelligence and emergence of Technology

Our survey of the possibilities for life in our Solar System has left us with the strong likelihood that ours is the only world with intelligent species, the only one on which at least one species has acquired technology, and quite possibly the only planet orbiting our Sun that hosts organisms of great size, complexity, and diversity. We will now turn to the question of why life with those characteristics arose on Earth, and what the explanation portends for the possibility that complex, intelligent, technologically competent life will be found on any given world.

## 12.1 Evolution of size, complexity, and biodiversity

In addition to a vast microbial biomass, the evolution of life on Earth has produced organisms of large size and complexity. It has done so by enabling the multiplication of the microscopic unit of life, the single cell, a billion times over into integrated colonies that produce living organisms of large size. Furthermore, by introducing cellular differentiation and specialization, it has generated organisms of great complexity. Because they are complex in many different ways, they have evolved into many different forms. Thus, the biosphere on Earth is highly diverse, consisting of life both large and small, both simple and complicated [1].

We don't know if beings of comparable size and complexity exist anywhere else in the Solar System. We have no evidence that they do, but until we can peer into the lava tube caves on Mars or beneath the icy crust of Europa, we can't rule them out. They almost surely do exist somewhere, on some other worlds. But why are we able to say that? In fact, why are we able to consider the possibility of multicellular, macroscopic forms of life on our neighboring worlds? By looking at a few specific factors that appear to have enabled the emergence of large size, complexity, and diversity on Earth, we should gain insight into where and under what circumstances we expect the same to happen on other worlds.

L.N. Irwin and D. Schulze-Makuch, *Cosmic Biology: How Life Could Evolve on Other Worlds*, Springer Praxis Books, DOI 10.1007/978-1-4419-1647-1_12,
© Springer Science+Business Media, LLC 2011

### 12.1.1 Energy

Organisms grow large when and if the biomass to support them is even larger. Biomass, in turn, is ultimately a function of the amount of energy available. Given what we learned about biomass pyramids in chapter 3, it stands to reason that consumers at the top of the pyramid can grow large only if a greater biomass of lower-level consumers and producers provides sufficient nutrients (hence energy) to make the organisms at the top of the pyramid big.

Complex food webs and intricate ecosystems likewise depend on an abundant supply of energy. This will support a large and diverse mass of producers, which in turn can feed large and diverse groups of consumers. Earth's essentially inexhaustible supply of energy from sunlight, one of the most efficient and effective means of powering living systems, makes possible the biodiversity found on the planet.

Earth also harbors a few macrobiotic members of ecosystems based on chemoautotrophs that derive their energy directly from chemicals in the environment. These examples provide analogs for what we suspect is the most likely energy source on worlds where life may exist in dark, subsurface habitats. Noteworthy, however, is the fact that such cases on Earth – where organisms reach sizable dimensions at all – are limited to deep ocean vents, for which the ecosystems supported by them do not extend very far. All other forms of life dependent on chemoautotrophy at the producer level are microbial in size. That isn't to say that large size based on any of the other energy sources we have surveyed in this book is impossible; just that we have no precedent for it on Earth other than in the limited case of deep ocean vents.

The high metabolic rates that support the maintenance of large, integrated organisms and high levels of activity are facilitated by atmospheric-biospheric interactions that make oxidation-reduction cycles not only feasible, but self-sustaining. Earth's high atmospheric content of oxygen makes oxidation of reduced carbon compounds (like glucose) an effective means of harvesting biologically useful energy, while photosynthesis turns the carbon dioxide thereby produced back into reduced carbon compounds for nutrients and replenishes the atmospheric supply of oxygen. Similar cycles (not necessarily of oxidation and reduction, though those seem most likely) would appear to be necessary in any ecosystem on any world where significant size and complexity are to be achieved. It is thought that the Cambrian Explosion, which led to the first large organisms on Earth, occurred primarily because oxygen finally reached a critical level in the atmosphere.

### 12.1.2 Temperature

The average surface temperature on Earth is about $15°C$, well below the point at which most proteins become inactivated (around $45–50°C$) but well above the melting point ($0°C$) of water. This means that water-borne life can operate in most of the habitats on Earth above ground and wherever bodies of water are liquid. With space to expand in the air, on land, and in the water, no physical limitations constrain the size of organisms. This is not true for forms of life sequestered in the soil, or encased in ice inclusions, for example.

Seasonal thermal cycles promote the evolution of organisms specialized to take advantage of extreme cold, extreme heat, moderate temperatures year round, and everything in between. Specialization of body parts and organ systems inevitably means more complexity.

Life is effectively restricted to temperatures at which its major solvent can exist in liquid form. On worlds where the solvent is water, thermal habitats can vary over the entire range at which water is liquid, from its freezing to its boiling point. While extremophiles at either end of the range are microbial, between about 5°C and 45°C, multicellular organisms can thrive, and natural selection on Earth has given rise to a full range of adaptations at every thermal point in that range. A great variety of adaptations means a great variety of organisms.

Does a narrower thermal range, as for a solvent other than water, preclude large size and complexity? Not necessarily; but in theory, the challenges are greater. One can imagine, for instance, the evolution of large, bottom-dwelling organisms in lakes of ethane and methane on Titan. The narrower thermal range in such an environment simply means that the "degrees of freedom" for evolution of different specializations are going to be fewer. The biochemical versatility of non-aqueous solvents might also be a good bit less as well, though we don't know that for sure.

### 12.1.3 Mobility
Space in which to move about, and temperatures that allow for dynamic activity, enable many organisms to be mobile. For macroorganisms, this requires muscular specializations and a means of coordinating incoming information with outgoing motor commands. On Earth, this has meant the evolution of nervous systems in animals. And complex nervous systems are associated with complex, and usually larger, animals.

The biological utility of mobility is the ability to chase prey, escape from predators, find mates (often involving a complicated behavioral repertoire required for mate recognition and consummation), spread out to optimize resources, and seek shelter. Each of these can be accomplished in many different ways, each of which requires different types of anatomical and physiological specialization.

### 12.1.4 Time
Considering all the features of Earth favorable for the evolution of large, complex, and diverse organisms, it should be noted that life evolved for nearly two billion years before becoming macroscopic. For at least 1.5 billion years, it was unicellular and prokaryotic – the simplest of cellular organizations. A testimony to life's tendency to remain stable and unchanging if optimally adapted lies in the fact that, while life arose fairly soon after it could, it then remained simple and not that diverse for over half the span of its entire existence.

The one case of life we know does not give us enough information to generalize about how rapidly life can emerge or how fast it can evolve and diversify under ideal conditions. We are left with the empirical evidence from our single sample that, even under the most favorable conditions, organisms may require a considerable amount of time to achieve much size or complexity.

The larger and more complex the organism, the greater the opportunity for diversification. But if evolution requires a considerable amount of time to produce large and complex organisms in the first place, the full development of a diverse biosphere takes even more time. Again, we don't know whether there are inherent limits on how fast biospheres can diversify. Intuitively, it would seem that the rate limiting factor is most likely the rate at which planetary conditions are changing.

### 12.1.5 Habitat fractionation

Habitats can be fractionated, not only on the basis of prevailing or oscillating temperature, but by topography, altitude, moisture, incident sunlight, substrate (rock, soil, sand) or lack thereof (air, water). On a global scale, the different biomes, such as polar, taiga, tundra, alpine, prairie, woodland, desert, and tropical forest, to name the major ones, represent habitat differentiation on a large scale. Within each of these, fractionation occurs from the community level down to the microhabitat, on a highly differentiated planet like ours.

Wherever different habitats abut or grade into one another, interfaces or gradients between them produce yet more habitats. On planets like Earth, which consist of ample bodies of liquid (water) and substrate (land) in contact with one another at and below the surface, the interfaces are extensive. The formation of novel habitats at these points of contact between land and water, land and air, and water and air provides the opportunity for even greater biodiversity. Consider the rich variety of life found in tidal pools at the sea shore, along river banks, and in ascending layers of canopy in a rain forest. Where the climate and topography are much more homogeneous, as in polar regions, biodiversity is much reduced.

### 12.1.6 Planetary history

Because evolution is essentially an irreversible process, the course of geophysical history on a planetary body affects the sequence of organisms produced by the evolutionary trajectory on that particular world. To the extent that planetary history has been complicated, the resulting biosphere will reflect its historical twists and turns. On Earth, for example, abundant energy and mild temperatures through most of the Mesozoic enabled reptiles, including their dominant forms, the dinosaurs, to occupy and dominate nearly every niche on land, in the air, and in the seas. With the catastrophic end of the Mesozoic and extinction of the dinosaurs, the surviving heterothermic reptiles were superseded by the homeothermic mammals and birds, as the world turned colder and the vegetation changed. Thus, mammals, preadapted with homeothermy and internal gestation, and birds, preadapted with homeothermy and the ability to fly, became the

**Figure 12.1** Biodiversity in a coral reef. The interface between solid substrate and water provides an opportunity for a great diversity of organisms (Photograph by Carol Irwin).

dominant vertebrates in Earth's biosphere. Absent the event that led to demise of the dinosaurs, the Earth might still be dominated by them.

## 12.2 Evolution of intelligence

Intelligence can be defined as the ability to integrate experience and anticipate the future. Once mobile forms of life – organisms that seek out their food by grazing, collection, or capture – have evolved, the nervous systems (or whatever system provides integration and coordination) would appear to be capable of developing sufficient complexity for intelligence to evolve. Indeed, intelligence has emerged independently in several major groups of animals on Earth, appearing first among the cephalopods over 400 million years ago. It has not, however, turned out to be a frequent or conspicuous product of evolution.

Technology can be defined as the use of energy, tools, material, and information by a living organism to amplify its impact on the environment. With some limited exceptions, only the human species has acquired significant technological capabilities on Earth. Below we consider the factors necessary for the evolution of intelligence and technology, and (in light of the fact that intelligence and technology are rarer than might be expected) why those factors are not necessarily sufficient to lead inevitably to either intelligence or technology.

### 12.2.1 What is intelligence?

Intelligence is a graded phenomenon and notoriously difficult to assess from a purely human perspective. Nonetheless, several instances of conspicuous intelligence have appeared during the course of evolution on Earth, as has at least one case of meta-intelligence (Figure 12.2). An examination of these cases may provide some insight into the main factors that appear to be important for its emergence.

Cephalopods have achieved the pinnacle of intelligence among all the invertebrates [2]. They diverged from other mollusks in the late Cambrian, about 500 Mya, became numerous and diverse in the succeeding periods until suffering a cataclysmic decline during the Permian crisis, 250 Mya, with only the octopi, squids, cuttlefish and a nautiloid surviving to the present day. Those forms, however, are active benthic foragers and predators, with highly developed tactile and visual sensory abilities, and elaborate motor systems for the control of jet-like propulsion, complex mouth part movements, and fine manipulation of each of their 8–10 appendages. The high degree of sensorimotor coordination in cephalopods is managed by a large brain centralized into a conglomerate mass of lobes and lobules, with differentiated inputs and functions, arranged in essentially a hierarchical fashion. Four levels of control can be distinguished: (1) intermediate motor centers that control discrete but unpatterned movements; (2) higher motor centers that generate coordinated motor patterns; (3) receptor analysis and memory stores, where pathways from gustatory, olfactory, and visual sensory receptors are brought together; and (4) lobes which appear to control motivation and assessment of rewards. Workers familiar with octopus behavior claim to detect "personality" changes in animals subjected to injury or surgical removal of the subfrontal lobes.

Insects diversified to fill every niche in the coal forests of the Carboniferous Period, around 300 Mya. By about 150 Mya, the social insects – ants, wasps, and bees – had evolved from their ancestral forms. Biologists do not rate individuals among the social insects as intelligent in the conventional sense, but in the aggregate, they display some of the features that would suggest intelligence, were they a single organism. They build elaborate housing, divide labor, communicate symbolically (in the case of bees), radically modify their microenvironment, grow food (in the case of fungal cultivating ants), domesticate other species, wage war, and cooperate for the good of the whole [3]. As such, they represent a case of meta-intelligence. The success of the social insects is testimony to the adaptive utility of this form of intelligence, and suggests that it could be just as relevant as individual intelligence for studying the generic fate of intelligent-like capabilities.

Among the birds that evolved from flying reptiles in the late Cretaceous, about 70 Mya, were the parrots and their relatives. Macaws appeared in the New World after the rise of the Andes about 28 Mya, and modern cockatoos date from about 20 Mya. The parrots, macaws, and cockatoos are members of the Psittaciformes, birds known for a level of intelligence claimed to be comparable to that of some non-human primates [4]. They have brain to body weight ratios

**Figure 12.2** Intelligent species. Conspicuous intelligence has arisen on Earth in relatively few animals. Among them are (a) cephalopods like octopi, (b) psitaccine birds like macaws, (c) cetaceans like dolphins, and (d) primates like orangutans. Social insects like (e) honeybees have evolved a high degree of meta-intelligence (Drawing by Ernst Haeckel (a) and photographs by Beatrice Murch (b), Piotr Konieczny (c), Julie Langford (d), and Stephanie Garnett (e)).

as high as chimpanzees. Most species are highly social. The African grey parrot can learn vocabularies of several hundred words, in multiple dialects. They can associate words with meaning, and understand simple syntax (sentence structure). There is some question of how far their cognitive ability extends beyond a clever capacity for mimicry, but that capacity alone reflects considerable brain power.

Some other birds show evidence of the capacity for symbolic learning. The gregarious Corvidae – the family of birds consisting of crows, jays, and ravens – hide food that they return to later, use automobile traffic to crack open nuts, and show other evidence of causal reasoning, flexibility, imagination, and planning on a par with that of chimpanzees [5].

Whales, dolphins, and porpoises constitute the mammalian Order Cetacea. They probably diverged from their terrestrial ancestors near the start of the Cenozoic 65 Mya, becoming increasingly amphibious, where the buoyancy of the ocean enabled them to become the largest animals on Earth. They also have the largest brains that have ever evolved, the brain of the blue whale measuring nine times the size of the human brain [6]. By the Miocene ($\sim 20$ Mya), cetacean brains had achieved essentially their modern size. Most of the enlargement of the brain in cetaceans reflects a huge elaboration of the neocortex beyond the sensorimotor primary projection areas. It appears that their extraordinary reliance on hearing, along with the neural specializations for that sense, evolved rather rapidly once they were totally marine. Other anatomical changes have enabled the sound production for a sophisticated echolocating capability and a communication system whose full complexity is not yet known. Their vocalizations are complex; they are good at learning and imitating distinct vocal signals, and appear to be able to recognize and address one another as individuals [7]. They are playful and anticipatory. They are highly social and unquestionably altruistic. Data from field studies have even been interpreted as evidence for cultural transmission in dolphins.

Elephants have the largest brains, and the largest neocortical volume, of all terrestrial animals. While they are generally regarded as intelligent, their performance on short-term cognitive tests including evidence of cause-effect learning is unimpressive. Where they do excel is in long-term spatiotemporal memories and in complex social interactions [8]. They show evidence of self-recognition in mirror tests, and are particularly sensitive to disabled and deceased companions.

Primates diverged from ancestral insectivores early in the great mammalian radiation at the start of the Cenozoic, ~65 Mya. Simians (monkeys and apes) split from other primates ~40 Mya, and began to evolve larger brains. By the middle Miocene, ~16 Mya, apes were distinct from old world monkeys, and the common ancestor of chimpanzees and humans diverged from ancestral gorillas ~6 Mya. By 4.4 Mya, humans had split from chimpanzees, and begun to diversify into a number of species [9]. *Homo sapiens* is the sole survivor of several competing human lineages, achieving modern morphology and brain size ~200,000 years ago. The evolution of hominids shows a relentless increase in brain size, characterized mainly by expansion of the neocortex. The evolutionary acceleration in brain size occurred in the anthropoids much more recently than in the cetaceans—the expansion of the human brain over that of the chimpanzee occurring within the last 6 My, while neocortical expansion in the Cetacea exceeded that of humans probably 20 My earlier [10]. The acceleration of neural complexity in these two distantly related mammals has thus been a completely independent event. With increasing brain size, a steady increase in cognitive and behavioral complexity can be traced through the history of primate evolution, from the onset of social grooming in monkeys 30 Mya, to a capacity for anticipation in orangutans 16 Mya, and evidence of creativity in gorillas 7.5 Mya. The chimpanzee/human ancestor of 7.5 Mya presumably had the social hierarchies and interactions, capacity for playfulness, altruism and deceit, and ability to form images and abstract thoughts seen in modern forms [12]. Though the capacity for symbolic language must have been present, since modern chimpanzees can master a modest vocabulary and syntax with sign language, a full-blown symbolic verbal language evolved only in humans, probably less than 2 Mya.

### 12.2.2 Under what circumstances does intelligence arise?

Each of the examples cited above is an independent evolutionary event. The relationship between cephalopods, social insects, and the vertebrates is so distant that their brains show little resemblance to one another. Among the vertebrates, mammals diverged from the reptilian lineage that would lead to birds well before much intelligence had evolved in either group. The last common ancestor between cetaceans and primates probably lived at least 40 million years ago, well before brain enlargement in either group accelerated. Elephants and cetaceans may be more closely related than either is to primates, but their divergence was still tens of millions of years ago, before either had been subjected to the selective pressures that led to exceptional brain enlargement. In each case, brain enlargement and enhanced cognitive capacity emerged from different starting points.

Another noteworthy point is the relative rarity with which conspicuous intelligence has evolved. No single ecological theme ties the groups together – marine, terrestrial, and arboreal flying habitats are all represented. Are there other circumstances, therefore, that are particularly associated with the evolution of intelligence? The following appear to have been critical.

**Size.** In all the cases of individual intelligence, the animals displaying it have tended to have larger body sizes than average for their taxonomic group. Cephalopods, whales, and elephants are particularly notable in this regard. Note that the rule does not apply to meta-intelligent forms.

**Activity level.** An active life style requires more sensory processing and motor control than one that is sedentary. Active organisms move through an ever-changing environment, which requires them to analyze features that sedentary organisms don't have to cope with, like balance and acceleration, depth perception, distinguishing foreground from background, feature extraction (is the looming object a friend or foe?), spatial orientation, and memory for places and situations experienced previously. Activity also obviously requires a system to command and coordinate muscle activity in the appropriate sequence, to the proper degree. This requires another type of learning. The larger the animal and the faster it moves about, the more elaborate and precise its sensory input and motor control have to be. All of these factors place pressure on the evolution of more neural elements and more communication among those elements.

**Cortical neural architecture.** Expansion of cortical brain structures is a common feature of intelligent species. Cortical architecture allows columns of nerve cells to be stacked vertically in high density, and a greater number of neurons translates into greater information processing capacity. The other notable feature of cortical architecture is its topographical representation of the external environment, such that information from a particular point in the outside world projects to a specific point in the brain. The expansion of cortical area more than volume within the brain (which is the reason for the crumpling of the cerebral cortex in mammals) increases the number of points in the environment that can be represented in a two-dimensional expanse of brain tissue. In the language of visual imaging, this increases the *resolution* of detail, which in turn, increases the complexity of information handled.

**High sensory resolution.** High intelligence invariably goes with detailed sensory processing. Both visual and tactile information are processed in great detail by cephalopods and primates. Birds have high resolution for vision and audition. Note that parrots and primates both evolved in visually complex arboreal environments. The hearing ability of the marine mammals is among the greatest of any vertebrate, and their echolocating ability is probably matched only by bats.

**Fine motor control.** The counterpart of high sensory resolution is fine motor control. Octopi have the challenge of controlling eight arms, each with a combination of delicate and complex movements, along with complicated mouth movements. The ability of parrots to mimic sounds is facilitated by motor control of a complex vocal apparatus. Primates likewise have precise control over subtle muscle movements to control vocalization (especially in humans), as well as hand and finger manipulations and multifaceted facial expressions important in social communication. Detailed coordination of gross limb movements, like those for grasping and throwing objects, has also been an important adaptive advantage for hominids. They evolved from an arboreal lifestyle which favors development of grasping forearms and good eye-hand coordination, and places a premium on depth perception and acuity. This would have favored integration of visual inputs with other modalities and a corresponding expansion of associative areas of the neocortex. It would also have preadapted them for the grooming that is the hallmark of primate social interactions.

**Social behavior.** All the vertebrate examples of intelligent animals above are highly social. This requires the ability for sophisticated communication – whether by behavior, vocalization, or facial expression. It also frequently involves hierarchical and territorial awareness, as well as accurate social memory. Thus, a significant degree of learning and abstraction, in addition to high sensory resolution and fine motor control, are required for appropriate social behavior.

This interplay between sensory-motor resolution and cognition is seen nicely in the hominid example. A long established assumption has been that the ancestral apes that would later lead to hominids moved from the forest onto the newly emerging savannas in the early Miocene (over 20 Mya). Increasing body size, partially upright posture with increased ability to use extractive tools, and predation on the larger animals of the plains would have further increased selective pressure for excellent eye-hand coordination, a keen sense of time, more complex social organization and behavior, and greater coordination of group behavior.

More recent insights provided by documentation that two contemporary characteristics of modern humans – bipedality and reduced male canine size – were present in *Ardipithicus ramidus*, which still followed at least a partial woodlands lifestyle 4.4 Mya, suggests that enhanced cognition may have had more to do with reduced intermale aggressiveness, food-for-sex tradeoffs leading to monogamy, and associated selective pressure for increased paternal invest-ment in child care, at early stages of hominid evolution [11]. These social innovations, which are fairly rare in mammals, could also have spurred, or at least been conducive to, the increased encephalization associated with the evolution of high intelligence. In either formulation, as humans competed with other hominids and one another, and dealt with the challenges of the Pleistocene ice ages, their now enlarged neocortex was preadapted for the development of a spoken language, the more sophisticated use of tools, and the attainment of culture [12].

**Trophic Level.** Other things being equal, animals at higher levels of the food web tend to be more intelligent than those at the lower levels on which they prey. The brain to body weight ratios are typically higher for carnivores than for herbivores, for example [6]. This may reflect the fact that more anticipation and behavioral coordination are required to capture prey than to escape predators. Fear and instinct are more ancestral, hard-wired brain functions than are planning, anticipation, and learning, which require newer, more plastic brain circuitry.

### 12.2.3 Under what circumstances has meta-intelligence arisen?
The vast majority of social insects—the ants, wasps, and bees—are restricted to the Order Hymenoptera (Class Insecta, Phylum Arthropoda). Socialization has arisen independently in numerous lineages among this group, and except for the termites, in no other insect order. This has been attributed to the evolution of haplodiploidy, whereby one sex has one set of chromosomes (is haploid) and the other has two sets (is diploid). In this case, genetic differences dictate different roles, leading to a caste division of labor [3]. This entails a different repertoire of behavior that appears to be highly rigid and instinctive. While honeybees can communicate precise information about direction and distance, their communication system is highly stylized and apparently quite literal. Thus, the behavior of individuals shows no evidence of the plasticity normally associated with intelligent behavior, but the aggregate effect of many individuals serving different roles results in outcomes that, were they conducted by a single individual, would appear quite intelligent. This case of meta-intelligence has astrobiological significance because it could be a different way for life to arrive at "intelligent" ends on another world without the necessity of evolving individual intelligence as we recognize it on Earth.

### 12.2.4 Why has intelligence arisen so rarely on Earth?
It would seem intuitively obvious that intelligence imparts a selective advantage, which therefore ought to be a favored evolutionary outcome. If that were really true, intelligence ought to be more common than it is. To be sure, life on Earth has shown a steady progression in the evolution of increasingly intelligent organisms – sword fish are smarter than jellyfish, reptiles show more behavioral plasticity than amphibians, and primates are more intelligent than any representative of the insectivores from which they evolved. But for most

animals, a qualitative elevation of intelligence is seen in only a restricted number of species within a given lineage. And the vast majority of the biomass on Earth is made up of plants and microbes, which lack intelligence entirely. Our intuitive premise must therefore be wrong.

Intelligence is an emergent property of biological features, such as brain mass, neural architecture, and circuitry. While neocortical expansion clearly makes intelligence possible, the initial expansion was probably driven by the need to expand brain capacity to better integrate high-resolution sensory information and coordinate fine-tuned motor activity, not to provide more intelligent behavior. Once the underlying components for intelligence were in place, however, natural selection could then act on them to favor intelligence for its own sake.

Consider the case of two mammals, the gazelle and the lion. Evolution has favored the development of a large brain with an expanded forebrain for processing complex sensory information and a high level of motor control in both. The outcome has been slightly different in prey and predator, however – the gazelle is an exquisitely sensitive and highly mobile but less intelligent animal than the less agile but more intelligent lion that chases it.

The question then arises, why did intelligence evolve in the lion but not in the gazelle? A partial answer lies in the fact that the gazelle doesn't need to be smart, just fast. The validity of this strategy is supported by the fact that 11 times out of 12, the gazelle escapes. But when two or more lions hunt together – a behavior requiring more coordination and "intelligence" – their success markedly rises. The evolution of prey and predator is an ongoing balancing act, in which natural selection optimizes different strategies – one for escape, the other for capture. Starting from a baseline of different advantages, selection will favor making the gazelle fast more than smart, and the lion smart more than fast. What the actual outcome of evolution is telling us in this example is that being fast and being smart are equally advantageous.

Is there more to it than that, however? Is it possible that intelligence is maladaptive under some circumstances? Neuroscientists don't understand intelligence and the factors that bring it about well enough to answer that. It may be that the neural changes required to make information processing, integration, and learning possible take away from other primary capacities in some way. It could also be that intelligence leads to social and behavioral complexities that are counterproductive; simpler behavior and less complex social structures may actually promote survival of the group more often than not. This may explain why intelligence actually appears to be underutilized in some cases [12]. Finally, intelligence could be self-limiting – a possibility more evident when we consider the consequences of technology, for which intelligence is a prerequisite.

**Figure 12.3** Intelligence as maladaptive. Curiosity, a corollary of intelligence, may be an evolutionary disadvantage more often than an advantage (©Rick London and Rich Diesslin, www.londonstimes.us).

## 12.3 Emergence of technology

The origin and fate of technology has clear astrobiological relevance for two reasons. First, any detailed information we are ever going to get about life on other worlds, and any information that life on other worlds is likely to get about us, will require some sophisticated technology for detecting, analyzing, and eventually traveling to, life on other worlds. Second, technology transcends biology, and as such could have a profound effect on the fate of its biological precursors (and/or their successors).

### 12.3.1 Under what conditions does technology arise?

The human species on Earth has developed technology to a substantial degree. Since this is the only case in the universe of which we are currently aware, offering a generic hypothesis about the origins of technology is problematic, at best. We can, however, note what appear to have been critical factors in the development of our particular case of technology (Figure 12.4).

The rapid enlargement of the forebrain was essentially completed in humans by 200,000 years ago. This enabled not only the fine motor control required for tool use and vocalization, but provided expanded brain areas for planning and abstract thought. These abilities spurred the development of symbolic spoken language, which facilitated increasingly complex social interactions culminating in the development of rudimentary hunter-gatherer cultures.

The first step away from total subservience to environmental conditions was probably the discovery of how to start and harness fire. This provided humans with a concentrated and controllable source of energy. It gave them respite from cold, expanded their diet through cooking, and probably afforded them protection from wild animals. Much later, it would become an important aid to fabrication.

The next critical step down the road toward technology was probably the domestication of wild animals – especially dogs and several hoofed animals. Over the long time it took to develop language, acquire and improve tools, and learn how to start and control fire, humans made their living by hunting and gathering. As they did so, they also learned more about plants and animals, finally reaching the point where they could grow plants and breed animals, thereby giving birth to agriculture.

Once agriculture became a fixed part of human culture, more food could be produced by the labor of one individual than was needed to keep that individual alive. This led to divisions of labor, then to the social stratification, elaborate cultural structures, and highly coordinated social behavior that we collectively refer to as civilization.

With civilization came construction of buildings and weaponry, which

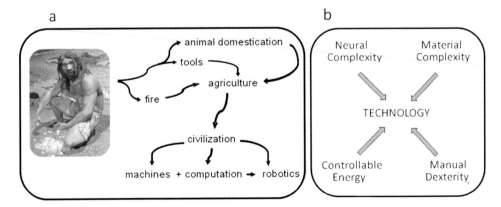

**Figure 12.4** Path to technology. The development of technology by humans depended on the emergence of civilization and a confluence of circumstances. (a) Sequence of events leading to civilization, including the potential for robotic succession of organic forms. (b) Minimal requirements for the development of technology (Line drawing by Louis Irwin, with art courtesy the National Museum, Smithsonian Institution).

depended on smelting ores to create metal objects. This skill extended the more ancestral crafts of creating pottery, paper, cloth, and other objects from natural materials. Using fabricated objects and domesticated animals, humans began to control and impact their environment to an extent far exceeding that of any other species. At this point, technology had arrived.

Technological advances were spurred by mathematics. First agriculture, then travel and navigation, turned ancient astrology into astronomy, creating the need to understand geometry and other forms of computation. As the utility of mathematics became evident, its use spread throughout nearly all human endeavors involving physical manipulations. Machines were created with mathematical precision, science enlarged understanding of the natural world, and new forms of energy were harnessed, giving humans the power to affect their environments in ever greater ways. Eventually machines were built for the purpose of computation, and computers brought information processing to a level competitive in speed (though not yet in complexity) with the human brain.

Today, computers and machines have combined to create robotics, thus embarking on a path toward creating mechanized forms with some of the essential features of living organisms.

## 12.3.2 Why has technology arisen more rarely than intelligence on Earth?

*Neural complexity*, the substrate for intelligence, is only one of several prerequisites for the development of technology. The cognitive capacity to evaluate options, envision structures and functions not yet realized, anticipate consequences, and compute quantitative relationships is necessary for fabricating and controlling the instruments of technology. However, at least three other factors have to be present for technology to arise (Figure 12.4b).

*Material complexity.* The resources for construction and fabrication must be available. In the human case this meant, first, stone, wood, and animal products like fur, gut, and bones. Then later, metal ores that could be extracted and forged became essential.

*Controllable energy.* Harnessing fire was critical in the progression toward technology for humans. In addition to its role in protection and partial independence from local environmental conditions, it was essential for the smelting of metals. Much later, other forms of controllable energy, like machines to capture wind and water power, and ultimately electricity, provided the supply of concentrated energy necessary for fabrication.

*Manual dexterity.* On a planet like Earth, where materials and energy are abundant, an intelligent being still needs to be able to manipulate them in precise ways. This calls for fine manual dexterity of the type that appendages with individually controllable digits like fingers enable.

In reviewing the limited cases in which conspicuous intelligence has evolved, we can see why the development of technology has been even more rare.

Two of our examples are marine animals. Cetaceans (whales, dolphins) have high intelligence, exquisite sensory abilities, and complex communicative skills, but lack manual dexterity. Cephalopods (octopi and squids) have both complex nervous systems and manual dexterity. But even if cephalopods were intelligent enough for abstract thinking, and cetaceans had some capacity for object manipulation, it is questionable whether the material resources for the fabrication of tools and machines could be gathered from the relatively homogeneous marine environment. Controllable, concentrated energy would be another problem. Fire is not possible underwater. While thermal vents do supply concentrated heat energy, it isn't portable or controllable. Thus, the marine environment presents challenges for the development of technology that have not been overcome in four billion years of evolution.

The terrestrial environment does provide material complexity; but, of the highly intelligent animals, only primates have the manual dexterity to manipulate objects and materials with precision. Parrots have fine manipulative skills with their beaks, and elephants have considerable dexterity with their large trunks. Neither of these two animals, however, is capable of the type of micromanipulation required to thread a needle. The limb structure of primates that enables them to throw objects, like stones or spears, may also be a very important form of object manipulation which represented an important competitive advantage.

The meta-intelligent example is interesting, in that the social insects show the ability to fabricate moderately complex structures – honeycombs and elaborate

Thanks to the kitty-cam, curiosity no longer kills the cat.

**Figure 12.5** Technology as an amplification of intelligence. Technology and intelligence build on one another. Whether the ultimate outcome will be survival or extinction is unclear (© Mike Baldwin, www.CartoonStock.com).

termite hills, for instance. Though individually much smaller in size than the animals that display individual intelligence, the social insects provide a hint of what collective meta-intelligence might be capable of. That a mature, flexible technology has not evolved from this group of meta-intelligent organisms on Earth may just reflect that, like intelligence for many lineages, it hasn't been necessary for evolutionary success. It would be premature to assume that technology cannot arise from meta-intelligent forms of life, provided the other requirements are met.

## 12.4 Where are they? Dealing with the Fermi Paradox

Enrico Fermi, the great nuclear physicist, was having coffee one day in the summer of 1950 with his fellow workers at Los Alamos, New Mexico, when suddenly he looked up and said, "Where are they?" When asked to elaborate, he pressed the question: if technologically competent aliens existed, why had they not visited Earth, or if they have, why haven't we known about it? Thus, the paradox attributed to Fermi is that, on statistical grounds, aliens with sufficiently advanced technology must exist, yet we have no evidence that they do. If they really do exist, would they not have the capability of knowing about life on Earth, and making themselves known to us by now?

Many ways of resolving the Fermi paradox have been offered. Most notably, Stephen Webb [13] proposed 50 solutions, none of which were satisfactory enough for him to avoid the conclusion that space-traveling aliens probably do not exist! We think that this degree of pessimism is unwarranted, but agree that the paradox is real. We're not prepared to offer a definitive solution, but would like to make the following observations.

### 12.4.1 The improbability of discovery

There are statistical grounds for arguing that the probability of contact by extraterrestrial forms of intelligent life (ETI) is quite low. The logic is based on the fact that the probability of contact is the product of a number of factors, each of which in itself has a low probability. Those factors include:

*Remoteness.* Simultaneously existing ETIs are likely to be far apart. There are about 130 stars within 20 light years of our Solar System. Unless the chances are greater than 1/130 that one of those stars includes planetary bodies on which an ETI has arisen, the nearest ETI (on statistical grounds) is 20 or more light years away. If stars extend at this density from our Sun in a flat plane out to a distance of 100 light years, the chances only have to be better than 1/32,500 to predict an ETI within a hundred light years of us. While these seem like good odds for another ETI, 100 light years is still a formidable distance.

Another ETI 100 light years from Earth would put it at the very limits of just now being able to detect the first human-generated radio signals (since they are barely 100 years old), assuming its radio antennas were pointed right at us. Since our closest ETI neighbor could lie in any direction, the chances of randomly picking up our electronic noise, or of happening upon us in the course of a mission for another purpose, are correspondingly smaller. Add to this the fact that Earth is not a particularly conspicuous planet, even in our own Solar System, the chances of an accidental find are smaller still. Earth is a tiny target in a vast ocean of space.

*Small time window.* If an ETI on another world is relying, as our own Search for Extraterrestrial Intelligence (SETI) project does, on radio transmissions of obvious technological origin, they are just now entering the time window within which they can detect us, unless they are closer than a hundred light years away. But let's suppose they have known about us by some means or another long enough to beam an electronic signal directly at us. We have only been able to detect such signals for about a hundred years ourselves. So what if they decided to just show up, and had the capability to do so? They could plop down in the middle of the rift valley of Kenya in spaceships lit by flashing neon lights, and no living being would know what to make of them unless they timed their landing within the past two or three million years, when humans smart enough, maybe, to distinguish them from flying birds would have been present. And had they decided to land in central Kansas, they would need to have done it in the last 10,000 years or so to be appreciated. Considering that life has existed on Earth for at least 3.5 billion years, and may have existed much longer on the planet from which the ETI mission was launched, hitting precisely within the time window of 2 million years or 10,000 years (0.06% and 0.0003% of life's existence, respectively) would be highly improbable.

*Communicability.* Even if by chance, an ETI were to stumble on Earth's physical presence, or pick up curious electronic signals from our remote corner of the universe, our alien observers would have to be able to make sense of the information and recognize it as being generated by some form of life. Likewise, for us to detect any information of technological origin from outer space, it would have to be in the form of a language that we could understand. There are generic ways of presenting information that make sense in our own minds, as embedded in the plaques and recordings that have been sent into space aboard our spacecraft, but this is information meaningful to the human brain, which may or may not make any sense at all to an intelligence of a different kind.

### 12.4.2 The disincentive for contact

Even if an ETI knew of our existence and was capable of visiting us, it may have chosen not to do so for various reasons. If technologically advanced relative to us, it may have no more interest in us than we have in ant hills. Perhaps it knows from experience that contact between intelligent beings from different origins is likely to end badly for the technologically inferior forms (as human history on Earth has repeatedly confirmed). Or, if our potential visitors are based on an organic chemistry anything like ours, they may be aware of the danger of infectious diseases for both of us.

Humans show no signs of disincentive for contacting ETIs. Our curiosity has served us well in general. Coupled with our superior learning ability, it has enabled us to absorb information about our world rapidly, and adapt to it quickly. In this case, however, our curiosity could be naïve, and possibly fatal. Any ETI capable of reaching us in the foreseeable future is going to be technologically well advanced over us. At best, we will not be in control of the situation. If our potential visitors are generously benign, they will probably know that it's best to keep us ignorant of their existence.

### 12.4.3 The possibility of past or present visitations

Two versions of this position are typically advanced. The first is that we have in the past been visited by aliens, and periodically continue to be visited in the present, but that the alien ETIs keep their distance, avoiding contact for any of the reasons mentioned above. A variant of this version is that ETI beings actually have made contact in various forms, including human abductions, animal desecrations, and other bizarre events. The second version is that alien ETIs are in fact living among us, disguised as normal humans or other forms of life.

Evidence allegedly consistent with the possibility of past and ongoing visitations by alien ETIs consists of a steady stream of reports of unidentified flying objects (UFOs), and of historical events attributed retroactively to UFOs. The vast majority of UFOs have a conventional explanation on close examination. Furthermore, no material evidence of UFO spacecraft or remnants of exotic forms of life have been produced which are widely accepted as fact by a scientific community strongly motivated to make such a discovery. Thus, while the possibility that UFOs represent actual alien spacecraft sent by or carrying some form of ETI has its passionate adherents, until objective evidence is presented which garners widespread scientific agreement, this has to remain an unproven hypothesis.

One consideration that tempers our temptation to dismiss entirely the notion that we are being visited by alien ETIs is that the reported characteristics of UFOs are indeed what one would expect of technologically advanced spacecraft – unusual or unrecognizable shapes, the ability to hover at will, and the tendency to accelerate at physically improbable speeds, with or without sound.

The second version of the notion that we have been visited already, consists of the staple science-fiction plot that aliens are living among us, indistinguishably from other forms of life, including humans. This possibility has been expressed with art and humor in the movie, *Men in Black*. Aside from the fact that motivation on the part of alien ETIs would be unclear (unless they are just as recklessly curious as we are), no biologically or mechanically exotic remnants plausibly attributable to such covert beings have been found. Despite the fervor of its adherents, the self-indulgent conceit that aliens far above us in technological competence and intelligence would wish to become one of us instead of peremptorily consuming us or blotting us out has no evidentiary support.

### 12.4.4 The possibility that technology is self-limiting

While pure speculation, it is not inconceivable that technology is inherently self-limiting, and therefore incapable of mounting sustained exploration. It could be, in other words, that any ETI that manages to make the transition to technology inevitably proceeds to its own destruction fairly quickly. This may be because it renders its home planet uninhabitable before survivors can colonize another world, or because the technological power to compete and destroy accelerates faster than natural selection or programmed evolution can mitigate the destructive behavior. The counter argument is that, even if true in most cases, surely exceptions occur somewhere in the vast expanse of the universe. This makes sense, but alone is not able to overcome the improbabilities listed in section 12.4.1 above. Indeed, if technology *is* self-limiting most of the time, this serves to narrow the time-window within which a comprehending and technologically competent form of life can be the recipient of alien contact.

### 12.4.5 Argument by analogy: the discovery and fate of the Hawaiian Islands
A volcanic hot spot in the Pacific Ocean gave rise to the major Hawaiian Islands between 0.5 and 5 million years ago. The youngest, the Big Island of Hawaii, rose from the ocean about 500,000 years ago. Life gradually took hold on the Island (from immigrant species) and evolved in isolation along a unique trajectory. Humans arrived for the first time probably between 300 and 500 years BCE, providing a species intelligent enough to establish a stone age culture that persisted to modern times.

Well before humans appeared in Hawaii, civilization had spread throughout most of the rest of the world, reaching a technological level and social complexity in the Eastern Hemisphere far beyond that of the Hawaiian islanders. Not until 1778, however, did any human from the civilized nations of the East make contact with the human inhabitants of Hawaii. As far as any Hawaiian knew in 1777, theirs was the only form of intelligence in the world.

Due to its remoteness and the small target it presented, Hawaii was not discovered by a technologically superior civilization until 2000 years after humans had inhabited the island, and half a million years after it had risen from the ocean. By the time the rudimentary technology of a stone age culture had appeared in Hawaii, civilization had advanced to the iron age in the East. Yet 2000 years would elapse before contact was made, and even this came 300 years after the knowledge and technology for open ocean voyages had developed in the Eastern hemisphere.

To push the analogy further, the contact between Hawaiian and European went well enough in the beginning, but didn't last for long. Conflict soon arose, superimposed upon the ongoing inter-island warfare among the natives, and European diseases decimated the native population. Within a hundred years, settlers from Europe, Asia, and North America had effectively wrested control of the islands from their indigenous inhabitants, and established a new, alien culture and technology. For two thousand years, Hawaiians were kept safe from aggressive and technologically superior aliens by their remoteness and inability to advertise their existence in the vast space of the Pacific Ocean. Once they were discovered, they succumbed quickly.

## 12.5 Chapter Summary

The history of life on Earth shows that certain factors favor a net increase in size and complexity of individual organisms over time. Those factors include an abundant supply of energy, temperatures that enable life on the surface, in the air, and in large volumes of liquids, a mobile lifestyle, and ample evolutionary time. Biodiversity is also more likely to occur on worlds where habitats are highly fragmented, geophysical conditions favor seasonal and global variations in temperature, time for evolution has been ample, and planetary history has been complex.

The evolution of intelligence has occurred on Earth several times but in each case as independent evolutionary events. Conspicuous intelligence, however, has not emerged that often – cephalopods, marine mammals, certain birds, elephants, and primates representing the prime examples. The social insects, though not intelligent as individual organisms, display a striking degree of meta-intelligence in their behavior as a group. Factors that appear to have favored the evolution of intelligence include large size, an active lifestyle, appropriate (pre-adapted) neural structures, high degrees of sensory resolution, fine motor control, and complex social behavior.

Intelligence is necessary but not sufficient for the development of technology. Beyond the requisite cognitive capacity, an intelligent organism must have manual dexterity, access to concentrated and controllable sources of energy, and the appropriate material resources.

Even assuming that intelligent, technological civilizations have the motivation to explore beyond their home planets, the chance that they will ever encounter another such civilization is made remote by the sheer volume of space and expanse of time that separates them, in all probability. Thus, the failure of humans to contact, or be contacted by, an alien intelligence is neither surprising nor evidence that they do not exist.

## 12.6 References and further reading

1   Pimm, S. L. and Jenkins, C. 2005. Sustaining the variety of life. *Sci Am* **293**: 66–73.
2   Young, J. 1964. *A Model of the Brain.* London: Oxford Univ. Press.
3   Wilson, E. 1980. *Sociobiology.* Cambridge: Harvard Univ. Press.
4   Pepperberg, I. M. 2002. In search of king Solomon's ring: cognitive and communicative studies of Grey parrots (*Psittacus erithacus*). *Brain Behav Evol* 59: 54–67.
5   Emery, N. J. and Clayton, N. S. 2004. The mentality of crows: convergent evolution of intelligence in corvids and apes. *Science* **306**: 1903–7.
6   Jerison, H. 1973. *Evolution of the Brain and Intelligence.* London: Academic Press;
7   Marino, L. 2004. Dolphin cognition. *Curr Biol* **14**: R910–1.

8   Hart, B. L., Hart, L. A. and Pinter-Wollman, N. 2008. Large brains and cognition: where do elephants fit in? *Neurosci Biobehav Rev* **32**: 86–98.
9   Wood, B. 1996. Human evolution. *Bioessays* **18**: 945–54.
10  Marino, L. 2002. Convergence of complex cognitive abilities in cetaceans and primates. *Brain Behav Evol* **59**: 21–326.
11  Lovejoy, C. O. 2009. Reexamining human origins in light of *Ardipithecus ramidus*. *Science* **326**: 741–8.
12  Byrne, R. 1995. *The Thinking Ape: Evolutionary Origins of Intelligence*. New York: Oxford Univ. Press.
13  Webb S. 2002. *Where Is Everybody? Fifty Solutions to the Fermi Paradox and the Problem of Extraterrestrial Life*. Copernicus, New York.

http://www.eoearth.org/article/Biodiversity
A brief overview of factors that promote biodiversity, with links to further reading

http://www.biodiversityhotspots.org/Pages/default.aspx
Interactive website showing hotspots around the globe where biodiversity is threatened

http://serendip.brynmawr.edu/bb/kinser/Size1.html
Interactive website exploring the relationship between body mass and brain size

http://setiradio.blogspot.com/2010/02/dolphins-and-evolution-of-intelligence.html Entertaining and informative website about the evolution of intelligence on Earth and in the universe

# 13 Anticipating the Future

## *The generic fate of living systems*

Our material universe, as best we can tell, is around 13.7 billion years old. Our Solar System has existed for about a third of that time. Thus, the possibility that life has had much longer than we have had to evolve in other regions of the cosmos is obvious. But has it, and is it still there?

For any specific case, on any given world, this two part question requires a two part answer. First, we seek to know whether and under what circumstances life *could have* emerged and then evolved. Second, we need to ask, under what circumstances, and for how long could it have *lasted*. All the previous chapters of this book have been devoted primarily to the first issue. Now we turn to the second, not only because it piques our imaginations, but because it has practical implications for how long we, ourselves, can expect to be around; and when we go, how and why we will disappear.

## 13.1 Three prospects for any form of life

Theory predicts, and the history of life on Earth illustrates, that there are three basic fates for all forms of life, once they have diverged from their ancestral forms sufficiently to be regarded as distinct [1]. First and most commonly, new forms emerge rather rapidly (on geological timescales), acquire optimal adaptive characteristics for the environment in which they have evolved, then settle into an evolutionary *plateau* in which form and function are kept more or less constant over long periods by stabilizing selection (see Figure 3.6 in chapter 3). Second and inevitably, most forms eventually succumb to *extinction*, because the biotic or abiotic environment changes faster than the organisms can adapt to the changes. Third and rarely, a lineage may acquire such a favorable adaptation that it undergoes dramatic *transition* to a whole new set of descendants that radiate into the new opportunities opened by evolutionary innovation or environmental change.

### 13.1.1 Plateau

Most of life at any given point in time is in a plateau stage of evolution. That is because stabilizing selection predominates over directional selection – placing a premium on keeping organisms the way they are, over changing them – as long as the environment is stable. Major geophysical changes, like climate and

L.N. Irwin and D. Schulze-Makuch, *Cosmic Biology: How Life Could Evolve on Other Worlds*,
Springer Praxis Books, DOI 10.1007/978-1-4419-1647-1_13,
© Springer Science+Business Media, LLC 2011

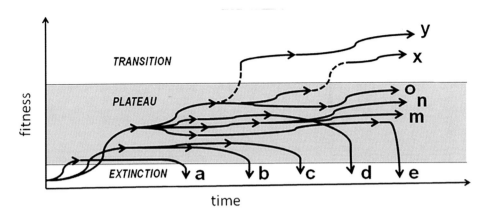

**Figure 13.1** Generalized evolutionary trajectories. Different forms of life radiate into new niches over time as natural selection increases the fitness of each to its particular niche. Most organisms settle into an evolutionary plateau for different lengths of time (a-o) after their initial divergent evolution, due to the constraining pressure of stabilizing selection. Eventually, environmental changes, natural disasters, or superior competitors lead to extinction (a-e) before they have time to evolve adjustments. In rare cases, the appearance of a new niche for which an organism may be pre-adapted, or a particular innovation that conveys a marked new advantage (dashed lines), results in a major transition (x,y) to a new form, that will in turn radiate and renew the cycle of stabilization, extinction, and transition.

topography, are usually very gradual; and ecosystems tend to settle into either a steady state equilibrium or predictable cycles, unless they are perturbed. Thus, both the physical and biological aspects of the environment are stable in the short-run for species (if not individuals) as a whole; so any mutations or other evolutionary mechanisms that would tend to change the way organisms are adapted to their environment, are selected against.

Over the long term, of course, environments and ecosystems change, and so must the organisms that occupy them if they are to avoid extinction. Nonetheless, examples abound of life that remains relatively unchanged over long periods of time. The most striking examples come from the microbial world, where the simple unicellular plan for the two earliest domains of life on Earth – the Archaea and the Eubacteria – is basically unchanged from their ancestral forms. Bacteria are evolving all the time to evade new drugs or newly-mutated adversaries, but fundamentally they remain little different from the way they were at least 3.5 billion years ago. Parasitic forms are particularly prone to the plateau scenario, living as they do in relatively constant environments even as their hosts evolve. For instance, the diplomonads – binucleated parasitic cells like *Giardia* that infect vertebrate digestive systems and cause diarrhea – have remained relatively unchanged since their ancestral transition from prokaryotes to primitive eukaryotes [2]. And cyanobacterial descendants of the first photosynthetic organisms have retained their simple bacterial structure and function for over two billion years

Within the plant kingdom, representatives of one of the earliest organisms to invade the land, the leafless and stemless bryophytes (mosses), still carpet the floor of our cool, damp forests after 360 million years. Notable for their highly successful and therefore unchanging morphology among invertebrates are flatworms like planaria, arthropods like the horseshoe crab, and mollusks like the nautilus. Among vertebrates, sharks have not changed fundamentally in their morphology, physiology, or lifestyle since their appearance 400 million years ago. The highly successful defensive morphology of turtles has likewise remained the same for 200 million years.

### 13.1.2 Extinction

Notwithstanding the compelling examples of biological perseverance above, extinction is the ultimate fate of most groups of organisms. This is because optimal adaptations are essentially irreversible, and once the environment changes or disaster strikes, as it inevitably will, adjustments can't be made quickly enough. Extinctions are occurring all the time, one species at a time. Indeed, many biologists believe that Earth is in the midst of a significant extinction event right now, as many more species are dying out than are being generated by the evolutionary process.

To make our case more starkly, however, we need only cite a few of the better-known and more spectacular extinctions in the history of life on Earth. A group of organisms referred to as the Ediacaran fauna experimented with a variety of forms early in the history of animal evolution, but the Cambrian explosion 540 million years ago replaced the experimental forms with a host of more competitive and stable body plans, like the arthropods, bivalve mollusks, and eventually, vertebrates, which would come to dominate the animal kingdom. Armored fish called placoderms ruled the seas during the Devonian period, but were replaced by their more agile unarmored descendants by the end of that age. Two spectacular extinction events occurred 251 million and 65 million years ago, respectively. The first, presumably due to a natural disaster at the boundary between the Paleozoic and Mesozoic eras, wiped out over 80% of all animal species. The second, thought to have been caused by the impact of a huge meteor at the Cretaceous-Tertiary boundary, led to the rapid extinction of the dinosaurs and many other major terrestrial and marine groups.

### 13.1.3 Transition

From the seeds of disaster and hardship, transforming innovation sometimes arises. The diminutive and mostly nocturnal mammals were kept out of day-time niches for the 150 million years that dinosaurs ruled the land, so in order to stay active enough to find food in the safety of nighttime darkness, they developed homeothermy – the ability to maintain a high, constant internal temperature. This innovation preadapted them for a tremendous evolutionary radiation, once the dinosaurs were gone. The presence of airbladders in fish, used to make them more buoyant in water, provided a preadaptation that was transformed into

a. Plateau          b. Extinction          c. Transition

**Figure 13.2** Plateau, extinction, and transition. (a) Organisms in long evolutionary plateaus include cyanobacteria, horeseshoe crabs, and crocodilians. (b) Examples of animals that disappeared in the face of new competition or environmental alterations include early squids, dinosaurs and trilobites. (c) Major transitions in the history of life have resulted from the evolution of jaws in fish, the innovation of flight among insects and reptiles, and homeothermy in early mammals (Art by Heinrich Harder (crab), Gerhard Boeggemann (dinosaur), Dmitry Bogdanov (jaws), Nobu Tamura (mammal), and Ballista (flight) and photographs by Valdosta State University Virtual Fossil Museum (trilobite) and Eric Christensen (squid)).

lungs, when lungfish had to start migrating in air from one desiccated pool of water to another. The evolution of lungs opened up a whole new way of life for amphibians and their successors, once they had access to the much higher oxygen content of air than of water.

Other innovations have come about more gradually, but with equally momentous consequences. The gradual enlargement and articulated movement of forward gill arches in fish that turned into jaws transformed the jawless, filter-feeding fish into fearsome predators, with consequential changes in musculature and nervous systems to control and coordinate a bigger body and more active life style. The evolution of feathers (quite possibly for thermoregulation in the beginning) and limb morphology changed arms into wings and enabled the reptilian ancestors of birds to take to the air. In each of these cases – the jawed fish, air-breathing animals, and flying reptiles – transformation to a new way of life led to a hugely successful radiation of descendant forms into new niches and life-styles previously unoccupied.

## 13.2 Thoughts on the relative frequency of different forms of life

Our ability to speculate on the extent of biodiversity in our Solar System and beyond is severely limited by our lack of knowledge about the *flexibility* of life; that is, the extent to which it can thrive under conditions unlike those prevailing on Earth. It makes a huge difference whether life can be based on building blocks other than carbon, borne by solvents other than water, and fueled by energy other than light and chemical reactions.

We have previously noted the advantageous nature of carbon-based, water-borne, light-driven (and a few chemically-supported) forms of life. The richness and complexity of the biosphere on Earth is consistent with those theoretical expectations. It could even be the fact that life can't be any other way. We will call this the **familiar** scenario, since it assumes a biochemical basis for life more or less as we know it. Such a scenario will obviously limit the existence of life to those worlds where carbon-based polymers, liquid water, and appropriate levels of light and chemical energy are available.

We have been insistent on arguing as well that other building blocks, solvents, and energy sources could support forms of life unfamiliar to us. We will call this the **exotic** scenario, since it would entail forms of life based on biochemistry totally foreign to our experience. This scenario clearly expands the variety of worlds on which life could arise and persist.

### 13.2.1 Biodiversity in the Solar System under the familiar scenario
Assuming that life can only exist if made of carbon-based polymers in aqueous solutions, it still should have been able to emerge early in the planetary histories of Venus, Earth, Mars, Europa, possibly other of the larger water-ice satellites (Ganymede, Enceladus, Titania), Titan, and possibly Triton. Because of the requirement that water be liquid, however, life would have perished in any habitat that became too cold to sustain water in liquid form. On Venus, the only way to have survived would be in sulfuric acid droplets in the clouds, and there only in microbial form.

All the conditions for giving rise to large, complex organisms, in addition to maintaining a vast microbial biosphere, are met today on Earth, including (obviously) circumstances allowing for the evolution of intelligence and technology.

Of the other planets still capable of sustaining some biosphere under the familiar scenario, all require sequestration beneath the surface. On Mars, this means within the soil or in caverns, where mobility would be severely limited. This, plus the absence of robust sources of energy, likely restrict any life remaining on Mars to being small, not very mobile, and mostly microbial.

On Europa and possibly two or three of the larger water-ice satellites, life could persist under the familiar scenario, but only beneath the icy covering at the surface. With large volumes of liquid water in which to move about, organisms would not be restricted from becoming large and complex in principle, provided energy would be ample enough to support the ecosystems that give rise to such organisms.

### 13.2.2 Biodiversity in the Solar System under the exotic scenario

When the possibility of life in solvents other than water is expanded under the exotic scenario, extremely cold satellites like Titan and Triton become distinct possibilities for hosting some forms of life, even under contemporary conditions. On Titan, the chemical building blocks would almost surely be carbon-based, because of the abundance of organic compounds there. On Triton, silane polymers conceivably could give rise to complex chemical building blocks in liquid nitrogen. In both cases, all forms of life would have to be subsurface. Absent knowledge of the volume of liquid reservoirs beneath the surfaces of Titan and Triton, or of the nature of the biochemistry possible within them, speculating on the size and level of complexity of organisms that could have evolved there is very difficult. Given the extremely cold temperatures, high levels of mobility are unlikely for any forms of life. Thus, the evolution of intelligence, even under the exotic scenario, seems unlikely – leaving Earth the only planet in the Solar System expected on theoretical grounds to be capable of sustaining intelligent forms of life.

### 13.2.3 Biodiversity in the cosmos under the familiar scenario

In extrapolating what we think the possibilities are for life in our own Solar System to all the worlds beyond our local neighborhood, we are further hampered by a lack of knowledge of how typical our Solar System is of all the others. This may not be as serious as it appears, however, since our Solar System includes a variety of examples of what is likely to be found anywhere. We have rocky planets orbiting close to the Sun, gas giants in orbits further out, greenhouse ovens, frozen deserts, airless snowballs, convulsive volcanic moons, atmospheres of various combinations, hydrocarbon plains and petrolakes, and captured satellites in neighborhoods where they don't belong.

The one type of world missing from our sample is the giant planet orbiting close to its central star. We know they exist because they constitute the majority of exoplanets discovered thus far. While no planet like that is found near us, we can deduce its properties and estimate the probability that it could be a home for some sort of life, using the principles applied in this book.

The so-called "habitable zone" is frequently assumed to delineate the range of distances from a central star within which worlds could exist that support life under the *familiar scenario*, because those are the worlds on which water could be liquid. In practice, the tacit assumption is that water must be liquid on the surface. In that sense, the habitable zone could define the range within which Earth-like worlds exist or could have existed at an earlier stage of planetary history. Our own Solar System, however, illustrates the fact that water may exist as a liquid or liquid solution beneath the surface of bodies all the way to its outer limits, provided internal sources of energy are available. Thus we find the concept of the habitable zone to have little utility.

We did argue in the previous chapter that complicated ecosystems giving rise to large and complex forms of life are more likely to arise on worlds with highly fragmented habitats, like those on which life is able to thrive on the surface. In that sense, the concept of a habitable zone (renamed, perhaps, to something like a "zone of macrobiotic complexity") would be defensible.

Our Solar System is probably typical in having many planets and moons within which liquid water can exist, but only under conditions that cannot give rise to large and complex organisms. Appropriate conditions and historical circumstances for the evolution of intelligence and technology are surely much rarer still. If these assumptions are true, it follows that water-based life of some sort could be fairly common, but that large, complex life would be less frequent, and intelligent life would be very rare.

### 13.2.4 Biodiversity in the cosmos under the exotic scenario

Under the exotic scenario, the habitable zone concept has even less utility, since it would differ for every different combination of conditions, depending on the nature of the life in question. In the case of our Solar System, every single planet and most of their major satellites, with the possible exception of Mercury and our Moon, could conceivably meet the requirements for either the carbon-based, water-borne life we are familiar with, *or* exotic forms of life we've not yet met.

Based on the assumption above, we have to conclude that life in the universe under the exotic scenario is fairly pervasive. We don't mean that every solar system has to have some form of life. Even under the exotic scenario, planetary history and other considerations may preclude the origin and persistence of life more often than not. We simply mean that worlds that harbor life must be scattered throughout the vastness of the universe in substantial number. Even if uncommon on a percentage basis of worlds that theoretically could harbor life, the total number of such worlds must be very large.

What is more difficult to estimate is the proportion of worlds on which life would have to be exotic, as opposed to familiar, from our point of view. To the extent that our Solar System is a typical sample, well over half the worlds either have or may have had liquid water, so life or its remnants on them would probably be of a form that we could recognize. But the assumption that our Solar System is typical, is an assumption that can't be verified at this time. To the extent that worlds where water could be a solvent are rare, and the exotic scenario is simply not found anywhere, our assumption that life in the universe is pervasive would have to be tempered.

### 13.2.5 Revisiting the Rare Earth Hypothesis

We began this book by contrasting two models for life on any planet. The first was the *Rare Earth* model, expressed most explicitly and cogently in the book of the same name by Peter Ward and David Brownlee [3]. This model argues that complex macro organisms evolve very rarely because the number of conditions required to support carbon-based, water-borne life on an Earth-like planet is so large that they very seldom coincide in their totality. Consequently, Earth-like planets are very rare, and therefore, so is life of the type that has evolved on Earth. The *Rare Earth* model requires what we have called the *familiar scenario.*

While arguing the case for the *Rare Earth* model, Ward and Brownlee acknowledged that life in microbial, sequestered form could be much more common. We have formalized this as an alternative but not incompatible model which we refer to as "Life Unseen."

Taken together, the Rare Earth and Life Unseen models most likely depict accurately the nature of life throughout the universe. Notwithstanding this fine-tuning of possibilities, even within the restrictions of the *Rare Earth*-familiar scenario, it is near inconceivable to us that the evolution of conspicuously intelligent, technologically advanced forms of life have *never* evolved anywhere else in the universe. On the contrary, we believe that this must have happened many times in many places.

## 13.3 The fate and future of life on Earth

Life on Earth will ultimately come to an end, because the planet itself will be consumed in the dying embers of a bloated Sun in the final throes of its natural lifespan. Like all stars as massive as our Sun, the hydrogen that fuses into helium at its core gradually increases the density (mass/volume) of the star, causing it to burn more brightly. For our Sun, energy output increases by roughly 10% every billion years. This means that the distance from the Sun within which planetary bodies and their satellites are exposed to intolerable heat and radiation expands progressively outward.

The Sun will exhaust the supply of hydrogen fuel in its core in about five billion years from now. The center will collapse under its own gravity, eventually becoming hot enough to fuse helium into carbon and oxygen in the core. Meanwhile, its atmosphere becomes unstable and starts to expand, transforming the Sun into an enormous red giant star. As the helium supply runs out, an inert mass of carbon and oxygen builds up at its center. The star then sheds its outer layers in a series of shells, leaving behind this core, which forms a remnant white dwarf – a hot, compact star the size of the Earth, composed of carbon and oxygen, which has exhausted its sources of fuel for fusion. From this point on the Sun will gradually fade away, becoming dimmer and dimmer until its light is finally extinguished. Earlier, as the red giant Sun expands to over 100 times its original size, Mercury and Venus will be consumed; perhaps Earth too. Long before that happens, five or more billion years from now, Earth will probably become uninhabitable, as it continues to follow the trajectory of Venus toward a runaway greenhouse world. Before that happens, however, how can we expect the biosphere to change?

There is no reason to assume that the generalized fates of plateau, extinction, and transition described in section 13.1 will not continue to play out as long as conditions on Earth remain tolerable for life. At the moment, mammals, reptiles, and, amphibians especially, appear to be declining in biodiversity, while birds seem to be ascendant. Among the invertebrates, insects are maintaining their lock on the biodiversity championship, while roundworms in the soil maintain their overwhelming but unobtrusive and unseen lead in biomass for organisms above the microbial level. As climate gradually changes, some species will disappear while a few new ones emerge, but for the most part, the major groups of organisms will probably persist.

Plants, being exquisitely sensitive to temperature, moisture, nutrients, and animal agents of dispersal, will likely change or stay the same, in proportion to changes in local climate, animal life, and their own parasites.

At this point in the history of life on Earth, human intervention through technology is beginning to have a major impact. Plant communities increasingly will be affected by genetically engineered plants. The same might be said of domesticated animals. Genetic manipulation, either by the direct insertion or deletion of genes, or by selective breeding, can be expected to alter domesticated animal stocks according to human intentions. Since these hereditary changes can be brought about fairly rapidly, they can promote the survival of animals favored by humans beyond the point where natural selection might have led to their extinction.

Thus, at this point, the trajectory of evolution for certain organisms will be directed by technology in accordance with human intent. This "directed selection" will then join Darwinian evolution by natural selection as a potent determinant in the fate of any given species. For some species, like humans and domesticated plants and animals, directed selection will become the dominant force in their evolution. Since changes in any given species could affect other species by a domino effect, directed selection could indirectly affect a much broader portion of the biosphere.

In addition to redirecting the course of organic evolution, advances in robotic technology will continue to drive a form of mechanical evolution toward ever more sophisticated life-like machines, with the potential to impinge upon or replace living organisms, including humans. Add to this the human capacity to affect large scale climatic changes, whether intentionally or inadvertently, and the potential impact of humans on the biosphere will continue to accelerate [4].

### 13.3.1 Fate of human life
In considering the future and fate of humans, the parallel evolution of humans and machines needs to be taken into account. Biomedical technology has already reached the point where the human body can be augmented by fabricated parts – artificial limbs, pacemakers, and hearing aids, for example. Computerized mechanical supplements that can enable stroke victims to communicate, the paralyzed to move about, and even the blind to see, are well advanced in development. From the moment that humans used their first tool,

in fact, they have been "natural born cyborgs," co-evolving with mechanical devices to amplify their ability to manipulate their environment [5]. The fate of humans thus is inseparable from the fate of their supplementary machines and devices. The time may come when humans cross over from being organic beings with mechanical supplements, to mechanical beings with organic residues. For clarity, we will distinguish between the fate of humans as organic beings and the fate of their possible mechanical successors.

### 13.3.1.1 As organic forms

Peter Ward doubts that any single catastrophe could totally extinguish human life, because of its ability to insulate itself from the adverse conditions that could doom other forms of life lacking the protective potential of technology [6]. However, as a species with unexceptional physical abilities but highly specialized physiological needs, like homeothermy, high metabolic rate, and prolonged internal gestation, humans are vulnerable to drastic changes in climate. The ability of humans to survive will depend on their capacity to protect themselves from an increasingly hostile environment. In the face of catastrophic events or cataclysmic changes, this would likely mean sequestration in fortress-like structures or underground, where the greatest challenge would be maintaining a flow of energy for food production.

In the final analysis, humans could find the challenge of biological survival on Earth too costly or risky, making emigration to another world a desirable alternative. While there would be formidable challenges to such a move, humans would be able to establish and control their own artificial environment – and possibly over the long run, terraform their new world to an environment more compatible with their biological needs.

As organic forms of life, disease might be a continuing challenge of unknown magnitude for humans, even on another world. There is certainly no reason to assume that microbes will retreat in the face of any threat short of sterilizing temperatures or planetary destruction. Humans have made great inroads toward conquering infectious diseases, but two hundred years of experience with inoculation and a hundred years of drug development have not eradicated the threat of a global pandemic that could devastate the human species. The cardiovascular diseases will ultimately be overcome, in all likelihood, by mechanical devices, but a cure for cancer and neurodegenerative diseases remains elusive. Indeed, an ever growing inventory of carcinogens and unknown triggers could continue to keep these diseases at the forefront of challenges to the health of human beings.

An alternative to defeating disease and avoiding biological degradation would lie in a transition to a mechanical form of life.

### 13.3.1.2 As mechanical forms

Mechanical beings, whether as intelligent machines or unintelligent robots, could clearly be constructed as unitary, bounded individuals capable of maintaining their organization and activity by taking energy in from their environment. Once they achieved the ability to create self-perpetuating copies of themselves from raw materials, they would fulfill all the criteria we have used for being alive. In that sense, we can legitimately speak of mechanical life.

A number of advantages would be available to mechanical over organic forms of life. They would be free of disease, much less vulnerable to environmental perturbations, more durable, and more energy efficient. They could be repaired more easily (except, perhaps, for their control and cognitive systems), controlled more precisely, specialized to a higher degree, and inactivated or kept active indefinitely. While some of these characteristics might not seem positive from a contemporary human perspective, from the standpoint of natural or directed selection, they could be highly favored. That being the case, at what point would mechanical humans become more highly fit than organic humans; and once they did, what would happen to the latter?

In the short run, robots have so much potential for extending the reach of technology that organic humans are motivated to accelerate the development and production of them. There is no reason to doubt that their sophistication will continue to increase, making them ever more versatile and pervasive. Whether their cognitive abilities can ever match those of the brains of the more intelligent species *is* a matter of controversy. While brains and computers do use different strategies for problem solving [7], there appears to be no barrier in principle to the attainment of computer intelligence sufficient for programming anticipation and intention – at which point the distinction between organic and mechanical human cognition may become negligible.

If and when robots attain human levels of intelligence and become self-replicating, they will have the capacity to occupy the same niche as humans, and therefore become direct competitors. The competitive exclusion principle is a long-standing theory in biology usually stated as "complete competitors cannot coexist." If two forms of life are occupying exactly the same niche, natural selection will ultimately favor one over the other, however slightly, and the other will become extinct or adapt to a new niche. It is possible that before robots become truly competitive with humans, they will be programmed by their creators to be perpetually subservient (for instance, by keeping them "sterile", or incapable of self-replication) or sufficiently different to avoid competing with humans. However, if, as the history of technology indicates, robots will be made increasingly sophisticated, intelligent, and capable – just because they can be – a transition from an organic to a mechanical form of life may occur in the niche currently occupied by the only species heavily impacting its environment through technology.

### 13.3.2 Fate of insects

Whether humans survive as an organic form of life, or create their own mechanical replacement, the insects are the one form of organic life on Earth most likely to be among the last multicellular forms to perish from the planet. More species of insects – close to a million – are known than all other species of life combined; and it has been estimated that 10 times that number might exist.

They occupy every terrestrial niche known, including fresh water. A small number of species are even found in the ocean, though their arthropod relatives, the crustaceans (shrimp, crabs, lobsters) dominate the marine environment.

Insects are known from the early Devonian, 400 Mya. They radiated into the large number of niches made available by proliferation of the carboniferous forests 350 Mya. Their arthropod heritage of a hard exoskeleton, jointed appendages, small size, mobility, multistage life cycle, and prolific reproductive capacity have served to make them extremely durable. They have survived at least two global catastrophes, and remain undefeated by all the technology that humans can throw at them, including the strongest poisons known. Given the extreme diversity their evolution has generated over 400 million years, it seems highly likely that they will be able to evolve to meet whatever challenges a dying planet presents, up till almost the point when all of life perishes.

### 13.3.3 Fate of everything else

Absent major chemical changes in the oceans, or in the atmosphere that would affect the oceans, marine life should continue largely in the plateau state, subject to individual adjustments by natural selection, as long as photosynthetic producers survive to support the food web. Major crises, like the one at the Permian-Triassic boundary that wiped out a majority of animal species then in existence, could cause widespread extinction, but representatives of the major groups will likely survive, as they did then. Fish have lived in the ocean for over 400 million years, and mollusks have been there since the early Cambrian, nearly 540 Mya. Bony fish, bivalve mollusks (clams, oysters), and probably crustaceans will likely persist until the oceans dry away.

Peter Ward's prognosis for the future of animal life envisions humans, domesticated animals, small animals that can co-exist with them, and birds constituting the majority of terrestrial vertebrates likely to survive the massive extinction currently underway [6]. Domesticated animals will survive because technology will be developed that enables them to endure all but the most drastic environmental challenges. Birds are the analogues of insects in the vertebrate world; they will continue to be successful because of their diversity and adaptability, so long as plants continue to be available at the base of their food chain.

Plants will adapt, as they always have, with the changing times. Their energy source will only increase in abundance, as the Sun burns brighter with aging. But as the planet warms severely and water becomes scarcer on land, the continental masses will gradually become drier, retaining moisture only at the poles, and eventually not even there. Once plant life on land is gone, all animal and fungal life will follow.

Microbes will retreat underground where they will last as long as they can sequester any moisture and manufacture food by some means other than photosynthesis.

### 13.3.4 Summary of the fate of life on Earth

Over half a billion years ago, multicellular life began to blossom on Earth, joining the microbial biosphere already thriving for at least three billion years, to give rise to the greatest biodiversity in the Solar System. Life on Earth was sufficiently varied and resilient to survive several global catastrophes and repeated ice ages. Highly diverse groups like insects, mollusks, birds, and fish, subject to Darwinian evolution, are likely to see some of their representatives endure as long as plant life can survive and the planet stays habitable.

Through the aid of technology, humans and their descendants (or their mechanical successors), and a few domesticated species for which they provide protection, might endure until the planet becomes uninhabitable, if they do not leave the planet earlier. What the biosphere will look like as it approaches the point of crisis for its survival is extremely difficult to predict. We are in the middle of an extinction period, driven largely by the human struggle to achieve technological supremacy over the environment and its own fate. Genetic engineering is already altering the flow of evolution through directed selection. This trend is likely to continue with very unpredictable, and likely unintended, consequences.

## 13.4 The fate and future of life on other worlds

Using the history and fate of life on Earth as a model, we can generalize the fate of life wherever it occurs as follows: Once it has emerged, life will adapt as well and with as little change as possible to prevailing planetary conditions. When those conditions change, the engine of evolution will kick in to bring about change in the biosphere. If there is a great deal of habitat fractionation, and if a variety of environments capable of hosting liquid-borne chemistry are present, life will become diverse, thereby improving the chances that some forms will survive as long as the planetary body itself survives. As conditions become more restrictive, however, the number of survivors will go down. If there is environmental stability – even of a very harsh and restrictive nature – some life, albeit not very diverse, will survive.

On worlds where conditions for the evolution of intelligence and technology are good, the potential for altering or countering natural trajectories in climate, and dealing with other natural catastrophes, will be present. Also, the possibility of a transition from organic to mechanical forms of life must be considered. Once the potential for travel through space also becomes a real possibility for beings on such a world, emigration to other worlds becomes an alternative fate for those beings, whether organic or mechanical, that have the means to do so.

## 13.5 Chapter summary

Once life arises, it shows remarkable durability and persistence. This is because stabilizing selection dominates directional selection over the short run, keeping organisms optimally fit for their environments. At any point in time, most forms of life are therefore in a *plateau* phase of evolution. Over the long term, as environments change or catastrophes strike, evolutionary mechanisms cannot respond quickly enough to the altered reality, so *extinction* eventually occurs. On rare occasions, new innovations arise, or circumstances change to make a previously inconsequential characteristic suddenly much more advantageous, leading to a major evolutionary *transition*, which serves to establish a new baseline from which descendant forms can radiate.

Our Solar System provides a good variety of the types of worlds that are likely to be found in any solar system, though the full variety of worlds is obviously unknown to us. Hopefully, it provides a good proxy for the probability that life exists anywhere else in the universe.

Carbon-based, water-borne forms of life have some clear biochemical and physicochemical advantages over other building blocks and solvents, respectively. If the conservative assumption is made that this form of life is the only form of life possible (the *familiar* scenario), we nonetheless could expect to find life, or remnants of past life, on at least three of the four rocky planets (Venus, Earth, and Mars), and on satellites of at least three of the four gas giants (Jupiter, Saturn, and Uranus). We therefore conclude that life, both past and present, has been widespread throughout the universe.

Under this conservative assumption, however, we are certain that complex macro organisms arose in only one case. They may have existed at one time on Venus and Mars, as well as Earth, but today are likely to be non-existent on Venus and relegated to caves or underground reservoirs on Mars. A fairly complex ecosystem in theory could exist in the subsurface oceans of icy satellites – notably Europa, and possibly Enceladus. However large or complex organisms may have become at any time in their history, they are now totally sequestered on every other world in our Solar System.

If a more liberal assumption is made about the conditions under which life could arise (the *exotic* scenario), the worlds on which it could exist today in our Solar System would have to be enlarged to include Titan and Triton. This expands the range of worlds on which life could be found throughout the cosmos, but doesn't change the fundamental fact that in the vast majority of cases, life is going to be sequestered in subsurface habitats, and most likely small to microbial in size.

Once life has arisen on a given world, especially on those where the biosphere has become diverse, life is likely to trace the planetary history of that world – remaining stable as long as the environments are stable, and changing to meet the challenges of new environments as they emerge. Life in general is likely to persist as long as conditions on that world allow any form of life to exist there. Ultimately, however, as the central stars of solar systems burn out, then die in a final convulsive output of energy, the planets close to them will be incinerated, and those far away will be deprived of whatever life-giving energy their sun had provided. Either way, the extinction of life on worlds thusly affected is a high probability, and the ultimate fate of life wherever it occurs.

The emergence of complex ecosystems populated by large organisms with significant information-processing powers – what we'll refer to as intelligence – would appear to be rare. The development of technology appears to be rarer still. But it is highly likely to have happened, and to be happening recurrently, in isolated outposts of the universe.

The advent of technology generates a slightly amended scenario for the evolution of life. Technology provides the capacity not only to radically impact the environment, but even create mechanical beings with the same essential features as their organic creators, thereby setting the stage for competition and possible replacement. On those worlds where intelligent forms of organic, and possibly mechanical, life can arise, the prospect of space travel provides an alternative end point to extinction.

## 13.6 References and further reading

1   Schulze-Makuch D. and Irwin L. N. (2008) *Life in the Universe: Expectations and Constraints*. Springer-Verlag, 2nd ed.
2   Adam, R. M. 1991. The biology of *Giardia* spp. *Microbiol. Rev.* **55**: 706–732.
3   Ward P. D. and Brownlee D. (2000) *Rare Earth: Why Complex Life Is Uncommon in the Universe*. Springer-Verlag, New York.
4   Woodruff, D. S. 2001. Declines of biomes and biotas and the future of evolution. *Proc Natl Acad Sci U.S.A.* **98**: 5471–6.
5   Clark, A. 2003. *Natural-Born Cyborgs: Minds, Technologies, and the Future of Human Intelligence*. New York: Oxford Univ. Press.
6   Ward, P. 2001. *Future Evolution*. W. H. Freeman.
7   Hawkins, J. and Blakeslee, S. 2005. *On Intelligence*. New York: Henry Holt.

http://imagine.gsfc.nasa.gov/docs/ask_astro/answers/980218c.html
Nice essay on the fate of the inner planets as the Sun progresses to a red giant.

http://specevolution.wordpress.com
An interesting blog about speculative evolution.

http://www.web4health.info/en/aux/homo-sapiens-future.html
Good discussion of the future of human evolution, with links to other sources.

# Glossary

**abiotic** Pertaining to any process not attributable to a living organism.

**absorption** Uptake of gas or liquid by liquid or solid material.

**acidophile** Any form of life, though most commonly a microorganism, that can grow in an acidic (low pH) environment.

**accretion** Growth of a massive object, such as a protoplanet, by gravitationally attracting more matter.

**ADP** Adenosine diphosphate, a nucleotide. ADP is the end-product that results from the loss of the terminal phosphate from ATP with the release of energy.

**aquifer** Layer of permeable rock or sediment in which all pore spaces are filled with water.

**albedo** A measure of how strongly an object reflects light from light sources such as the Sun.

**alkaliphile** Any form of life, though most commonly a microorganism, that can grow in an alkaline (high pH) environment.

**Amazonian** Youngest epoch of planetary history on Mars, from about 1.8 billion years ago to the present.

**amino acids** Molecules containing an amine group, a carboxylic acid group, a hydrogen atom, and a variable organic side chain bonded to a central carbon atom. Amino acids are the building blocks of proteins.

**ammonia** Compound of nitrogen and hydrogen with the formula $NH_3$. Ammonia is a gas under typical Earth environmental conditions.

**amphibian** Poikilothermic (variable body temperature) animal such as a frog, toad, salamander, newt, or caecilian, that reproduces in water and typically metamorphoses from a juvenile water-breathing form to an adult air-breathing form, often with retention of some juvenile characteristics. Amphibians evolved from lungfish and gave rise to reptiles.

**anaerobic** Pertaining to a process that occurs in the absence of, and does not require, free oxygen.

**Astronomical Unit (AU)** Mean distance from the Sun to the Earth (149.6 million kilometers).

L.N. Irwin and D. Schulze-Makuch, *Cosmic Biology: How Life Could Evolve on Other Worlds*,
Springer Praxis Books, DOI 10.1007/978-1-4419-1647-1,
© Springer Science+Business Media, LLC 2011

**animal** Any multicellular, eukaryotic, heterotrophic organism that digests its food internally within body cavities.

**Archaea** One of the three domains of life on Earth, consisting of single-celled microorganisms lacking a cell nucleus and other organelles with cell membranes made of lipids with ether linkages. Archaea often occur and grow in extreme environments.

**Archean** An eon of planetary history on Earth, from 3.8 to 2.5 billion years ago.

**arthropod** Invertebrate animal with an external skeleton, segmented body, and jointed appendages; prominent groups include insects, arachnids, and crustaceans. The number of arthropod species exceeds the total number of all other animal species combined.

**ATP** Adenosine-5'-triphosphate, a multifunctional nucleotide, most commonly used to store chemical energy for use in biological processes.

**autotroph** Organism that synthesizes its own high-energy compounds from inorganic substrates.

**benthic** Ecological region at the lowest level of a body of water such as an ocean or a lake, including the sediment surface and some sub-surface layers.

**binucleated** Pertaining to organisms with two cell nuclei.

**biofilm** Aggregate of microorganisms in which cells adhere to each other and/ or to a surface. These adherent cells are frequently embedded within a self-produced matrix of extracellular polymeric substance.

**biosignature** A chemical or physical marker produced by or characteristic of living organisms.

**bipedal** Form of locomotion in which an organism moves by means of its two rear limbs, or legs.

**bird** Winged, bipedal, homeothermic (constant body temperature), egg-laying, vertebrate animal with about 10,000 living species. Birds are living descendants of dinosaurs.

**bivalve** Marine and freshwater mollusks that have a shell consisting of two rounded plates (valves) joined at one edge by a flexible ligament (hinge), including mussels and clams.

**bolide** Term used by geologists to describe any large natural object that impacts on a planet or moon forming a crater.

**buoyancy** An upward acting force that keeps bodies afloat because the density of the liquid is greater than the density of the body.

**bryophytes** Nonvascular land plants that lack true stems, leaves, or flowers, and reproduce with spores rather than seeds. Major groups include the mosses, liverworts, and hornworts.

**budding** A form of asexual reproduction, in which new individuals are produced by outgrowth and separation from the parent. In single-celled organisms it differs from binary fission in that the two resulting cells are not of equal size. Budding also occurs in multicellular organisms such as sponges.

**Cambrian** A period of planetary history on Earth, from about 545 to 500 million years ago.

**Cambrian Explosion** Relatively rapid appearance, over a period of several million years, of most major groups of complex animals around 530 million years ago, following a Snowball Earth event which ended about 50 million years earlier.

**Carboniferous** A period of planetary history on Earth, from about 360 to 290 million years ago.

**carnivore** An organism that derives its energy and nutrient requirements from a diet consisting mainly or exclusively of animal tissue, whether through predation or scavenging.

**catalyst** Compound that promotes a chemical reaction but which itself remains unchanged.

**Cenozoic** An era of planetary history on Earth, from about 65 million years ago to the present.

**cephalopod** Mollusks characterized by bilateral body symmetry, a prominent head, and a modification of the foot into arms or tentacles. Major groups include octopi, squids, and cuttlefish. Cephalopods are generally considered to have the highest intelligence of all the invertebrates.

**cerebral cortex** Outermost layer of the cerebrum of the vertebrate brain. It plays a key role in memory, attention, perceptual awareness, motor control, thought, language, and consciousness.

**Cetacea** Order of mammals fully adapted to marine habitats, with large brains, exquisite hearing, and complex social behavior. Prominent groups include whales, dolphins, and porpoises.

**chemoautotroph** An organism that obtains energy through metabolic transformations of inorganic substrates and uses carbon dioxide as a carbon source.

**ciliate** Group of protozoans characterized by the presence of hair-like organelles (cilia), which are identical in structure to flagella but are typically shorter and more numerous with a different undulating pattern.

**clathrate** Chemical substance consisting of a cage structure of one type of molecule enclosing a second type of molecule, often referring to a water ice structure trapping a gas.

**cognition** Analysis and integration of information, generally associated with purposeful behavior.

**commensalism** Symbiotic relationship in which an occupant organism benefits from a host which is neither helped nor harmed.

**comet** Loose collections of ice, dust, and small rocky particles, ranging from a few hundred meters to tens of kilometers across, often displaying a characteristic tail when heated by solar radiation.

**condensation** Change from the gaseous phase of an element or chemical into liquid droplets or solid grains of the same element or chemical.

**conjecture** A proposition that is unproven but appears correct and has not been disproven.

**consciousness** Perceptual awareness of and attention to specific aspects of the environment, including causal relationships and the temporal sequence of current and past events.

**consumer (primary)** In ecology, any organism that derives its energy by feeding on producers (autotrophs).

**consumer (secondary)** In ecology, any organism that derives its energy by feeding on primary consumers.

**consumer (tertiary)** In ecology, any organism that derives its energy by feeding on secondary consumers.

**cortex** Outer layer of any subdivision of the brain; prominent in the brains of all vertebrates, but restricted to just a few invertebrates.

**Corvidae** Family of birds, notable for complex social behavior and an apparently high level of cognitive function, consisting of crows, jays, and ravens.

**covalent bond** Chemical bond in which electrons from different atoms are shared more or less equally.

**Cretaceous** A period of planetary history on Earth, from about 145 to 65 million years ago.

**cryovolcano** Eruption of water-based liquids into a freezing environment.

**cryptobiotic** Pertaining to a dormant state characterized by a severe but reversible depression of metabolism.

**cryptoendoliths** Cryptobiotic organisms living in rocks.

**cyanobacteria** Photosynthetic, oxygen generating bacteria.

**desecration** Mutilation of a dead organism.

**detritus** Remnants of dead organisms.

**detrivore** Organism that feeds on detritus.

**deuterium** Heavy hydrogen; hydrogen with a neutron in addition to a proton in the nucleus.

**Devonian** A period of planetary history on Earth, from about 410 to 360 million years ago.

**dexterity** Facility for finely controlled manipulation with appendages such as tentacles or fingers and hands.

**diapir** Intrusion in which a more mobile and ductile-deformable material is forced into brittle overlying rocks.

**diffusion** Process by which molecules spread from areas of high concentration to areas of low concentration.

**diplomonad** Binucleated protozoan, frequently parasitic, such as *Giardia*.

**ecosystem** All the organisms in a given environment, and the abiotic factors with which they interact; the physical components and biological aspects of an environment considered collectively.

**Ediacaran** Pertaining to a group of soft-bodied fossils preceding those from the Cambrian Explosion, originally discovered in the Ediacara Gorge in Australia; also used to refer to the timeframe of Earth's history from about 630 to 545 million years ago.

**elasmobranchs** Cartilagenous fishes, like sharks, skates, and rays. Elasmo-branchs occupy the highest trophic level in most marine ecosystems, but are seldom found in fresh water.

**electroconductive** Pertaining to a material that carries an electric current.

**electronegativity** Degree to which an atom attracts and holds bonding electrons.

**encephalization** Evolutionary enlargement of brain structures at the anterior end of a central nervous system, most commonly including cortical lamination, in association with increased central processing of sensory information and motor control.

**endolith** Organism that lives inside rock, coral, animal shells, or in the pores between mineral grains of a rock.

**ephemeral** Lasting for a fixed period of time, or recurring on a seasonal basis.

**ethane** Hydrocarbon with the chemical formula, $C_2H_6$; the major component of precipitation and surface liquid on Titan.

**eukaryote** Organisms whose cells contain complex, membrane-enclosed structures (organelles), including a cell nucleus or nuclear envelope, within which the genetic material is carried.

**evolution (Darwinian)** Change in the form and/or function of organisms over time, attributable mainly to natural selection acting on variations that have differential survival value within a population, but sometimes due to random factors like genetic drift.

**exoplanet** A planet within a solar system other than ours

**extremophile** Any organism that grows under environmental conditions not tolerated by most forms of life; used especially to characterize microorganisms that survive at temperatures, pH levels, and solute concentrations well outside the range normally encountered on Earth.

**fish** Aquatic vertebrate, most commonly poikilothermic (variable body temperature), covered with scales, equipped with two sets of paired and several unpaired fins, and gills for extracting dissolved oxygen from the water.

**flagella** Cellular appendages specialized for locomotion, consisting of a central pair of microtubules encircled by nine doublet microtubular filaments. Motion is achieved by sliding the outer filaments against the central pair.

**fumarole** Opening in a planet's crust, which emits hot steam and gases such as carbon dioxide, sulfur dioxide, hydrochloric acid, and hydrogen sulfide.

**fungus** Multicellular, eukaryotic, heterotroph that digests its food externally, then absorbs it directly into cells.

**genetic drift** Evolution resulting from differential reproduction within a subpopulation whose gene pool deviates by chance from that of the larger parent population; considered to be the only random process for gradual evolution acting at the population level.

**Giardia** Genus of binucleated, parasitic protists lacking some organelles; thought to be deeply rooted in the evolution of eukaryotes.

**gustatory** Pertaining to taste.

**habitat** The natural environment in which an organism lives.

**Hadean** Oldest eon of planetary history on Earth, from its accretion approximately 4.56 billion years ago, to 3.8 billion years ago.

**halophile** Any form of life that can grow in a highly saline environment. While many multicellular organisms are halophilic, the term is used most commonly with microorganisms.

**haplodiploploidy** Condition in which one gender has one set of chromosomes (is haploid) and the other has two sets (is diploid).

**herbivore** Organism which consumes principally plants, algae or photosynthesizing bacteria.

**Hesperian** An epoch of planetary history on Mars, from about 3.5 to 1.8 billion years ago.

**heterotroph** Organism that derives its metabolic energy from the breakdown of high-energy organic compounds, generally coming from other organisms.

**hibernation** A state of inactivity and metabolic depression in animals, characterized by lower body temperature, slower breathing, and lower metabolic rate.

**homeothermy** Maintenance of a constant internal body temperature.

**hydrocarbon** Molecule made up of atoms of carbon and hydrogen, typically with high potential energy content; the major component of lipids.

**hydrogen bond** Relatively weak chemical bond arising from attraction between the partial positive charge of a hydrogen covalently bound to an electronegative atom of one molecule and the partial negative charge of an electrophyllic atom of another molecule; the partial positive charge of hydrogen bridges the gap between the two electronegative atoms (most typically O, N, or P).

**hydrophilic** Capable of easily dispersing and dissolving in a polar solvent like water, but able to dissolve little if at all in a non-polar solvent.

**hydrophopic** Capable of easily dispersing and dissolving in a non-polar solvent, but able to dissolve little if at all in a polar solvent like water.

**hygroscopic** Ability of a substance such as honey, sugar and some salts to attract water molecules from the surrounding environment through either absorption or adsorption.

**hypolithic** Organism that grows underneath rocks.

**hypothesis** An unconfirmed but testable assumption based on accepted facts and logical inferences.

**intelligence** Capacity for integrating experience and anticipating the future.

**invertebrate** Animal without a backbone, which includes 95% of all animal species (excluding vertebrates only).

**ion** Atom or molecule in which the total number of electrons is not equal to the total number of protons, giving it a net positive or negative electrical charge.

**ionotroph** Hypothesized organism that uses ionic gradients as its primary energy source for metabolism.

**Jurassic** A period of planetary history on Earth, from about 210 to 145 million years ago

**kinetic energy** The additional energy an object possesses because of its motion.

**kinetroph** Hypothesized organism that uses kinetic energy as its primary energy source for metabolism.

**Kuiper Belt** A disk-shaped region composed of small icy bodies and planetesimals orbiting the Sun at a distance of about 30 to 55 AU; the source of many short-period comets.

**Late Heavy Bombardment** A period approximately 4.1 to 3.8 billion years ago during which a large number of impact craters are believed to have formed on the Moon, and by inference on Earth as well.

**ligand** An ion or molecule that binds to a molecular site or receptor.

**lipophilic** Capable of easily dispersing and dissolving in lipid solvents; essentially equivalent to hydrophobic.

**lipophobic** Incapable of dispersing and dissolving in lipid solvents; essentially equivalent to hydrophilic.

**liposphere** Global collection of habitats in which the liquids are lipophilic, as in the ethane/methane lakes of Titan.

**litho(auto)troph** Phototroph or chemoautotroph that lives in rocks.

**macrobiota** All organisms large enough to be perceived by the human eye without microscopic magnification; informally, all multicellular organisms.

**macromolecule** Very large molecule such as a protein, nucleic acid, conjugated lipid, or carbohydrate polymer, synthesized by joining monomeric units through covalent bonds.

**magnetar** Neutron star with an extremely strong magnetic field.

**magnetotroph** Hypothesized organism that uses magnetic fields as its primary energy source for metabolism.

**mantle** Semi-molten rock layer within a planetary body between the core and the crust.

**Mesozoic** An era of planetary history on Earth, from about 245 to 65 million years ago, beginning with the Triassic and ending with the Cretaceous periods.

**methane** Carbon compound with the chemical formula $CH_4$; the simplest saturated organic compound. Because methane is readily broken down by solar radiation, its presence is considered an indication of ongoing generation, possibly from biotic sources.

**methanogen** Microorganism that produces methane as a metabolic byproduct under anaerobic environmental conditions.

**methanogenesis** The formation of methane as a metabolic by-product by methanogens.

**metabolism** Collection of all the biochemical interactions and associated transformations of energy and matter carried out by living organisms.

**microbialite** Sedimentary body formed on the bed of a lake from the remains or activity of microorganisms; usually includes communities of algae or cyanobacteria.

**Miocene** An epoch of planetary history on Earth, from about 23 million to 5 million years ago.

**mollusk** Invertebrate animal composed ancestrally of three main body parts: a muscular foot, visceral mass with a dispersed rather than centralized nervous system, and mantle that secretes a calcium carbonate exoskeleton; consists of many diverse members, including some highly derivative from the ancestral form. Major groups include gastropods like snails, bivalves like clams and oysters, and cephalopods like squid and octopi.

**monomer** Molecules with a consistent but often variable chemical structure that can be polymerized through covalent bonds to form larger molecules Monomer/polymer examples include amino acids/proteins, sugars/polysaccharides, and fatty acids/triglycerides.

**mons** Designation of a mountain on a planet or moon.

**motility** Ability of an organism to move in its environment.

**multicellular** Consisting of more than one cell, with differentiated cells that perform specialized functions.

**mutation** Change in the base sequence of a cell's genome, caused by environmental factors such as radiation, errors in replication, or by other cellular processes.

**nautiloid** Complex, highly derived group of marine mollusks with a rigid, conical exoskeleton. Nautiloids flourished during the early Paleozoic, and were the major marine predators of the late Cambrian. Now mostly extinct, *Nautilus* is a modern survivor.

**neocortex** Laminar cortical expansion of the outer cerebrum; most recently evolved region of the (primarily) avian and mammalian forebrain.

**natural selection** Change in the mean value or range of variability for a biological characteristic, due to differential survival of organisms displaying more favorable variants of the characteristic; considered the most important mechanism for bringing about evolutionary change in populations over time.

**nitrile** Organic molecule containing nitrogen.

**Noachian** Oldest epoch of planetary history of Mars, from the accretion of Mars, about 4.5 billion years ago, to 3.5 billion years ago.

**nonpolar** Pertaining to a molecule whose electrons are distributed symmetrically so that no part of the molecule carries a partial positive or negative charge.

**normophilic** Pertaining to an organism that lives within the normal temperature range for life zones on Earth, as opposed to thermophilic and psychrophilic extremophiles.

**nucleic acid** Macromolecule composed of chains of monomeric nucleotides, which can carry genetic information.

**nucleotide** Molecule composed of a heterocyclic nitrogenous base, a five-carbon sugar, and one to three phosphate groups. When joined together, nucleotides make up the structural units of RNA and DNA.

**obliquity** Angle between an object's rotational axis, and the line perpendicular to its orbital plane.

**olfactory** Pertaining to smell.

**oligotrophic** Pertaining to an aquatic habitat low in nutrients and plant life but high in dissolved oxygen content.

**omnivore** Organism that eats both plants and animals.

**Ordovician** A period of planetary history on Earth, from about 500 to 440 million years ago.

**organosilicon** Organic compound containing one or more atoms of silicon.

**osmosis** Spontaneous movement of water across a barrier like a semipermeable membrane from an area of lower solute concentration to an area of higher solute concentration.

**osmotroph** Hypothesized organism that uses osmotic gradients as a primary energy source for metabolism.

**Paleozoic** An era of planetary history on Earth, beginning with the Cambrian and ending with the Permian periods, from about 545 to 245 million years ago.

**panspermia** Theory that life can spread through outer space from one point or planetary body to another; as a corollary, life on different planets would have a shared common ancestry.

**parasitism** Symbiotic relationship in which an occupant organism benefits from the host organism at the expense of the host.

**pelagic** Open-sea environment; any water in the sea that is not close to the bottom or near to the shore.

**Permian** A period of planetary history on Earth, from about 290 to 245 million years ago.

**petrolake** Lake of petroleum or petroleum-like compounds.

**Phanerozoic** An eon of planetary history on Earth, from about 545 million years ago until today.

**phosphorylation** Addition of a phosphate group to another molecule; in biochemistry, this reaction usually adds potential energy to the phosphorylated compound.

**photoautotroph** Organism that harvests energy from sunlight in order to synthesize its own organic nutrients.

**photodissociation** Disintegration of a molecule due to light waves, such as a water molecule into hydrogen and oxygen.

**photopigment** Organic compound, typically complexed with a metal atom, that absorbs light as a means of transducing its energy into a biological process; chlorophyll is a prominent example.

**physiology** Study of the mechanisms and function of coordinated living processes.

**pigment** Any compound that absorbs discreet wavelengths of visible light, and therefore is perceived as colored.

**placoderm** Members of an extinct Class of fishes characterized by thick bony plates and the evolutionary origin of jaws.

**planaria** A group of fresh water flatworms.

**planetesimal** Solid objects thought to form in protoplanetary disks, but which haven't reached the mass equivalent of a large moon or planet.

**plant** Multicellular, eukaryotic, autotrophic organism that synthesizes its own high-energy carbon-based nutrients through photosynthesis.

**plate tectonics** Movement of pieces of the upper crust (plates) on top of a semi-molten interior that gives rise to volcanic activity, earthquakes, and mineral recycling, especially along plate boundaries.

**Pleistocene** An epoch of planetary history on Earth, from about 1.8 million to 12,000 years ago.

**polar** In planetary science, pertaining to the poles, at either the north or south end of the planetary axis of rotation. In chemistry, the attribute of a molecule with asymmetrically distributed charges, so that one portion of the molecule is more positively or negatively charged than another.

**polycyclic aromatic hydrocarbons (PAHs)** Chemical compounds that consist of fused aromatic rings and do not contain heteroatoms (atoms other than carbon or hydrogen) or incorporate substituents.

**polymer** Macromolecule composed of repeating structural units.

**polysilane** Elongated molecule of silicon and hydrogen with the general chemical formula $SiH_3[SiH_2]_nSiH_3$.

**primate** Order of mammals with large brain/body weight ratios, complex social behavior, and high intelligence; includes prosiminians (lemurs and tarsiers), monkeys, and apes, including hominids. Except for humans, most primates are restricted to tropical or sub-tropical habitats.

**producer** In ecology, any autotroph that serves as a food source for consumers; occupants of the lowest trophic level of the energy pyramid.

**prokaryote** Unicellular organism lacking a cell nucleus or any other membrane-bound organelles.

**protein** Macromolecule composed of amino acids arranged in a linear chain, forming either elongated strands, sheets, or folded into a globular form.

**Proterozoic** An eon of planetary history on Earth, from about 2.5 billion to 545 million years ago.

**protist** Diverse group of eukaryotic microorganisms, which are either single-celled or multicellular without specialized tissues; may be either autotrophic or heterotrophic. Protists constitute a Kingdom in the Domain Eukarya.

**protocell** A precursor of the first cell without all the attributes of life.

**protostellar disk** Collection of gas and dust flattened into a disk rotating around a central mass that has not yet compressed into a star.

**Psittaciformes** Order of birds that includes parrots and cockatoos, character-ized by brain/body weight ratios as high as primates; generally regarded along with corvids as the most intelligent of the birds.

**psychrophile** Organism that grows at low temperatures (usually below 5°C).

**Quaternary** A period of planetary history on Earth, from about 1.8 million years ago to the present.

**red dwarf** dM star; a star which is much less luminous than our Sun.

**reed** Grass-like plant growing in shallow water or on marshy ground.

**reproduction** A biological process by which new individual organisms are produced, either sexually or asexually (see also **budding**).

**reptile** Vertebrate with four limbs and scaly skin that breathes air and lays amniotic eggs (though a few forms are viviparous); major groups include crocodilians, lizards, snakes, turtles, and many extinct lineages including the dinosaurs. Surviving reptiles are poikilothermic (variable body temperature), but some dinosaurs may have been mostly homeothermic. Reptiles were the earliest vertebrates to be able to reproduce away from water.

**rift** Geological feature, usually linear, where the Earth's crust is being pulled apart by tectonic forces.

**sedentary** Lifestyle characterized by stationary attachment to a single location; lack of locomotion.

**selection (directional)** Form of natural selection that changes the mean value of a biological characteristic over time in a particular direction due to a directional modification in the environment or biological circumstance of a population.

**selection (disruptive)** Form of natural selection that changes a biological characteristic over time in two or more different directions due to environmental changes within the range of a single population that subjects different portions of the population to different selective pressures This form of natural selection commonly leads to allopatric (non-overlapping) speciation.

**selection (natural)** see **natural selection**

**selection (stabilizing)** Form of natural selection in stable environments that reduces the range of variability of a biological characteristic without changing its mean value over time. In unchanging environments, characteristics tend to evolve closer to the optimal adaptation for that environment.

**sensorimotor** Pertaining to physiological processes that integrate sensory input with motor output.

**silane** Molecule with the chemical formula $SiH_4$; the silicon analogue of methane. At average Earth temperatures, silane is a gas and undergoes spontaneous combustion in air.

**silicate** Molecule consisting of an ion in which one or more central silicon atoms are surrounded by electronegative ligands. Silicates are the largest and most important class of rock-forming minerals, constituting approximately 90% of the crust of the Earth.

**Silurian** A period of planetary history on Earth, from about 440 to 410 million years ago.

**Snowball Earth** An event when Earth's surface becomes nearly or entirely frozen over. The last event occurred between 650 and 750 million years ago, near the end of the Proterozoic eon, prior to the Cambrian Explosion.

**solute** Substance dissolved in a liquid.

**solvent** Liquid in which solutes are dissolved.

**spectrophotometry** Analytical technique for assessing the chemical composition and/or quantity of a substance based on the wavelengths of radiation absorbed or emitted by the substance.

**spore** Reproductive structure adapted for dispersal and survival for extended periods of time in unfavorable conditions with no stored food sources.

**stratigraphy** Application of the Law of Superposition to soil and geological layers to determine their relative ages.

**subduction** Process that takes place at certain types of plate boundaries, by which one tectonic plate moves under another one, sinking into the Earth's crust, as the plates converge.

**subterranean** Existing or operating below the surface of the Earth.

**supercritical** Physical state of a gas in which it remains liquid beyond its boiling point due to high pressure.

**symbiont** Any organism that lives inside or in close association with an organism of a different species.

**syngamy** Union of cells resulting in fertilization during sexual reproduction.

**thermosynthesis** Use of heat as an energy source for endergonic metabolic reactions.

**thermotroph** Hypothesized organism that uses thermal gradients as its primary source of energy for metabolism.

**tectonic** Movement of rock units.

**teleost** Largest and most dominant group of the bony fishes; the single most diverse group of all vertebrates. Teleosts include the vast majority of marine and freshwater, ray-finned fishes, such as perch, trout, and salmon, but not the cartilaginous fishes like sharks (see **elasmobranchs**).

**terra** Pertaining to the planet Earth.

**terracentric** Perspective based on conditions characteristic of the planet Earth.

**terraforming** Process of deliberately modifying the atmosphere, temperature, surface topography, or ecology of another planet or moon to be similar to that on Earth, to make it habitable for organisms from Earth.

**Tertiary** A period of planetary history on Earth, from about 65 to 1.8 million years ago.

**tholin** Heteropolymeric molecule formed by ultraviolet irradiation of simple organic compounds and nitriles. Tholins do not form naturally in Earth's atmosphere, but commonly occur on icy planetary bodies, like Titan, Iapetus, and probably Triton.

**thylakoid** Flattened membrane sac within the chloroplast of plants, where the energy from sunlight is converted to chemical energy.

**topology** Three dimensional spatial contour of a surface.

**transcription** Synthesis of an RNA strand from a DNA template – the first step in reading out genetic information.

**translation (genetic)** Mechanism by which the proper alignment of amino acids into proteins is arranged, based on information from an RNA strand.

**Triassic** A period of planetary history on Earth, from about 245 to 210 million years ago.

**tributary** Channel that carries water or other liquids to points of convergence with other channels, leading to larger but fewer channels downstream.

**trophic level** Hierarchical position of an organism in the flow of energy within an ecosystem, defining what it eats and what it is eaten by.

**vertebrate** Bilaterally symmetrical animal with a cartilaginous or solid backbone enclosing a dorsal nerve cord.

**viscoelastic** Pertaining to topographical distortions in the surface of a body due to tidal forces.

# Index

Subjects featured in figures or tables are shown in bold